DORMANCY IN PLANTS
From Whole Plant Behaviour to Cellular Control

DORMANCY IN PLANTS
From Whole Plant Behaviour to Cellular Control

Edited by

J.-D. Viémont
UMR-SAGAH
INRA-INH-Université d'Angers
Angers
France

and

J. Crabbé
Morphogenèse Végétale Appliquée
Faculté des Sciences Agronomiques
Gembloux
Belgium

CABI *Publishing*

CABI *Publishing* is a division of CAB *International*

CABI Publishing
CAB International
Wallingford
Oxon OX10 8DE
UK

Tel: +44 (0)1491 832111
Fax: +44 (0)1491 833508
Email: cabi@cabi.org
Web site: http://www.cabi.org

CABI Publishing
10 E 40th Street
Suite 3203
New York, NY 10016
USA

Tel: +1 212 481 7018
Fax: +1 212 686 7993
Email: cabi-nao@cabi.org

A catalogue record for this book is available from the British Library, London, UK.

Library of Congress Cataloging-in-Publication Data
Dormancy in plants : from whole plant behaviour to cellular control / edited by J.-D. Viémont and J. Crabbé.
p. cm.
Includes bibliographical references.
ISBN 0-85199-447-4 (alk. paper)
1. Dormancy in plants--Congresses. I. Viémont, J.-D. (Jean-Daniel) II. Crabbé, J. (Jacques)

QK761 .D67 2000
571.7'82--dc21 00-023170

ISBN 0 85199 447 4

Typeset by AMA DataSet Ltd, UK.
Printed and bound in the UK by the University Press, Cambridge.

Contents

Contributors ix

Preface xiii

Part I: Whole Plant and Organ Physiology

1 Growth Cycle and Dormancy in Plants 1
H. Okubo

2 Bud Dormancy: a Suggestion for the Control Mechanism and its Evolution 23
A. Erez

3 The Distinct Controlling of Dormancy Release and Germination Commencement in Seeds 35
N.V. Obroucheva and O.V. Antipova

4 Sugar Metabolism in Apple Embryos 47
S. Lewak, R. Bogatek and B. Żarska-Maciejewska

5 Population Responses to Temperature for Loss and Induction of Seed Dormancy and Consequences for Predictive Empirical Modelling 57
A.J. Murdoch, L. Sonko and E. Kebreab

6 Pre-treatment at Controlled Seed Moisture Content as an Effective Means to Break Dormancy in Tree Seeds 69
M.P.M. Derkx

Part II: Water Relations and Stress

7 Organismic and Environmental Controls of Bud Growth in Tropical Trees 87
R. Borchert

8 Water Relations in Winter: Effect on Bud Break of Walnut Tree 109
T. Améglio, A. Guilliot, A. Lacointe, J.L. Julien, G. Alves, V. Valentin and G. Pétel

9 Morphological Structure and Water Status in Tulip Bulbs during their Transition from Dormancy to Active Growth: Visualization by Magnetic Resonance Imaging 121
R. Kamenetsky, H. Zemah, A. van der Toorn, H. Van As and P. Bendel

10 Seed Dormancy as One of the Survival Strategies in Annual Plant Species Occurring in Deserts 139
Y. Gutterman

11 Seasonally Regulated Proteins in Peach (*Prunus persica* L. Batsch): What Are They and What Do They Do? 161
M. Wisniewski and R. Arora

Part III: Abscisic Acid and Hormonal Control

12 Aspects of ABA and Fusicoccin Signal Transduction in Barley Grains: I Effect of Fusicoccin on ABA-induced Gene Expression in Embryo and Aleurone; II Spatial and Temporal Expression of 14-3-3 Proteins in Dormant Embryos 173
C. Testerink, R.M. van der Meulen and M. Wang

13 Dormancy of Cereal Seeds as Related to Embryo Sensitivity to ABA and Water Potential 183
F. Corbineau and D. Côme

14 Abscisic Acid Control of Seed Dormancy Expression in *Nicotiana plumbaginifolia* and *Arabidopsis thaliana* 195
M. Jullien, D. Bouinot, S. Ali-Rachedi, B. Sotta and P. Grappin

15 The Role of Endogenous Hormones in Potato Tuber Dormancy 211
J.C. Suttle

16 Role of ABA, Gibberellins and Auxin in Dormancy and Dormancy Release of Tulip Bulbs 227
M. Saniewski, L. Kawa-Miszczak, E. Wegrzynowicz-Lesiak and H. Okubo

Part IV: Biochemical and Cellular Aspects

17 ATP, ADP and NTP Contents in Vegetative and Floral Peach Buds During Winter: Are They Useful for Characterizing the Type of Dormancy? 245
M. Bonhomme, R. Rageau and M. Gendraud

18 Alcohols that Break Seed Dormancy: the Anaesthetic Hypothesis, Dead or Alive? 259
M.A. Cohn and H.W.M. Hilhorst

19 Are Cellular Membranes Involved in the Control of Seed Dormancy? 275
H.W.M. Hilhorst and M.A. Cohn

20 Intercellular Communication Channels and Intracellular Calcium Levels Involved in the Dormancy Development of Poplar (*Populus deltoides*) Plants 291
L.-C. Jian, J.-H. Li, P.H. Li and T.H.H. Chen

21 The Developmental Basis of Bud Dormancy in 1-year-old *Picea* and *Pseudotsuga* Seedlings 313
J.E. MacDonald

Part V: Genetics and Molecular Biology

22 Molecular Genetic Analysis of Bud Dormancy-related Traits in *Populus* 319
T.H.H. Chen, J. Davis, B.E. Frewen, G.T. Howe and H.D. Bradshaw, Jr

23 Regulation of Growth and Dormancy in Pea Axillary Buds 331
J.P. Stafstrom

24 Expression of Endo-β-mannanase and SNF-related Protein Kinase Genes in True Potato Seeds in Relation to Dormancy, Gibberellin and Abscisic Acid 347
V. Alvarado, H. Nonogaki and K.J. Bradford

25 The Genetics of Seed Dormancy in *Arabidopsis thaliana* 365
M. Koornneef, C. Alonso-Blanco, L. Bentsink, H. Blankestijn-de Vries, I. Debeaujon, C.J. Hanhart, K.M. Léon-Kloosterziel, A.J.M. Peeters and V. Raz

Index 375

Contributors

S. Ali-Rachedi, *UMR INRA-INA PG Biologie des Semences, Chaire de Physiologie Végétale, INA P-G, 16 rue Claude Bernard, F-75231 Paris cedex 05, France*

C. Alonso-Blanco, *Laboratory of Genetics, Wageningen University, Dreijenlaan 2, 6703 HA Wageningen, The Netherlands*

V. Alvarado, *Department of Vegetable Crops, University of California, One Shields Avenue, Davis, CA 95616-8631, USA*

G. Alves, *UA Bioclimatologie-PIAF, Université Blaise Pascal, 24 Avenue des Landais, F-63177, Aubière cedex, France*

T. Améglio, *INRA, UA Bioclimatologie-PIAF, Domaine de Crouelle, F-63039 Clermont-Ferrand cedex 2, France*

O.V. Antipova, *Russian Academy of Sciences, Institute of Plant Physiology, Botanical Street 35, Moscow 127276, Russian Federation*

R. Arora, *Department of Plant and Soil Science, University of West Virginia, Morgantown, WV, USA*

D. Bartels, *Friedrich-Willems Universität Bonn, Botanisches Institut, Kirschallee 1, 53115 Bonn, Germany*

P. Bendel, *Department of Chemical Services, MR Center, Weizmann Institute of Science, Rehovot, Israel*

L. Bentsink, *Laboratory of Genetics, Wageningen University, Dreijenlaan 2, 6703 HA Wageningen, The Netherlands*

J.H.W. Bergervoet, *CPRO-DLO, Center for Plant Breeding, PO Box 16, 6700 AA Wageningen, The Netherlands*

H. Blankestein-de Vries, *Laboratory of Genetics, Wageningen University, Dreijenlaan 2, 6703 HA Wageningen, The Netherlands*

R. Bogatek, *University of Warsaw, Institute of Plant Experimental Biology, ul. Pawińskiego 5a, 02-106 Warszawa, Poland*

M. Bonhomme, *INRA, UA Bioclimatologie-PIAF, Domaine de Crouelle, F-63039 Clermont-Ferrand cedex 2, France*

R. Borchert, *Biology Division, Haworth Hall, University of Kansas, Lawrence, KS 66045-2106, USA*

D. Bouinot, *UMR INRA-INA PG Biologie des Semences, Chaire de Physiologie Végétale, INA P-G, 16 Rue Claude Bernard, F-75231 Paris cedex 05, France*

K.J. Bradford, *Department of Vegetable Crops, University of California, One Shields Avenue, Davis, CA 95616-8631, USA*

H.D. Bradshaw, Jr, *College of Forest Resources, University of Washington, Seattle, WA 98195, USA*

T.H.H. Chen, *Department of Horticulture, Oregon State University, ALS 4017, Corvallis, OR 97331, USA*

M.A. Cohn, *Department of Plant Pathology and Crop Physiology, 302 Life Sciences Building, Louisiana State University, Baton Rouge, LA 70803, USA*

D. Côme, *Laboratoire de Physiologie Végétale Appliquée, Université Pierre et Marie Curie, Tour 53, 1er etage, 4 Place Jussieu, 75252 Paris cedex 05, France*

F. Corbineau, *Laboratoire de Physiologie Végétale Appliquée, Université Pierre et Marie Curie, Tour 53, 1er etage, 4 Place Jussieu, 75252 Paris cedex 05, France*

J. Davis, *Department of Horticulture, Oregon State University, ALS 4017, Corvallis, OR 97331, USA*

I. Debeaujon, *Laboratory of Genetics, Wageningen University, Dreijenlaan 2, 6703 HA Wageningen, The Netherlands*

M.P.M. Derkx, *Research Station for Nursery Stock, Rijneveld 153, PO Box 118, 2770 AC Boskoop, The Netherlands*

A. Erez, *ARO, The Volcani Center, Institute of Horticulture, PO Box 6, Bet Dagan 50-250, Israel*

M.E. Foley, *USDA-ARS Biosciences Research Laboratory, State University Station, Fargo, ND 58105-5674, USA*

B.E. Frewen, *College of Forest Resources, University of Washington, Seattle, WA 98195, USA*

M. Gendraud, *UA Bioclimatologie-PIAF, Université Blaise Pascal, 24 Avenue des Landais, F-63177 Aubière cedex, France*

P. Grappin, *UMR INRA-INA PG Biologie des Semences, Chaire de Physiologie Végétale, INA P-G, 16 rue Claude Bernard, F-75231 Paris cedex 05, France*

A. Guilliot, *UA Bioclimatologie-PIAF, Université Blaise Pascal, 24 Avenue des Landais, F-63177, Aubière cedex, France*

Y. Gutterman, *Jacob Blaustein Institute for Desert Research and Department of Life Sciences, Ben Gurion University of the Negev, Sede Boker Campus 84990, Israel*

C.J. Hanhart, *Laboratory of Genetics, Wageningen University, Dreijenlaan 2, 6703 HA Wageningen, The Netherlands*

H.W.M. Hilhorst, *Department of Plant Physiology, Wageningen Agricultural University, Arboretumlaan 4, 6703 BD Wageningen, The Netherlands*

M. Holdsworth, *IACR-Long Ashton, Department of Agricultural Sciences, University of Bristol, Long Ashton, Bristol BS41 9AF, UK*

G.T. Howe, *School of Natural Resources, The Ohio State University, OARDC, Wooster, OH 44691, USA*

L.-C. Jian, *Laboratory of Plant Hardiness, Department of Horticultural Science, University of Minnesota, St Paul, MN 55108, USA*

J.L. Julien, *UA Bioclimatologie-PIAF, Université Blaise Pascal, 24 Avenue des Landais, F-63177, Aubière cedex, France*

M. Jullien, *UMR INRA-INA PG, Biologie des Semences, Route de Saint-Cyr, 78026 Versailles cedex, France*

R. Kamenetsky, *ARO, Department of Ornamental Horticulture, The Volcani Center, PO Box 6, Bet Dagan, 50-250, Israel*

L. Kawa-Miszczak, *Research Institute of Pomology and Floriculture, Pomologiczna 18, 96-100 Skierniewice, Poland*

E. Kebreab, *Department of Agriculture, University of Reading, PO Box 236, Earley Gate, Reading RG6 6AT, UK*

M. Koornneef, *Laboratory of Genetics, Wageningen University, Dreijenlaan 2, 6703 HA Wageningen, The Netherlands*

A. Lacointe, *INRA, UA Bioclimatologie-PIAF, Domaine de Crouelle, 63039 Clermont-Ferrand cedex 2, France*

H.M. Léon-Kloosterziel, *Laboratory of Genetics, Wageningen University, Dreijenlaan 2, 6703 HA Wageningen, The Netherlands*

S. Lewak, *University of Warsaw, Institute of Plant Experimental Biology, ul. Pawińskiego 5a, 02-106 Warszawa, Poland*

J.-H. Li, *Laboratory of Plant Hardiness, Department of Horticultural Science, University of Minnesota, St Paul, MN 55108, USA*

P.H. Li, *Laboratory of Plant Hardiness, Department of Horticultural Science, University of Minnesota, St Paul, MN 55108, USA*

J.E. MacDonald, *Natural Resources Canada, Canadian Forest Service, PO Box 4000, Fredericton, New Brunswick, Canada, E3B 5P7*

A.J. Murdoch, *Department of Agriculture, University of Reading, PO Box 236, Earley Gate, Reading RG6 6AT, UK*

H. Nonogaki, *Department of Vegetable Crops, University of California, One Shields Avenue, Davis, CA 95616-8631, USA*

N.V. Obroucheva, *Russian Academy of Sciences, Institute of Plant Physiology, Botanical Street 35, Moscow 127276, Russian Federation*

H. Okubo, *Laboratory of Horticultural Science, Faculty of Agriculture, Kyushu University 46-01, Fukuoka 812-8581, Japan*

A.J.M. Peeters, *Laboratory of Genetics, Wageningen University, Dreijenlaan 2, 6703 HA Wageningen, The Netherlands*

G. Petel, *UA Bioclimatologie-PIAF, Université Blaise Pascal, 24 Avenue des Landais, F-63177, Aubière cedex, France*

R. Rageau, *INRA, UA Bioclimatologie-PIAF, Domaine de Crouelle, F-63039 Clermont-Ferrand cedex 2, France*

V. Raz, *Laboratory of Genetics, Wageningen University, Dreijenlaan 2, 6703 HA Wageningen, The Netherlands*

M. Saniewski, *Research Institute of Pomology and Floriculture, Pomologiczna 18, 96-100 Skierniewice, Poland*

L. Sonko, *Department of Agricultural Services, Agricultural Pest Management Unit, Yundum, The Gambia, West Africa*

B. Sotta, *Laboratoire de Physiologie du Développement des Plantes, UMR de Physiologie Cellulaire et Moléculaire des Plantes, Université Pierre et Marie Curie (PVI), Tour 53 (E5, casier 156), 4 Place Jussieu, F-75252 Paris cedex 05, France*

J.P. Stafstrom, *Department of Biological Sciences, Plant Molecular Biology Center, Northern Illinois University, De Kalb, IL 60115, USA*

J.C. Suttle, *USDA-ARS, Northern Crop Science Laboratory, PO Box 5677, Fargo, ND 58105-5677, USA*

C. Testerink, *Center for Phytotechnology, Leiden University, TNO Department of Plant Biotechnology, Wassenaarseweg 64, 2333 AL Leiden, The Netherlands*

V. Valentin, *UA Bioclimatologie-PIAF, Université Blaise Pascal, 24 Avenue des Landais, F-63177, Aubière cedex, France*

H. van As, *Department of Biomolecular Sciences, WNMRC, Wageningen, The Netherlands*

R.M. van der Meulen, *Center for Phytotechnology, Leiden University, TNO Department of Plant Biotechnology, Wassenaarseweg 64, 2333 AL Leiden, The Netherlands*

A. van der Toorn, *Department of Biomolecular Sciences, WNMRC, Wageningen, The Netherlands*

M.K. Walker-Simmons, *USDA-ARS, Wheat Research Unit, 209 Johnson Hall, Washington State University, Pullman, WA 99164-6420, USA*

M. Wang, *Center for Phytotechnology, Leiden University, TNO Department of Plant Biotechnology, Wassenaarseweg 64, 2333 AL Leiden, The Netherlands*

E. Wegrzynowicz-Lesiak, *Research Institute of Pomology and Floriculture, Pomologiczna 18, 96-100 Skierniewice, Poland*

M. Wisniewski, *USDA-ARS, Appalachian Fruit Research Station, 45 Wiltshire Road, Kearneysville, WV 25430-9425, USA*

B. Żarska-Maciejewska, *University of Warsaw, Institute of Plant Experimental Biology, ul. Pawińskiego 5a, 02-106 Warszawa, Poland*

H. Zemah, *ARO, Department of Ornamental Horticulture, The Volcani Center, PO Box 6, Bet Dagan, 50-250 Israel*

Preface

Dormancy is a multiple and complex phenomenon. Its multiplicity arises from the fact that it affects diverse plant structures, which themselves appeared at different times during evolution and evolved in a variety of biotopes and climates. Its complexity, moreover, results from the numerous anatomical and physiological relations of the properly dormant structure with neighbouring parts.

The bud was probably born some 400 million years ago, in the lower Devonian, in the very early land plants. The upward struggle for light among these plants demanded strict control to adjust growth and branching to nutrient availability; axillary branching and apical dominance, and also flushing growth, sprang up from these constraints. The seed, with a dormant embryo, appeared 100 million years later, in the upper Devonian, as the last step of an evolution bringing female gametophyte development, fertilization and embryogenesis free from external water. These improvements were completed when ancient gymnosperms left the humid, already forested areas, to invade and colonize still empty and drier uplands. The display of the whole range of vegetative propagules (tubers, bulbs, corms, etc.) had to wait for the explosive diversification of the angiosperms, during the Cretaceous era *c.* 130 million years ago. The beginnings of dormancy occurred in a rather uniform – warm and moist – environment; biotope diversity and climate changes happened only later. Dormancy thus had a morphogenetic function – spatial and temporal organization of the plant – before it became also a survival mechanism against an unfavourable season.

Nevertheless, the latter aspect became a major concern in plant physiology in relation to the control of germination, growth and reproduction of temperate species. Survival of an unfavourable season implies many manifestations interfering with dormancy: formation of protective structures, storage of reserve metabolites in appropriate tissues, resistance mechanisms, etc., all of these generating internal correlations and submitted to environmental influences. Since the early 1930s, physiologists have toiled to unravel the numerous and intricate dormancy syndromes and to try to find out a supposed 'common denominator' for them; the long evolution of dormancy and its multiple appearances does not exclude the possibility that some basic cellular processes would have been re-used several times. The modern approaches of molecular biology probably constitute very efficient tools to achieve this goal. However, the multiplicity and the complexity of the phenomenon requires that still more cases and particulars of dormancy are brought under scrutiny and investigated by traditional physiological and anatomical analyses.

Five years after the 'first' symposium organized in Corvallis (Oregon, USA) by Greg A. Lang and co-workers, it was thought opportune to sum up again the progress made in the experimental pursuit and eventual elucidation of the difficult and challenging subject of dormancy. The city of Angers (France) and its university proved a convenient and pleasant setting for this event.

For their participation in organizing the Symposium, we would like to thank individual members of the following institutions: the University of Angers, The National Institute for Agronomical Research (INRA-Angers), the National Institute of Horticulture (INH), the National Station for Seed Testing (SNES-GEVES) and the Unité Mixte de Recherche SAGAH. Together with the members of the scientific and local organizing committees, special gratitude is addressed to those people who kindly helped in the elaboration of the programme: Drs Anne Fennell, Marie-Thérèse Lepage, Henk Hilhorst, Greg Lang and Jeffrey Suttle.

Neither the Symposium, nor this book would have come to existence without financial support. We particularly prize the subventions granted by the French Ministry of Education, Research and Technology, the Regional Council of 'Pays de la Loire', Atlantech, the General Council of Maine et Loire, the City of Angers and the 'Groupe d'Etude de l'Arbre'. Our gratitude also goes to the chapter authors of this book, the chairpersons of the Symposium and its 140 participants, who all together contributed to its success. Finally, we appreciated very much Tim Hardwick's patience, willingness and experience throughout the book's development with CAB International.

Jean-Daniel Viémont
Angers, France
Jacques Crabbé
Brussels, Belgium

Growth Cycle and Dormancy in Plants

1

Hiroshi Okubo

Laboratory of Horticultural Science, Faculty of Agriculture, Kyushu University 46-01, Fukuoka 812-8581, Japan

Introduction

The longevity of plant species ranges from a few weeks, e.g. desert annuals, to thousands of years, e.g. *Dracaena draco*. Life spans for clonal species are longer and some may have persisted since the time of the last glacial period. Starting with the gamete to fertilized egg, the developmental history of a plant is traced through seed development, germination, juvenility, maturity, flowering and fruiting. At fruiting, the basic cycle of plant growth is complete. In annuals, fruiting is the beginning of senescence and death, whereas in perennials, the plant is prepared to recycle after a period of quiescence. One growth cycle, therefore, corresponds to one life cycle in annuals, whereas in perennials the life span consists of repeated growth cycles. What is one growth cycle in tomato plants? At lower nodes of a growing tomato plant, fruits are maturing, whereas at the upper nodes, flowering is occurring while the apical meristem is vegetative. It grows as an annual in temperate regions and a perennial in the tropics.

Great diversity of size, form, life span and growth behaviour in plants including dormancy makes it difficult to establish general and universal definitions for plant growth cycle. In order to discuss dormancy, it seems necessary to establish a concept and definition of plant growth cycle that can satisfactorily explain the various growth behaviours of almost all the higher plants by simple terms, because dormancy (if it exists) is only one of the phenomena and processes in the plant growth cycle.

A new universal terminology for dormancy has been proposed by Lang *et al.* (1987) and it seems to be well accepted. However, the definition that 'Dormancy is a temporary suspension of visible growth of any plant structure containing a meristem' (Lang *et al.*, 1987) does not refer to the commencement of temporary suspensions of growth. How can we determine when a plant is in the dormant phase? For example, when does the bulb dormancy commence: before, at, during or after bulb induction or development?

In this article, a new idea on plant growth cycle is initially presented. Then, the new concept of dormancy, particularly concerning the commencement of dormancy, in the growth cycle is discussed.

History of Plant Evolution: Reconfirmation

Before discussing the primary topic a brief history of plant evolution, in particular of angiosperms, will be reviewed and three generally important facts in the history of plant evolution will be reconfirmed.

During the Early Mesozoic era (60–185 million years ago), gymnosperms arose and rapidly became dominant. The first angiosperms appeared in the Middle Mesozoic, developed rapidly, and had become very abundant by the end of the era. The Early Tertiary period of the Cenozoic era (60 million years ago) had a humid and warm climate which supported a rich and widespread flora, with woody angiosperms dominant. During the Late Tertiary, the flora became restricted in distribution and segregated into climatic types, as cooling occurred and climatic zones became established. Extensive glaciation occurred in the Pleistocene (2 million years before present) in the northern hemisphere, which strongly affected the distribution of many plant species. Modern herbaceous angiosperms arose and evolved rapidly during this time. Their short life cycle and small size had considerable survival value under the conditions of advancing cooling.

What should be kept in mind is: (i) angiosperms originated in humid tropical regions where temperature, daylength and precipitation were stable all through the year; (ii) the trend of their evolution was from woody to herbaceous; and (iii) one of the major evolution forces to differentiate families, genera, species, etc. was environmental change.

Growth Behaviours: Comparative Observation and Experimental Facts

Another point to note is that, in plants, growth is restricted to certain embryonic regions, i.e. meristems. By having localized embryogenic regions, higher plants differ from animals, where growth typically occurs throughout the whole organism (Wareing and Phillips, 1981).

In this section, various examples of plant growth traits, some of which are beyond our understanding, are presented together with some experimental facts. We have also to realize that irregular or erratic growth traits of plants such as flowering out of season are normal and a natural plant response to irregular or erratic external factors such as a cold summer, warm winter, etc.

Annual plants

Annuals are not always annuals. It is generally recognized that the herbaceous annual plant is one that normally completes its entire life cycle during a single growing season. However, they can become woody in appropriate climatic conditions as illustrated in Fig. 1.1: a 'tree' salvia (*Salvia splendens*).

True herbaceous annuals can also grow to become woody perennials under special conditions. In a greenhouse at Kyushu University with a special treatment, *Brassica* × *napus* (an interspecific hybrid of cabbage (*Brassica oleracea*) × Chinese cabbage (*Brassica campestris*), a heading-type cruciferous vegetable) grew to become a 'tree'. Several 'cabbages' were formed on the branches of the 5-year-old tree (Fig. 1.2). A cross-section of the stem indicated woody characteristics, e.g. lignification and annual ring-like tissues (Fig. 1.3). This example indicates that the plant genetically preserves its ancestral growth trait but it is suppressed in normal growth conditions. In addition, it indicates that the growth cycle of tree and herbaceous plants must be explained by the same theory. Of course, since many or almost all herbaceous annuals have evolved as true herbaceous annuals, we are, up

Fig. 1.1. A 'tree' salvia on Bali Island, Indonesia. Photographed in October, 1990.

Fig. 1.2. A 5-year-old *Brassica* × *napus* tree (top) and many 'cabbage' formations on a woody branch (bottom).

to this time, unable to reverse their growth behaviour from herbaceous annuals to woody perennials.

In order to explain the different growth traits of annual plants growing as monocarpic (one growth cycle) and polycarpic (repeated growth cycles) plants by the same theory, we propose to define the growth cycle as the process starting with cell division of the zygote or with burst of a lateral bud

Fig. 1.3. Cross-section of the stem of a 5-year-old *Brassica* × *napus*. Annual ring-like tissues and lignification are present (courtesy of S. Uemoto).

to flowering/fruiting on the top end of the shoot through vegetative growth with leaf development ('one bud – one growth cycle theory'). Plants consist of the accumulation of one to numerous growth cycles (one to several growth cycles in one year in annuals and numerous growth cycles continuing year by year in perennials). This definition must also apply to bulbous plants and true trees, as will be verified later in this chapter. In fact, the idea has been established not only from the above-mentioned phenomena but also from various growth traits and the experiments described below.

Perennial plants

Perennials from temperate regions have their origin in the tropics. Differences in the plants of the same family or genus distributed in the tropics and temperate climate regions will be reviewed in this section.

Giant bamboo (*Dendrocalamus giganteus,* family *Gramineae,* subfamily *Bambusoideae*) is of tropical origin. It forms a bush (Fig. 1.4, top) without developing rhizomes during growth. Moso bamboo (*Phyllostachys heterocycla,* same subfamily), a native of south China and well distributed in Japan, has edible young sprouts which emerge and become bamboo woods in spring from widespread rhizomes (Fig. 1.4, bottom).

The tropical water lily (*Nymphaea* sp.) grows by forming shrubs with the petioles radiating from the root crown and reduced development

Fig. 1.4. Bamboo plants in the tropics and temperate regions. Giant bamboo (*Dendrocalamus giganteus*) forming a bush without developing rhizomes (top) in Peradeniya, Sri Lanka (c. 8° N), photographed in January, 1987. Moso bamboo (*Phyllostachys heterocycla*) forming a bamboo forest by the numbers of shoots emerged from widespread rhizomes (bottom) in Nagoya, Japan (c. 35° N), photographed in October, 1997.

of rhizomes. In contrast, the temperate water lily develops and spreads rhizomes with each node producing new shoots. The shoots, therefore, emerge randomly on the water surface. *Zoysia* spp. are dwarf and creeping in temperate regions and used as a turf plant. Another *Zoysia* sp. found in the tropics stands erect and has reduced development of rhizomes.

These facts indicate that plants have acquired rhizomes during the evolution from tropical to temperate regions in order to survive unfavourable environments, e.g. low temperatures in winter.

Bulbous plants

Bulbous plants have the ability to change their growth habit from a bulbous to an evergreen one (Okubo and Uemoto, 1981). Three to four gram bulblets of Dutch iris (*Iris hollandica*) cv. 'Dominator' were kept at 20°C for 3 months after lifting. They were then grown at 25°C in a greenhouse for a month and subsequently either grown at 25°C or transferred to 15°C in a phytotron. Bulblets stored at 20°C and grown at 15°C developed only three or four leaves and the vegetative apex produced a new bulb (Fig. 1.5). This is a normal growth pattern for this bulblet weight class and the new bulb became dormant. Bulblets stored at 20°C and grown at 25°C, however, continued to grow and developed ten or more leaves without forming new bulbs over a 12-month period. Growth of the lateral shoots was also observed (Fig. 1.6). Endogenous abscisic acid (ABA), a dormancy-controlling hormone, increased only in plants grown under bulb-forming conditions (15°C). These data indicate that under certain environmental conditions bulbous iris can grow continuously without bulb formation and dormancy, and that ABA increases only in the bulb-forming plant.

Similar results were obtained from *in vitro* culture experiments with hyacinth and lily. Bach (1992) developed a method for *in vitro* mass

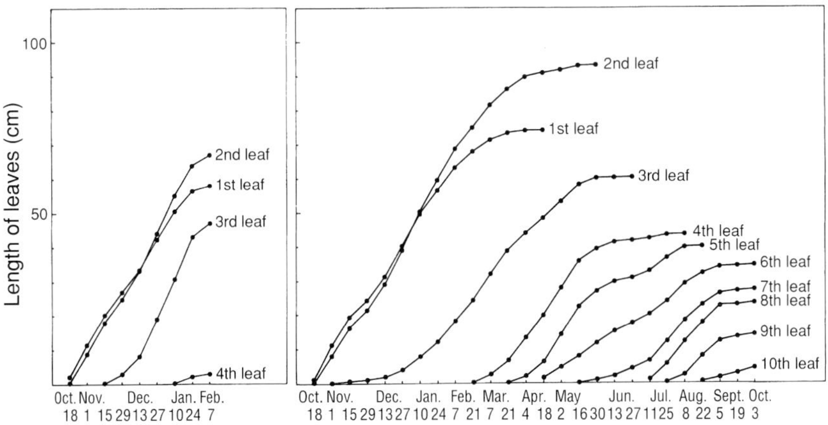

Fig. 1.5. Leaf development of 3–4 g iris bulbs grown at 15 (left) or 25°C (right) after storage at 20°C (reprinted from Okubo and Uemoto (1981) with permission from the Japanese Society of Plant Physiologists).

Fig. 1.6. Two lateral shoots (arrows) emerging from a mother plant of bulbous iris. Normally, the lateral meristems grow to form bulbs. The main shoot consists of more than ten leaves.

propagation of hyacinth (*Hyacinthus orientalis*). Initial explants prepared from the basal parts of immature leaves of low temperature-treated (2°C for 3–4 months) mature bulbs, developed shoots after being cultured on Murashige and Skoog medium (Murashige and Skoog, 1962) with 1 mg l^{-1} indole-3-butyric acid (IBA) at 23°C and a 16 h photoperiod. The epidermal and subepidermal cell layers, as secondary explants taken from the regenerated shoots in initial culture, regenerated shoots and did not regenerate bulblets when cultured at 23°C in subculture; those cultured at 4°C for 8 weeks followed by 23°C developed bulblets. Further experiments were carried out using Bach's system (K. Ii *et al.*, unpublished data). Addition of ABA to the medium or soaking the explants in ABA before culture led to bulblet formation under non-bulb-forming conditions (25°C), whereas addition of fluridone (an ABA biosynthesis inhibitor) prevented bulb formation of the explants in the bulb-forming condition (5°C, then 25°C). Endogenous ABA increased in the latter explants but not in the former explants during culture.

In *Lilium speciosum*, the addition of ABA blocked leaf formation and was required for the induction of dormancy, whereas fluridone blocked bulb formation and prevented the development of dormancy (Gerrits and de Klerk, 1992).

In the tulip, a species with an annual bulb replacement, the bulb has to reach a critical weight to be able to produce a flower bud. As long as the bulb has not reached that critical size, all the buds are vegetative and are in apical and lateral positions (in the axil of the scales). When the bulb reaches the critical weight, its apical meristem produces a flower bud while the lateral buds remain vegetative. The tulip bulb is thus a 'cluster' of buds that, according to its size and/or their position in the bulb, will either produce a flower stem or daughter bulbs. Low temperature simultaneously induces a stimulation of stem elongation and bulbing (transformation of vegetative buds into bulbs). Due to the characteristics of the tulip bulb the number of growth cycles between the initiation of one bud and its anthesis greatly varies with the type of bulb and/or the bud position in the bulb. Two different situations can be distinguished:

1. In large bulbs, the innermost vegetative bud is initiated in summer (in the axil of the flower bud); it produces scale primordia that, after induction by winter low temperatures, give rise to a bulb (following spring) whose apical meristem will produce a flower bud (summer). Its elongation, induced by low winter temperatures, leads to anthesis that is observed in the following spring; in the case of large bulbs two growth cycles are thus necessary from the initiation of the innermost vegetative bud to its anthesis.
2. In the case of small bulbs, several growth cycles are necessary from bud initiation to anthesis. The bulbs produced from seeds constitute an extreme situation as they need five to six growth cycles to reach a flowering size.

The behaviour of the axillary vegetative buds of large bulbs varies according to their size and is thus intermediate between the two extreme situations illustrated above.

In the tulip as in bulbous iris (Okubo and Uemoto, 1981), bulb formation (scale primordia enlargement) can be prevented by maintaining the bulbs, or plants, at temperatures higher than 20°C (Le Nard and Cohat, 1968; Le Nard, 1983). Under that condition, scale primordia tend to change into leaves and sometimes it is even possible to observe a floral initiation from lateral buds.

Ancestors of bulbous plants should not have needed to evolve the phenomenon of bulb formation if the environments had been reasonably constant. Some unfavourable environmental factors must have forced the ancestor plants to form bulbs for their survival. Numerous facts confirm and generalize this idea: (i) low temperature induces bulb formation in tulip (Le Nard and Cohat, 1968), garlic (Aoba, 1971), bulbous oxalis (Aoba, 1972) and bulbous iris (Aoba, 1974); (ii) the induction of dormancy is caused by low temperature (Le Nard, 1983); (iii) bulbous iris does not have a dormant period without bulb formation (Okubo and Uemoto, 1981); (iv) ABA is assumed to control dormancy in various organs of plants and it declines at the time of dormancy release in iris bulbs (Tsukamoto and Ando, 1973), *Gladiolus* corms (Tsukamoto, 1974), seeds of *Acer saccharum* (Webb *et al.*, 1973), etc.; and (v) ABA increases when the plants received low temperature in bulbous iris (Okubo and Uemoto,

1981) and hyacinth (K. Ii *et al.*, unpublished data). In conclusion, the induction of bulb formation and bulb dormancy are the same phenomenon. Delvallée *et al.* (1990) have found that the development of dormancy in lily bulblets corresponds to a switch in the development of the primordium. Other experimental data also show that, in tulip and bulbous iris, bulbing corresponds to a stop in the meristem activity in the buds, and that nullifying or preventing bulbing induction by high temperature treatments allows the continuation of meristem activity and organ differentiation in the buds (Okubo and Uemoto, 1981; Le Nard, 1983; Le Nard and De Hertogh, 1993).

Determinate versus indeterminate growth

In determinate plants, the main axis terminates in a floral bud, e.g. maize, and in indeterminate plants, it remains vegetative and the flowers form in an axillary position, e.g. cucumber. Are the two growth habits really different? Is the indeterminate growth habit contradictory to the proposed theory of growth cycle?

The growth habit is known to be genetically controlled. For example, in *Phaseolus vulgaris*, a single gene difference separates the determinate from the indeterminate growth habit (Yarnell, 1965). Recently, Bradley *et al.* (1996) isolated the *cen* gene which shifts indeterminate flowering of snapdragon (*Antirrhinum majus*) to determinate. However, there are examples of environment-dependent changes of the growth habit in some plant species. Determinate cowpea (*Vigna unguiculata*) elongated and became indeterminate at a night temperature of 24°C under a 12 h photoperiod (Summerfield and Wien, 1980). The determinate growth habit of lablab bean (*Lablab purpureus*) was changed to indeterminate by exposure to a daylength of 13 h or longer at 25°C or to a daylength of 10–11 h or longer at 30°C, but there was no daylength effect at 20°C (Fig. 1.7) (Kim and Okubo, 1995). Hormonal control of the growth habit in this species has also been proposed with auxin accelerating indeterminate growth and with cytokinin inducing determinate growth (Kim and Okubo, 1996). Judging from the plasticity observed, we may hypothesize that the branching of so-called indeterminate plants is in fact determinate, the lateral shoot growing vigorously assuming a terminal position and the initially terminal reproductive meristem appearing in a lateral position. The indeterminate growth is only an outward appearance due to strong vigour of lateral growth and is only a morphological variation of determinate growth.

Hippeastrum (*Amaryllidaceae*), a tropical bulbous plant, has a sympodial branching system (determinate growth habit) (Okubo, 1993). At flower initiation, a lateral growing point is formed on the side of the apex, and it develops four leaves and a multiflowered inflorescence. The inflorescence emergence from the bulb is delayed so that it appears lateral to the

Fig. 1.7. Effects of temperature and daylength on growth of determinate lablab bean (reprinted from Kim and Okubo (1995), with permission from Elsevier Science).

leaves of the next unit of the branching system that initiated later, but had already emerged (Fig. 1.8). Flowers in asparagus (*Asparagus officinalis*) are usually initiated two per node, one on each side of a lateral branch or cladophyll (Fig. 1.9, top). Under normal conditions in temperate regions, more than one year is required for the first flowering from the seed. However, when the seed is treated with carbamate, after only 1 month flowering occurs on the apex of the shoot (Fig. 1.9, bottom). This indicates that when growth of the flowering shoot is faster and more vigorous than vegetative growth, the plant exhibits a determinate flowering habit. The true growth habit appears thus masked by the extreme vigour of vegetative growth.

When an indeterminate cultivar of cucumber (*Cucumis sativus*) 'Kaga-fushinari' was cultured *in vitro* from seeds, it became dwarf with shortened internodes. The plant shape became zigzag (Fig. 1.10) (Suzuki, 1990), with

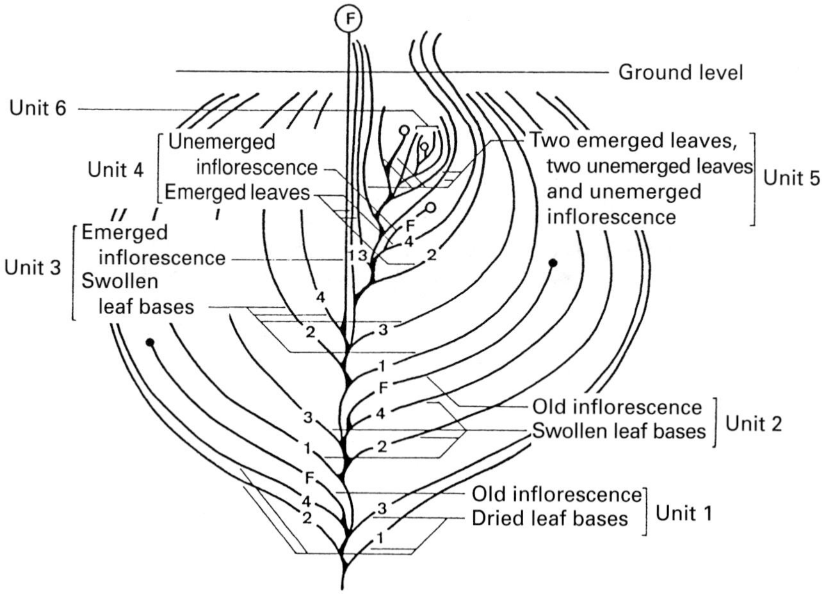

Fig. 1.8. Schematic drawing of a branching system of a mature *Hippeastrum* bulb at flowering (reprinted from Le Nard and De Hertogh (1993), with permission from Elsevier Science).

flower buds on the apex of each unit of the zigzag, and subsequent lateral vegetative growth forming a next unit of the zigzag also terminated by a flower bud. Thus, the two distinct types of flowering can sometimes be only one with different appearances, the determinate growth habit predominating in particular conditions.

Tree plants

There are four types of periodicity in the flowering of tropical plants: ever-flowering, non-seasonal flowering, gregarious flowering and seasonal flowering (Kramer and Kozlowski, 1979). The ever-flowering type must be an original and natural growth behaviour of the primitive woody angiosperms in ancient times when they evolved in stable warm and humid climates without fluctuation. Trees in tropical rain forest generally never stop growing; they continue growing by accumulating growth cycles. The reverse is also true. *Trachycarpus excelsa* is a palm of tropical origin, but it can survive in Japan where it flowers in late spring by sending out five to seven inflorescences at once from the axils. However, in the tropics the inflorescences appear after development of successive leaves with equal intervals, one by one, non-seasonally (Koriba, 1958).

Fig. 1.9. Natural (top) and carbamate-induced flowering of asparagus (*Asparagus officinalis*) (bottom). In the latter, flowering occurred on the apex of the shoots (courtesy of Y. Ozaki).

Pyracantha sp., a temperate *Rosaceae* native from southeast Europe to Turkey, flowers in early summer and its beautiful red fruit appears in autumn in temperate regions. In the tropics, however, flowers and matured red fruit are present at the same time on the same tree (Fig. 1.11). Neither synchronism in flowering nor in fruiting is found. *Magnolia* sp., in

Fig. 1.10. An *in vitro* grown cucumber. Flowering occurred on the apex of each internode. Each internode grew lateral to the flowers of its lower internode (courtesy of S. Uemoto).

Fig. 1.11. A *Pyracantha* tree with flowers and fruit developing at the same time; photographed in October, 1990 in Cibodas (c. 7° S), Indonesia.

temperate regions, flowers in spring before the development of leaves in the same year, whereas, grown in the tropics they have leaves below the open flower (Fig. 1.12). This difference in flowering is due to the fact that in the tropics the flower is formed terminally after developing several leaves. By contrast, in temperate conditions, a flower bud initiates at the

Fig. 1.12. *Magnolia liliflora* flowering in March without leaves in Fukuoka, Japan (top) and another *Magnolia* sp. flowering with leaves in Cibodas, Indonesia; photographed in October, 1990 (bottom).

end of the growth period, is induced into dormancy, and bursts in next spring. The mode of development, however, is identical.

Lemon (*Citrus limon*) and grapefruit (*Citrus* × *paradisi*) belong to subgenus *Archicitrus*; on the other hand, mandarin orange (*Citrus unshiu*) belongs to subgenus *Metacitrus*. The former species is more tropical and more primitive (ancestral) than the latter. Flowers of *Archicitrus* are grouped in inflorescences. In contrast, there are two flowering types in *Metacitrus*: leafy and leafless inflorescences (Fig. 1.13). On the peduncle of a leafless inflorescence in *Citrus unshiu*, leaf scars are observed. With a gibberellin treatment at the time of completion of flower initiation, the inflorescence developed leaves. Considering the evolutionary process from *Archicitrus* to *Metacitrus*, condensed stems with leaf scars appear as first sign of the acquisition of protection against cold winters (i.e. dormancy) by formation of winter buds.

Sealy (1958) describes the sequence of flower development in *Camellia* species (*Theaceae*). The flowers are formed in the axils of the lowest scales of the vegetative terminal and axillary buds. The scales which subtend the flowers abscise soon after the flower buds begin to develop and, since the flower buds rapidly exceed the parent vegetative bud in size, the flowers are often described as terminal or axillary. However, the flower bud (with contracted peduncle) of *Camellia japonica*, an evergreen tree of Japanese origin, actually represents the third growth cycle, the second being the vegetative

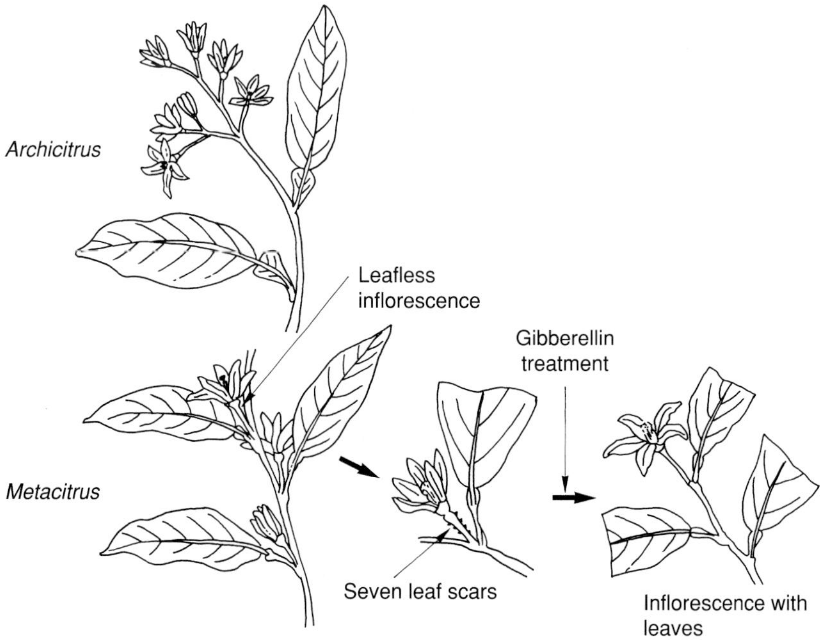

Fig. 1.13. Flowering traits in *Citrus* (from S. Uemoto).

truly axillary bud, and the first one the bearing axis (Fig. 1.14). Bud, leaf and flower initiation always occur in this sequence, but there are time lags and different growth rates of each organ. A *Camellia japonica* tree growing in Bogor, Indonesia is reported to flower almost year round. These phenomena found in *Camellia* are quite similar to those occurring in *Hippeastrum* (see previous section). The monopodial growth and seasonal flowering are really sympodial growth and non-seasonal flowering, respectively.

The majority of temperate woody plants exhibit a well-defined dormancy or resting phase during the annual growth cycle. It is usually accompanied by the development of resting buds which involve the bud scales (Wareing and Philips, 1981). *Camellia* species in the tropics and subtropics, e.g. *Camellia tsaii*, produce four or less scales in their leaf buds, whereas those in the temperate regions, e.g. *C. japonica*, produce 7–12 scales (Uemoto *et al.*, 1990). Within the same species, the number of scales in *C. japonica* var. *hozanensis* in southern-most Japan (24° N) is about five, whereas that of *C. japonica* var. *japonica* in Fukuoka, Japan (33° N) is about nine. In addition, the number of scales in one *C. japonica* var. *japonica* tree is

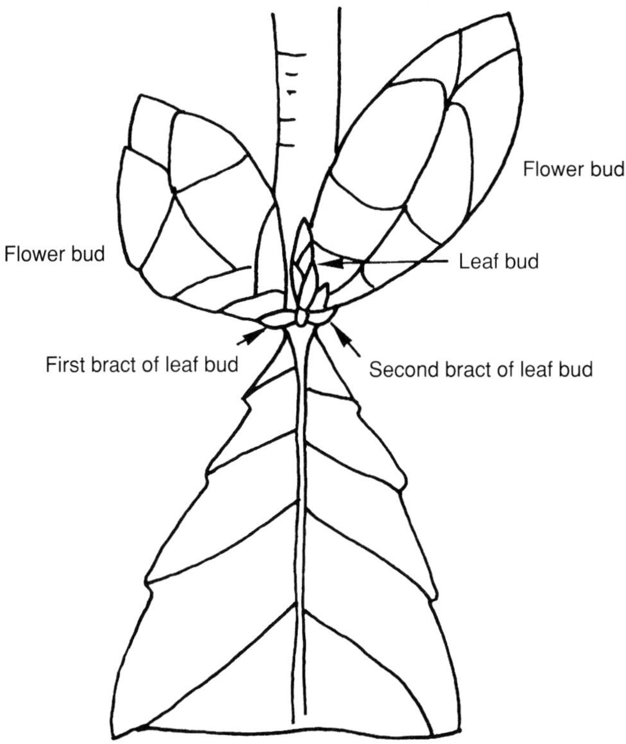

Fig. 1.14. A flower bud of *Camellia* (from S. Uemoto).

different when they are formed on spring shoots or on summer shoots, about 12 and seven, respectively (Uemoto *et al.*, 1990).

From these examples, the following conclusions can be made:

1. Under favourable environmental conditions, plants continue growing, and the growth of their buds is not synchronized because no factors force the buds to do so except for the apical dominance. In temperate regions with unfavourable conditions, climatic factors affect growth.
2. Trees can continue growing at any time, in nearly constant conditions. But, during their evolution from tropical regions, a mechanism to halt development was inserted in the developmental sequence at various stages, due to unfavourable conditions.
3. Rhizome development in herbaceous perennials and scale development in trees are similar as a result of the evolution from the tropics to temperate regions. Also, scale formation in bulbous plants, as well as in tree buds, is interpreted as the same phenomenon, i.e. the induction of dormancy.

New Hypothesis on Growth Cycle and Dormancy: Conclusion

A growth cycle consists of the process starting with the beginning of cell division in the zygote or with branching (initiation of lateral primordia) and ending with flowering/fruiting on the top end of the shoot through vegetative growth with leaf development. This 'one bud – one growth cycle theory' appears to be applicable to the diverse growth behaviours and morphologies described in various plant species. For example, in a tomato (*Lycopersicon esculentum*) plant starting from seed, after six to about 11 leaves, the initial axis terminates with an inflorescence (first growth cycle). Subsequent growth is initiated in the axil of the last initiated leaf and terminates in the second inflorescence after developing about three leaves (second growth cycle). Then the next growth cycle follows the same sequence. The sweet pepper (*Capsicum annuum*) plant develops a single stem with 8–15 leaves before the appearance of the terminal flower (first growth cycle). With the first flower bud development, the plant branches at the apex into two or more shoots (initiation of second growth cycle). Each shoot bears one or two leaves, terminates in a flower (fruit development follows) (termination of second growth cycle), and divides into two second-order branches (initiation of third growth cycle).

Figure 1.15 is a summary drawing of the growth cycles of annuals, perennials, bulbous plants and tree plants. The diversity of growth traits and morphology of various plant species displays homology that can be clearly explained and understood by the hypothesis. The hypothesis also states that plants are a colony of lateral buds regenerating asexually. It is a

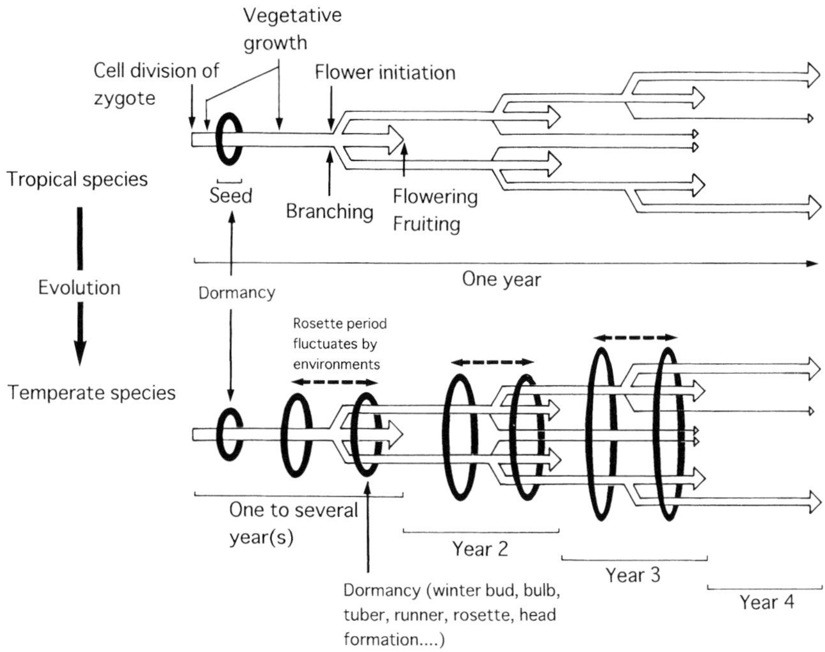

Fig. 1.15. Schematic drawing of plant growth cycle (adapted from Uemoto (1983)).

fact that seed formation was acquired later than asexual reproduction in the course of plant evolution.

Dormancy must also be explained clearly and simply and it must cover the various aspects of dormancy in almost all species of flowering plants, without inconsistency with the 'one bud – one growth cycle theory'. Dormancy is considered as the regulation imposed on the progressing growth processes at various stages (during vegetative growth, during flower formation and anthesis, etc.) with or without morphological modification as illustrated in Fig. 1.15. Induction of dormancy is, therefore, defined as 'the change of the primordia that cease growing for a while or that initiate special organs instead of producing shoots'. For example, dormancy induction of bulbs is, therefore, defined as 'the initiating process of bulb formation', i.e. the process whereby the meristem initiates scales instead of leaf and flower primordia. Thus, the swelling of bulbs (accumulation of reserves) after bulb initiation (= induction of dormancy) is the process of the bulbs that deepens their dormancy. It is said that *Hippeastrum* bulbs have no dormancy because they can sprout at any time under favourable climatic conditions. However, by the hypothesis proposed here it may be said that the dormancy of *Hippeastrum* is weak. Differences in lily and *Hippeastrum* are only a difference in depth of dormancy. As previously mentioned, and based on the proposed hypothesis, the induction of dormancy in bulbous plants and the induction of bulb formation are the same phenomenon.

Vegetative growth from seeds, flower initiation, bolting and flowering, and fruiting are the essentials for the plant to maintain the species eternally. Environmental changes (cold winter) brought about diverse growth phenomena to survive unfavourable conditions. Heading of cabbage, rosettes in radish, enlarged roots of dahlia, corms of gladiolus, bamboo and water lily rhizomes, strawberry runners, winter buds in trees, bulbs, tubers, etc., can all be conceived as an induction phenomenon of dormancy, as a form of strategy to survive, which their allied tropical species do not develop. Shape, position on plants, and the inducing factors are secondary differences. Dormancy is not only 'a temporary suspension of visible growth of any structure containing a meristem' (Lang *et al.*, 1987) as original or transformed organs, i.e. winter buds, bulbs, etc., but it also includes the 'dynamic' change of growth of the primordia to initiate and develop special dormant organs before the temporary suspension of visible growth. Without dormancy, all plants might show the same growth behaviour as tropical trees. Several growth cycles are continuing and overlapping in one plant in tropical trees. In contrast, only one growth cycle, in many positions of branches, occurs once in temperate plants (twice in some species). Dormancy lets one growth cycle occur in a year and limits the continuity of the growth cycle genetically and environmentally. It forces the next growth cycle to wait for the following growing season. Therefore, in the sense proposed above, the term dormancy only includes Lang *et al.*'s endodormancy (Lang *et al.*, 1987). Ecodormancy and paradormancy are only the forced cessation caused by environments and physiological factors outside the affected structure, respectively. As Junttila (1988) commented, '. . . occurred under conditions normally favourable for growth' should follow their definition when the dormancy is induced and appears as a survival mechanism.

Seeds achieve two major purposes in the life of plants (Osborne, 1981). One is the dispersal of new individuals of the same species with the potential for colonization of new habitats. The other is to maintain the survival of the species through adverse environmental conditions that are not conducive to active growth. Based on the second purpose and on the concept of dormancy proposed here, it is conceivable that the seed is one of the forms of dormancy as is the bulb. Seed development is, therefore, a process that is built into the growth cycle as a cessation period (Fig. 1.15). The seed of angiosperms is a specific organ formed to protect the developing embryo at the beginning stage of the growth cycle, after fertilization.

Acknowledgements

The concept of the growth cycle of plants and dormancy was first initiated by Professor Emeritus Dr Shunpei Uemoto, Kyushu University. The author is grateful to Professor Uemoto for his influence over many years of study, and for providing some of the photographs and figures in this chapter. The photographs in Fig. 1.9 are provided courtesy of Dr Yukio Ozaki of the same laboratory as the author. The author is also grateful to Professor August De Hertogh, North Carolina State University, Raleigh, USA, and Dr Marcel Le Nard, Station d'Amélioration de la Pomme de Terre et des Plantes à Bulbes, Institut National de la Recherche Agronomique, Ploudaniel, France, for reviewing the manuscript.

References

Aoba, T. (1971) Studies on the bulb formation in garlic plants II. On the effect of low temperature. *The Journal of the Yamagata Agriculture and Forestry Society* 28, 35–40.

Aoba, T. (1972) Effect of temperature on bulb- and tuber-formation in bulbous and tuberous plants II. On bulb formation in bulbous oxalis. *Journal of the Japanese Society for Horticultural Science* 41, 393–397.

Aoba, T. (1974) Effect of temperature on bulb- and tuber-formation in bulbous and tuberous crops VI. On the bulb formation in bulbous iris. *Journal of the Japanese Society for Horticultural Science* 43, 273–280.

Bach, A. (1992) Micropropagation of hyacinths (*Hyacinthus orientalis* L.). In: Bajaj, Y.P.S. (ed.) *Biotechnology in Agriculture and Forestry*, vol. 20, *High-Tech and Micropropagation IV.* Springer-Verlag, Heidelberg, pp. 144–159.

Bradley, D., Carpenter, R., Copsey, L., Vincent, C., Rothstein, S. and Coen, E. (1996) Control of inflorescence architecture in *Antirrhinum. Nature* 379, 791–797.

Delvallée, I., Paffen, A. and de Klerk, G.J. (1990) The development of dormancy in bulblets of *Lilium speciosum* generated *in vitro.* II. The effect of temperature. *Physiologia Plantarum* 80, 431–436.

Gerrits, M.M. and de Klerk, G.J. (1992) Dry-matter partitioning between bulbs and leaves in plantlets of *Lilium speciosum* regenerated *in vitro. Acta Botanica Neerlandica* 41, 461–468.

Junttila, O. (1988) To be or not to be dormant: some comments on the new dormancy nomenclature. *HortScience* 23, 805–806.

Kim, S.E. and Okubo, H. (1995) Control of growth habit in determinate lablab bean (*Lablab purpureus*) by temperature and photoperiod. *Scientia Horticulturae* 61, 147–155.

Kim, S.E. and Okubo, H. (1996) Hormonal control of growth habit in determinate lablab bean (*Lablab purpureus*). *Scientia Horticulturae* 65, 95–104.

Koriba, K. (1958) On the periodicity of tree-growth in the tropics, with difference to the mode of branching, the leaf-fall, and the formation of the resting bud. *Gardens' Bulletin Singapore* 17, 11–81.

Kramer, P.J. and Kozlowski, T.T. (1979) *Physiology of Woody Plants.* Academic Press, New York.

Lang, G.A., Early, J.D., Martin, G.C. and Darnell, R.L. (1987) Endo-, para-, and eco-dormancy: physiological terminology and classification for dormancy research. *HortScience* 22, 371–377.

Le Nard, M. (1983) Physiology and storage of bulbs: concepts and nature of dormancy in bulbs. In: Lieberman, M. (ed.) *Post-harvest Physiology and Crop Preservation. NATO Advanced Study Institute Series*; Series A *Life Science* 46. Plenum Press, New York, pp. 191–230.

Le Nard, M. and Cohat, J. (1968) Influence des températures de conservation des bulbes sur l'élongation, la floraison et la bulbification de la tulipe (*Tulipa gesneriana* L.). *Annales de l'Amélioration des Plantes* 18, 181–215.

Le Nard, M. and De Hertogh, A. (1993) Bulb growth and development and flowering. In: De Hertogh, A. and Le Nard, M. (eds) *The Physiology of Flower Bulbs.* Elsevier Science, Amsterdam, pp. 29–43.

Murashige, T. and Skoog, F. (1962) A revised medium for rapid growth and bioassays with tobacco tissue cultures. *Physiologia Plantarum* 15, 473–497.

Okubo, H. (1993) *Hippeastrum* (Amaryllis). In: De Hertogh, A. and Le Nard, M. (eds) *The Physiology of Flower Bulbs.* Elsevier Science, Amsterdam, pp. 321–334.

Okubo, H. and Uemoto, S. (1981) Changes in the endogenous growth regulators in bulbous iris in bulb-forming and nonbulb-forming aspects. *Plant and Cell Physiology* 22, 297–301.

Osborne, D. (1981) Dormancy as a survival stratagem. *Annals of Applied Biology* 98, 525–531.

Sealy, J.R. (1958) *A Revision of the Genus* Camellia. The Royal Horticultural Society, London.

Summerfield, R.J. and Wien, H.C. (1980) Effects of photoperiod and air temperature on growth and yield of economic legumes. In: Summerfield, R.J. and Bunting, A.H. (eds) *Advances in Legume Science.* HMSO, London, pp. 7–36.

Suzuki, F. (1990) Studies on the characteristics on dwarf, bush-type winged bean. MS thesis, Iwate University, Morioka, Japan.

Tsukamoto, Y. (1974) Changes in endogenous regulators and dormancy in bulbous plants. In: Antoszewski, L., Harrison, L. and Nowosielski, J. (eds) *Proceedings of the 19th International Horticultural Congress, Warsaw,* 19th edn. pp. 293–306.

Tsukamoto, Y. and Ando, T. (1973) The changes of amount of inhibitors inducing dormancy in the Dutch iris bulb. *Proceedings of the Japanese Academy* 49, 627–632.

Uemoto, S. (1983) Hana. In: Takahashi, N., Hirose, K., Sato, M., Saito, T. and Uemoto, S. (eds) *Shokubutsu Chouseibusshitsu no Engeiteki Riyou.* Seibundo-Shinkosha, Tokyo, Japan, pp. 238–287.

Uemoto, S., Okubo, H. and Nakakido, F. (1990) Relationships between the intensity of dormancy and numbers of bracts formed in leaf buds in *Camellia* plants. *Bulletin of Seibu Maizuru Botanical Institute* 5, 56–75.

Wareing, P.F. and Phillips, I.D.J. (1981) *Growth and Differentiation in Plants,* 3rd edn. Pergamon Press, Oxford.

Webb, D.P., van Staden, J. and Wareing, P.F. (1973) Seed dormancy in *Acer.* Changes in endogenous cytokinins, gibberellins and germination inhibitors during the breaking of dormancy in *Acer saccharum* Marsh. *Journal of Experimental Botany* 24, 105–116.

Yarnell, S.H. (1965) Cytogenetics of the vegetable crops. IV. Legumes (continued). *Botanical Review* 31, 247–330.

Bud Dormancy: a Suggestion for the Control Mechanism and its Evolution

2

Amnon Erez

ARO, The Volcani Center, Institute of Horticulture, PO Box 6, Bet Dagan 50-250, Israel

Introduction

The dormancy mechanism has puzzled researchers for decades. No convincing explanation is commonly accepted for the effects of temperature on dormancy release (Faust *et al.*, 1997). Various attempts have been made to explain how chilling overcomes dormancy, most commonly by relating it to growth inhibitors and promoters (Saure, 1985). This theory has not been proved by research work, in spite of the fact that growth retardants do induce dormancy and certain growth stimulators such as cytokinins and gibberellins do induce dormancy breaking (Crabbé, 1994). Other approaches dealt with the exchange of sink power between the bud initial and the surrounding tissues (Champagnat, 1989), and their changing growth potential (Petel *et al.*, 1992). Here, too, it was difficult to elucidate a mechanism of control. Another approach has concentrated on the change in the water status of the buds: Faust *et al.* (1991) have shown that during dormancy, water in the buds is found in a more bound state while towards release from dormancy, it becomes freer. They have suggested that the change in state of water controls induction of dormancy and later its release. More recent work failed to verify that hypothesis and indicated a closer connection of the water status of the buds with their cold resistance than with their dormancy (Erez *et al.*, 1997a; Parmentier *et al.*, 1998).

Membrane Phospholipids

One of the very few dynamic changes in dormant buds that are demonstrated to be correlated with dormancy is the increase in specific fatty acid contents in the phospholipids in cell membranes. This change, which was demonstrated in vegetative apple buds by Wang and Faust (1990) on a time basis, was further shown by Erez *et al.* (1997b) to be correlated with efficient chilling that breaks dormancy in peach vegetative and floral buds. The major change found was the increase in linoleic acid with chilling; this change seems to be required for enabling the buds to resume growth. Linoleic acid is produced by oleate desaturase (Lyons, 1973), a membrane-bound enzyme that is presumably activated by low temperatures. In order to express the growth resumption potential, a high temperature is needed to facilitate the further desaturation that occurs at high temperatures, namely desaturation of linoleate to linolenate in the phospholipid fraction by linoleate desaturase, also a membrane-bound enzyme. Wang and Faust (1990) found that during dormancy the linoleate level rose from less than 20% of total fatty acids in the phospholipids to 40% and above, while prior to initial bud break the linolenate in the phospholipids rose from around 40 to 55% or higher. These changes are by no means minor, and they represent a major change in the activity of the cell.

Furthermore, apart from this change, there is a major increase in the total level of phospholipids in the membranes between late autumn and the end of winter (Wang and Faust, 1990; Erez *et al.*, 1997b). The changes that occur in the cell membrane accentuate the fact that dormancy development is a local phenomenon, which may be overcome in one bud but not in another. There is little communication between buds during this period.

Characteristics of Dormancy Completion

Although we know rather little about the events that occur in the bud during dormancy, we know much more about the specific effects of temperature on dormancy development.

The basic elements are:

1. A slow accumulation of the chilling effect;
2. An optimum curve for the chilling effect at 6–8°C (Fig. 2.1);
3. A negation of the effect of chilling by high temperature in short cycles (Fig 2.2);
4. A fixation of the chilling effect (Fig. 2.3).

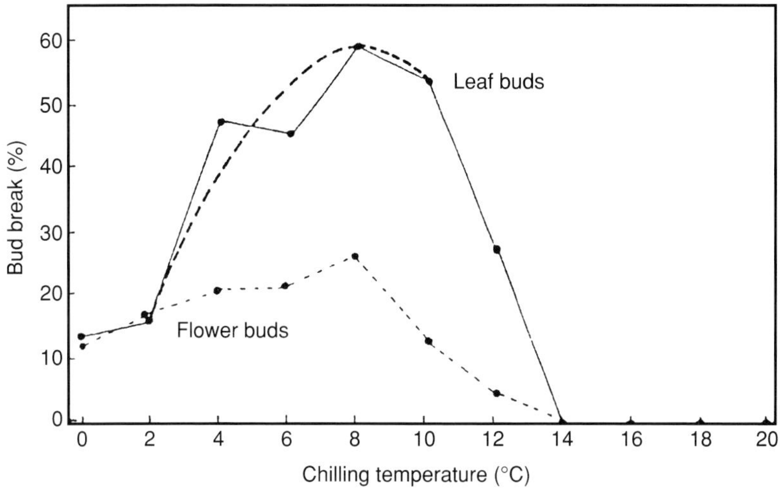

Fig. 2.1. Bud break of 'Redhaven' peach plants following exposure to 1200 h at various continuous temperatures (reprinted from Erez and Couvillon (1987) with permission).

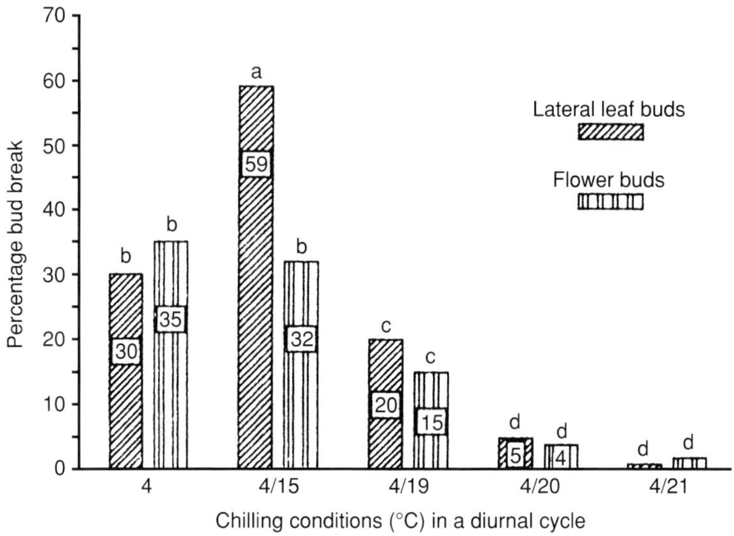

Fig. 2.2. The influence of diurnal temperature cycles on bud break in 'Redhaven' peach. Exposure to 4°C in the cyclic treatments for 16 h a day (reprinted from Couvillon and Erez (1985) with permission).

Basic Requirements of a Dormancy Control Mechanism

The following are, therefore, the basic characteristics of a dormancy control mechanism.

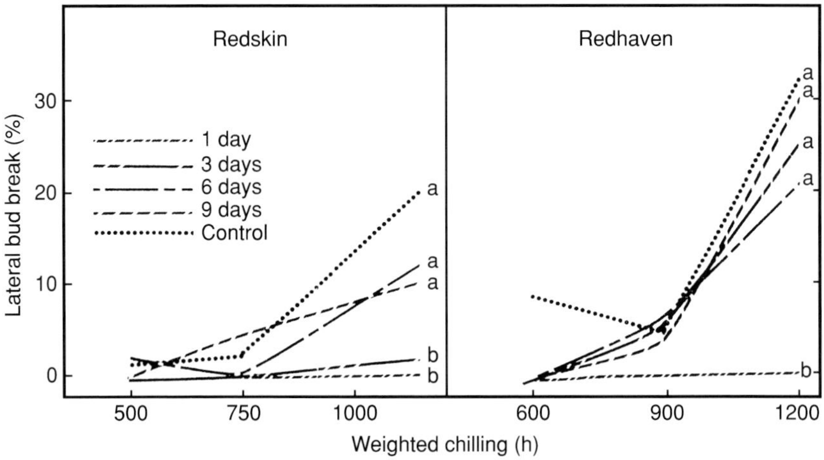

Fig. 2.3. The effect of temperature cycle duration on bud break of lateral leaf buds in two peach cultivars following exposure to increasing chilling. In all cyclic treatments two-thirds of cycle was maintained at low temperature (4–6°C) and one-third of the cycle at high temperature (24°C). Bud break after exposure to continuous 4°C (Control) was added for comparison (reprinted from Erez *et al.* (1979) with permission).

1. The changes occurring with time should be slow in order not to overwhelm the system after a relatively short exposure.
2. The mechanism must respond to chilling level and duration in a quantitative manner.
3. Its activity must match the optimum curve of temperature effect.
4. It must correctly include the effect of high temperatures in negating chilling and the fixation of the chilling effect.
5. The mechanism of control should reside in the bud itself, as connection between organs is disrupted during dormancy.
6. The changes occurring due to chilling must dispose the buds to grow when high temperatures return.

My suggestion is that changes in lipids in bud cell membranes conform with all these requirements. The basic mechanism is the activation of the two membrane-bound enzymes: oleate desaturase and linoleate desaturase.

What do we know about these two enzymes? We know they have to move within the membrane in order to operate; and we know that their mobility and, consequently, their activity are affected by temperature (Thompson, 1979). Under high temperature, oleate desaturase will not be activated, therefore, although its movement through the fluid membrane is easy, it is kept inactive. It is presumably activated by the increased viscosity of the membrane at lower temperatures (Thompson, 1979; Ferrante and Kates, 1986) but its activity at lower temperatures is probably low, because of the increased membranal viscosity and slow metabolic activity. The unique state of these enzymes is the requirement for their movement within the membrane to operate on a relatively stationary substrate. In order to reach a high level of the product, long periods of exposure to low temperatures are required. Thus, they fulfil the first requirement of slow accumulation of product.

To this biochemical effect we may add another physico-chemical element: an obligatory requirement for desaturase activity for oxygen. If we consider oxygen solubility in membrane lipids as a limiting factor for oleate desaturase activity, we may expect higher availability of oxygen at low temperatures according to Henry's law of gas solubility in liquids. As a result of that, a higher activity is expected at lower temperatures, where more oxygen is available. Furthermore, respiration may compete with desaturases for the available oxygen (Harris and James, 1969), further reducing oxygen availability under warmer conditions. Harris and James (1969) have pointed out that in non-photosynthesizing organs, which include all dormant organs, the activity of desaturases is strictly dependent on externally supplied oxygen, while in photosynthesizing organs oxygen is available as a product of the reaction. Hence the positive effect of low temperature in breaking dormancy. The unique dependence on oxygen, and the activation of oleate desaturase only at low temperatures lead to fulfilment of the second requirement of response to chilling.

A combined effect of a chemical and a physical factor can explain the optimum type curve of the effect of temperature on dormancy breaking (Fig. 2.1). The increasing activity of the oleate desaturase as the temperature falls from 14°C to 6°C presumably results from activation of the enzyme due to the increased viscosity of the membrane with reduced temperatures (Thompson, 1979; Ferrante and Kates, 1986) and from increasing solubility in the membrane of oxygen which is indispensable for the desaturase activity. The left side of the curve with temperatures lower than 6°C is affected by the reduced activity of the desaturase at excessively low temperatures. This fulfils the third requirement of conforming with the optimum curve of temperature effect.

The antagonistic effect of high temperature on the effects of previous low temperature (Fig. 2.2) is seen by the interaction of the two desaturases. No information is available in the literature regarding the difference in temperature response of these two enzymes, but from observation of the accumulation of the products, linoleate and linolenate, it can be deduced that linoleate desaturase activity must have a higher optimum temperature than that of oleate desaturase, since the linoleate content in dormant buds drops at high temperatures while that of linolenate increases (Fig. 2.4; Wang and Faust, 1990) (Fig. 2.5; Erez *et al.*, 1997b). While low temperature enhances the activity of oleate desaturase and hence linoleate build-up, higher temperatures will enhance the activation of linoleate desaturase to produce linolenate. If the total level of linoleate is low, its conversion to linolenate will reduce the level of the precursor without allowing the

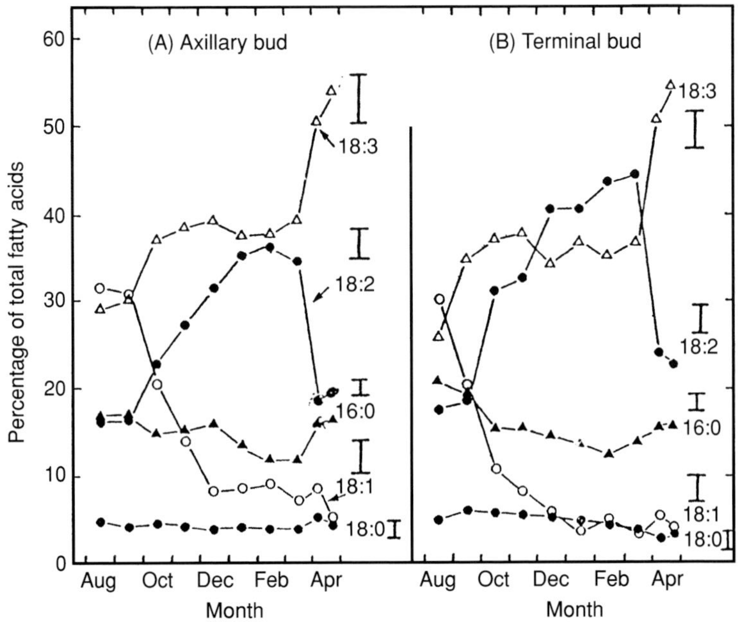

Fig. 2.4. Fatty acid composition (% of total) of phosphatidylcholine in 'Delicious' apple buds from August to April (reprinted from Wang and Faust (1990) with permission).

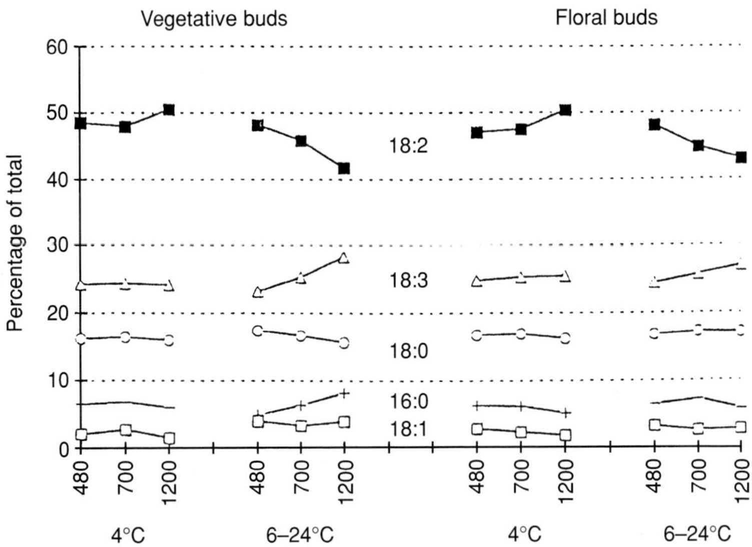

Fig. 2.5. Relative levels of fatty acids in phospholipids in vegetative and floral buds of 'Winblo' peach after exposure in darkness to continuous 4°C (efficient chilling) or a daily cycle of 6–24°C for 16:8 h (non-efficient chilling) (reprinted from Erez *et al.*, (1997b) with permission).

build-up of linolenate to a high enough level to allow growth resumption (Figs 2.5 and 2.6).

The reaction involved in chilling fixation is still unknown. We know that the chilling effect of a period of about 30 h at 6°C is fixed and cannot be negated by high temperatures (Fig. 2.3). It is tempting to hypothesize that there must be a minimal level of linoleate accumulation in the membrane that cannot be reversed by high temperature. High temperature will lead to a reduction rather than an increase in linoleate level. This will explain the susceptibility to high temperatures which occur prior to fixation. This satisfies the fourth requirement. The membrane-bound changes fulfil the fifth requirement that the chilling effect must be localized as the response is directly related to the cell temperature.

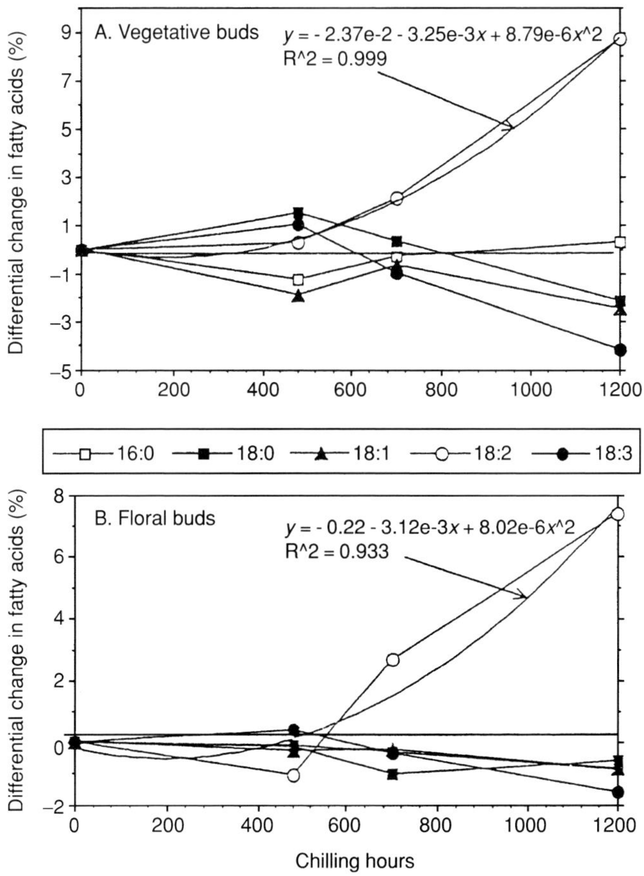

Fig. 2.6. Differential change of the fatty acid components in phospholipids in vegetative and floral buds of the 'Winblo' peach with exposure to chilling. Differential values calculated by subtracting values for 6–24°C treatment from those for 4°C treatment (reprinted with permission from Erez *et al.* (1997b)).

From the data presented it seems, indirectly, that these two desaturases have different temperature optima. Low temperatures favour the activity of oleate desaturase while higher temperatures favour that of linoleate desaturase. Together, these characteristics lead to fulfilment of the sixth requirement since the linoleate desaturase is activated at high temperatures but the level of linolenate, presumably needed for growth resumption, depends on the level of linoleate which accumulates only through long chilling exposure.

Stress Resistance

When dealing with changes in membranes, especially fatty acid changes, an association regarding cold stress resistance immediately comes to mind.

Tolerance to stress caused by non-freezing chilling temperatures is obtained by holding plants at near-chilling temperatures prior to exposure to cold (Wang, 1990). This acclimation leads to an increase in desaturation in membrane lipids (Wang, 1990). Lyons (1973) proposed early on that enrichment in polyunsaturated fatty acids (PUFA) may lower the melting point of membranes and hence, prevent chilling-induced membrane-lipid solidification (fluid-to-gel phase change) and an associated membrane dysfunction. This concept is supported by correlative studies showing that membrane-lipid enrichment in PUFA accompanies chilling acclimation in plants (Lynch and Thompson, 1984a,b; Yoshida, 1984; Lynch and Steponkus, 1987). Moreover, increased lipid desaturation, caused by genetic manipulation, was found to enhance chilling tolerance in chilling-sensitive cyanobacteria (Wada *et al.*, 1990) and plants (Ishizaki-Nishizawa *et al.*, 1996). Conversely, use of molecular methods to arrest lipid desaturation induced chilling susceptibility in chilling-tolerant plants (Murata *et al.*, 1992), further emphasizing the importance of lipid desaturation in cold acclimation.

Lipid desaturation is catalysed by a family of desaturases that require oxygen (Ferrante and Kates, 1986) and, in most cases, reduced pyridine nucleotides (Harwood, 1988). Thompson (1979) suggested that cold conditions may influence the activity of desaturases by changing the matrix properties of the enzyme microenvironment. In this view, temperature-dependent cold adaptation may arise from the activation of pre-existing enzymes (Skriver and Thompson, 1979).

Comparison between Stress Resistance and Dormancy Release

On the basis of the above it is interesting to compare the two phenomena. One has to consider the mechanism of stress resistance as a very basic mechanism that had to develop in order to allow the plant to survive a temporarily hostile environment. Only when plants managed to develop a mechanism of survival under stress could they proliferate and maintain the continuation of their species. The basic mechanism of stress resistance in plants involves desaturation of lipids, change of viscosity, hydration and, as a result, maintenance of membrane functionality in a hostile environment.

Once this mechanism was developed, I suggest that it has also been adopted for dormancy development. This basic system was adopted by plants for maintaining a state of reduced metabolic activity in the dormant bud. The plant adapted the stress-related development to the dormancy mechanism by using it in non-photosynthetic organs as a time–temperature measuring mechanism to obtain the information on when buds should respond to high temperature in order to resume growth. It seems that growth resumption is dependent on the changes that occur in membranes of the bud cells. As long as levels of linoleate are kept low, no bud break will occur, but under high temperatures, linoleate is desaturated to linolenate and its final level dictates bud break.

The following are supporting data for the theory. Sparing active oxygen enhances bud break: two potent dormancy breaking agents, thiourea and hydrogen cyanamide are strong catalase inhibitors (Brennan *et al.*, 1978; Shulman *et al.*, 1986), thus sparing active oxygen for other critical reactions. Exposure of dormant seeds to high oxygen can substitute for lack of chilling (Brennan *et al.*, 1978; Frenkel and Erez, unpublished data).

References

Brennan, T., Willemsen, R., Rudd, T. and Frenkel, C. (1978) Interaction of oxygen and ethylene in the release of ragweed seeds from dormancy. *Botanical Gazette* 139, 46–49.

Champagnat, P. (1989) Rest and activity in buds of trees. *Annales des Sciences Forestières* 46, 9–26.

Couvillon, G.A. and Erez, A. (1985) Effect of level and duration of high temperatures on rest in the peach. *Journal of the American Society for Horticultural Science* 110, 579–581.

Crabbé, J. (1994) Dormancy. *Encyclopedia Agricultural Science* 1, 597–611.

Erez, A. and Couvillon, G.A. (1987) Characterization of the influence of moderate temperatures on rest completion in peach. *Journal of the American Society for Horticultural Science* 112, 677–680.

Erez, A., Couvillon, G.A. and Hendershott, C.H. (1979) The effect of cycle length on chilling negation by high temperatures in dormant peach leaf buds. *Journal of the American Society for Horticultural Science* 104, 573–576.

Erez, A., Faust, M. and Line, M.J. (1997a) Changes in water status in peach buds on induction, development and release from dormancy. *Scientia Horticulturae* 73, 111–123.

Erez, A., Wang, S.Y. and Faust, M. (1997b) Lipids in peach buds during dormancy, a possible involvement in dormancy control. *Advances in Horticultural Science* 11, 128–132.

Faust, M., Liu D., Merle, M. and Stutte, G.W. (1991) Bound versus free water in dormant apple buds – a theory for endodormancy. *HortScience* 26, 887–890.

Faust, M., Erez, A., Rowland, L.J., Wang, S.Y. and Norman, H.A. (1997) Bud dormancy in perennial fruit trees: physiological basis for dormancy induction, maintenance and release. *HortScience* 32, 623–629.

Ferrante, G. and Kates, M. (1986) Characteristics of the oleoyl and linoleoyl-COA desaturase and hydrolase systems in cell fractions from soybean cell suspension cultures. *Biochimica et Biophysica Acta* 876, 429–437.

Harris, P. and James, A.T. (1969) The effect of low temperatures on fatty acid biosynthesis in plants. *Biochemistry Journal* 112, 325–330.

Harwood, J.L. (1988) Fatty acid metabolism. *Annual Review of Plant Physiology and Molecular Biology* 39, 101–138.

Ishizaki-Nishizawa, O., Fujii, T., Azuma, M., Sekiguchi, K., Murata, N., Ohtani, T. and Toguri, T. (1996) Low-temperature resistance of higher-plants is significantly enhanced by a nonspecific cyanobacterial desaturase. *Nature Biotechnology* 14, 1003–1006.

Lynch, D.V. and Steponkus, P.L. (1987) Plasma membrane lipid alteration associated with cold acclimation of winter rye seedlings (*Secale creale* L. cv Puma). *Plant Physiology* 83, 761–767.

Lynch, D.V. and Thompson, G.A., Jr (1984a) Microsomal phospholipid molecular species alterations during low temperature acclimation in Dunaliella. *Plant Physiology* 74, 193–197.

Lynch, D.V. and Thompson, G.A., Jr (1984b) Chloroplast phospholipid molecular species alterations during low temperature acclimation in Dunaliella. *Plant Physiology* 74, 198–203.

Lyons, J.M. (1973) Chilling injury in plants. *Annual Review of Plant Physiology* 24, 445–466.

Murata, N., Ishizaki-Nishizawa, O., Higashi, S., Hayashi, H., Tasaka, Y. and Nishida, I. (1992) Genetically engineered alteration in the chilling sensitivity of plants. *Nature* 356, 710–713.

Parmentier, C.M., Rowland, L.J. and Line, M.J. (1998) Water status in relation to maintenance and release from dormancy in blueberry flower buds. *Journal of the American Society of Horticultural Science* 123, 762–769.

Petel, G., Lafleuriel, J., Dauphin, G. and Gendraud, M. (1992) Cytoplasmic pH and plasmalemma ATPase activity of paranchyma cells during the release of dormancy of Jerusalem Artichoke tubers. *Plant Physiology and Biochemistry* 30, 379–382.

Saure, M. (1985) Dormancy release in deciduous fruit trees. *Horticultural Review* 7, 239–300.

Shulman, Y., Nir, G. and Lavee, S. (1986) Oxidative processes in bud dormancy and the use of hydrogen cyanamide in breaking dormancy. *Acta Horticulturae* 179, 141–145.

Skriver, L. and Thompson, G.A., Jr (1979) Temperature-induced changes in fatty acid unsaturation of Tetrahymena membranes do not require induced fatty acid desaturase synthesis. *Biochimica et Biophysica Acta* 572, 376–381.

Thompson, G.A., Jr (1979) Molecular control of membrane fluidity. In: Lyons, J.M., Grahas, D. and Raison, J.K. (eds) *Low Temperature Stress in Crop Plants.* Academic Press, New York, pp. 347–363.

Wada, H., Gombos, T. and Murata, N. (1990) Enhancement of chilling tolerance of cyanobacterium by genetic manipulation of fatty acids desaturation. *Nature* 347, 200–203.

Wang, C.Y. (1990) Alleviation of chilling injury of horticultural crops. In: Wang, C.Y. (ed.) *Chilling Injury of Horticultural Crops.* CRC Press, Boca Raton, Florida, pp. 281–302.

Wang, S.W. and Faust, M. (1990) Changes in membrane lipids in apple buds during dormancy and bud break. *Journal of the American Society for Horticultural Science* 115, 803–808.

Yoshida, S. (1984) Chemical and biophysical changes in the plasma membrane during cold acclimation of Mulberry bark cells (*Morus bombycis* Koidz. cv Goroji). *Plant Physiology* 76, 257–265.

The Distinct Controlling of Dormancy Release and Germination Commencement in Seeds

3

N.V. Obroucheva and O.V. Antipova

Russian Academy of Sciences, Institute of Plant Physiology, Botanical Street 35, Moscow 127276, Russian Federation

Seed transition from deep dormancy to visible germination is usually considered to be a single process tested by radicle protrusion through the seed coat. However, it may be subdivided into the periods of dormancy maintenance, dormancy release, preparation for cell elongation and growth initiation *per se*. To discriminate between these events, their sequence may be considered in the opposite order, beginning with growth initiation and ending with dormancy.

Growth commencement can be more closely examined in the quiescent or non-dormant seeds. An obligatory and primary process for growth commencement in the seeds is cell elongation in the axial organs of the embryo (Obroucheva, 1999), whereas cell division in the radicle meristem begins either simultaneously with the elongation in some seeds or some hours or even many hours later in others. Cell elongation provides rapid axis extension necessary not only for pushing the root tip through the seed coat, but for coming into contact with soil water as well. Thus, visible germination (radicle emergence) results from cell elongation beginning, and the processes preparing for growth initiation are those preparing for the commencement of elongation in the axial organs of embryos.

The scheme shown in Fig. 3.1 describes the events occurring in the axial organs of quiescent seeds; the ordinate being water content (% fresh wt) and the abscissa being time necessary to achieve the respective water contents (Obroucheva, 1997, 1999; Obroucheva and Antipova, 1997). The first period is metabolic activation; it starts at 20–23% water content, includes initial activation of respiration, commencement of amino acid transformations and operation of the Krebs cycle. These processes do not

(eds J.-D. Viémont and J. Crabbé)

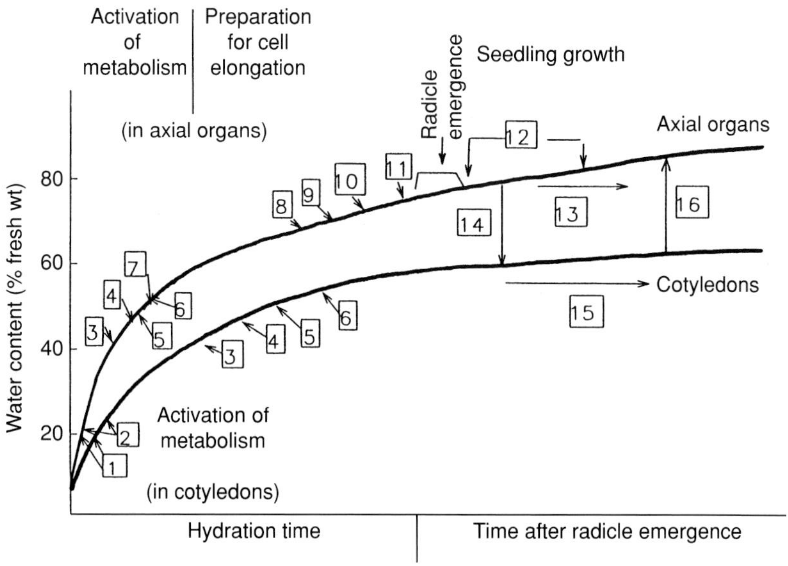

Fig. 3.1. Schematic representation of seed germination. 1. Primary activation of respiration; 2. Onset of amino acid metabolism; 3. Onset of mRNA synthesis; 4. Final activation of respiration; 5. Onset of starch degradation; 6. Onset of protein degradation; 7. Activation of protein synthesis; 8. Accumulation of osmotic solutes; 9. Enlargement of vacuole; 10. Activation of H^+-ATPase in plasmalemma; 11. Elongation start in the axial organs; 12. Beginning of mitotic activity; 13. Onset of *de novo* biosynthesis of hormones; 14. Induction of hydrolases by the axial organs; 15. Degradation of major reserves; 16. Import of degradation products into the axial organs.

start at hydration levels below 19–20%, but as soon as water content increases in imbibing axes up to these threshold values, they become active. This hydration level corresponds to the appearance of free water covering the protein surface as a monolayer and permitting the resumption of enzyme activities due to conformational mobility of enzyme molecules and delivery of substrates to their active centres (Vertucci, 1989).

At higher water content, that is within the range 42–52%, further metabolic activation commences: biogenesis of mitochondria is completed, syntheses of proteins and new mRNAs are resumed and degradation of starch and reserve proteins starts. These processes remain uninitiated for some time if water content remains below their threshold levels. Their activation needs not only sufficient hydration of the enzyme molecules, but much more complicated spatial structural reorganization as well. For example, triggering of a whole gene regulatory ensemble including chromatin conformational changes and RNA polymerase assembly needs additional water, to say nothing of the twofold higher hydration level required for nucleic acid molecules as compared with proteins. One more example is the rather high hydration level of densely packed reserve substances necessary for their solubilization prior to hydrolysis.

When the first pregerminative period is over, all major metabolic systems are brought into operation at a water content below 55–60%. This level of hydration can be achieved 'automatically' if water is available because the seed behaves as a porous body ramified by micro-capillaries and abundant in interfaces formed by diverse hydrophilic biopolymers. Water is absorbed mainly by matric forces up to this level, as follows from water content reached in the axes of heat-killed seeds.

It is necessary to mention that the data underlying this scheme were obtained by a number of authors with diverse seeds (soybean, pea, broad bean, french bean, maize, wheat, rape, lettuce and charlock) which were compared in terms of their threshold hydration levels (Obroucheva, 1997, 1999). However, in order to germinate, the seeds have to increase their moisture content to a higher level during the second pregerminative period marked by the processes preparing axial cells for germination (Antipova, 1997). They include, first of all, accumulation of solutes, mainly monosaccharides and potassium ions, i.e. additional osmotically active substances, providing the entry of new portions of water. The additional absorbed water not only fully hydrates the cytoplasm but accumulates in the enlarging vacuoles restored from the protein bodies after partial digestion of reserve proteins. The second important preparatory event is activation of proton-ATPase in plasmalemma resulting in acidification of cell walls, their loosening and increased extensibility, i.e. the events preparing for the commencement of acid-induced growth. Taken together, these processes lead to increased water inflow and cell wall extension under turgor pressure and culminate in the beginning of cell elongation. When the tissue hydration was stopped by polyethyleneglycol (PEG-6000) at the levels permitting solute accumulation and cell wall acidification, no radicle emergence occurred.

The above findings were obtained with broad bean (*Vicia faba minor* L.) seeds (Fig. 3.2), the germination of which proceeds only by elongation up to the length of 1 cm; cell division starts 18 h later than elongation (Antipova, 1997). The processes preparing cells for elongation in the broad bean were compared with those in wheat and horse chestnut seeds differing in dormancy state and cell growth events (Obroucheva and Antipova, 1999). Thus, we compared the quiescent seeds (broad bean), the seeds with dormancy released by dry storage (wheat) and the seeds possessing a deep dormancy (horse chestnut). Osmotic solutes (Table 3.1) accumulated in the axes at a hydration level exceeding that attained by matric forces. The seeds differed only in monosaccharide composition and the contribution ratio of sugars and potassium to osmotic pressure. Their accumulation resulted in a similar course of changes in the osmotic potential of cell sap (Fig. 3.3). The similarity in acidification commencement is also clearly seen

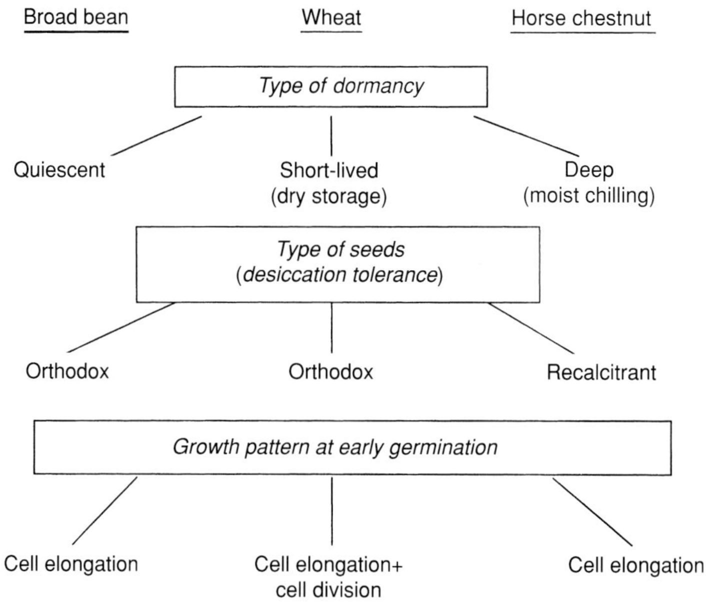

Fig. 3.2. Characteristics of plant material.

(Table 3.2); proton extrusion starts at clearly identified and very similar hydration levels.

Therefore, the accumulation of endogenous osmotica and acidification of cell walls are the prerequisites of elongation beginning in all seeds tested, independent of whether they belong to quiescent, dormant, orthodox or recalcitrant seeds. Their common pattern is the stepwise triggering of metabolic systems by inflowing water, overcoming corresponding hydration thresholds. Thus, hydration provides a controlling tool in a cell preparing for germination.

Now our attention must be turned to the transition from dormancy release to preparation for growth initiation. Among dormant seeds, horse chestnut (*Aesculus hippocastanum* L.) seeds were chosen because they germinate only by cell elongation up to the length of 3 cm. The freshly harvested seeds were subjected to moist chilling at 4°C, i.e. were kept under conditions preventing seed desiccation to avoid a fatal situation for the recalcitrant seeds. The horse chestnut seeds exhibit embryo dormancy in combination with coat-imposed dormancy. When tested for germinability in water at 27°C, the seeds exhibited the 4-week-long period of deep dormancy (Fig. 3.4) when only 30% of seeds slowly germinated for a month, and the following period of dormancy release lasted for about 10 weeks and was characterized by faster germination of almost all seeds. During the third period when the seeds were free of dormancy, they germinated uniformly at 27°C for 2–3 days (Obroucheva and Antipova, 1999). Only during the period of dormancy release were seeds sensitive to cytokinins (Fig. 3.4). Benzylaminopurine accelerated the germination twofold and other cytokinins, isopentenyladenine and zeatin-riboside, exerted a similar effect (Fig. 3.5). Other phytohormones (10^{-5} M indole-3-acetic acid and

Table 3.1. Amount of osmotic solutes (μg axis^{-1}) in cell sap from the axial organs of imbibing seeds.

Water content (% fresh wt)	Sucrose	Glucose	Fructose	K^+
Broad bean				
60	147	0	102	39
65	121	42	76	38
68	132	150	18	43
72[a]	162	226	21	59
80	168	383	12	99
Wheat				
53.1	19.2	0.3	5.4	3.5
60.2	—	0.2	15.2	3.3
63.0	18.0	0.1	16.1	2.6
65.2	19.8	0.1	16.3	3.4
73.4[a]	18.6	0.2	23.2	6.3
83.0	2.7	0.2	29.2	5.7
Horse chestnut				
64.0	3300	300	230	620
64.7	1600	390	1030	570
68.3	1200	370	1130	740
74.3[a]	300	1280	1640	1300
86.5	400	2210	2380	2040

[a]Radicle emergence.

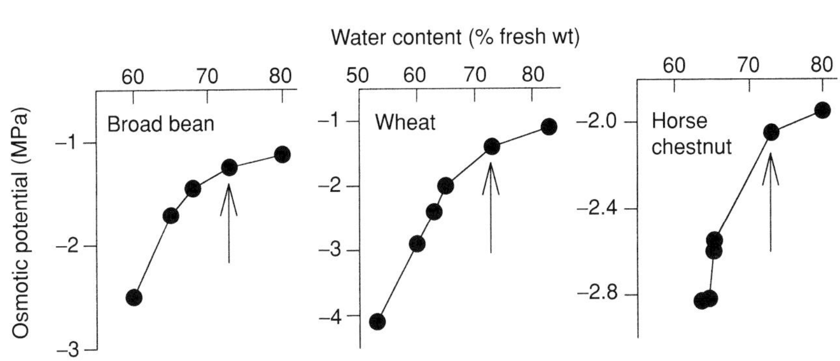

Fig. 3.3. Osmotic potential in the cell sap from seed axes at early germination. Arrows indicate visible germination.

Table 3.2. Acidification (Δ pH) of ambient solution, pH 6.3, by axial organs from the seeds imbibed to various water contents.

Water content (% fresh wt)	Δ pH
Broad bean axes	
60	0.01
65	0.09
68	0.34
75	0.65
85	0.59
Wheat germs	
53	0.24
60	0.27
65	0.72
73	0.77
83	0.90
Horse chestnut axes	
62	0.01
66	0.08
67	0.30
69	0.51
77	0.63

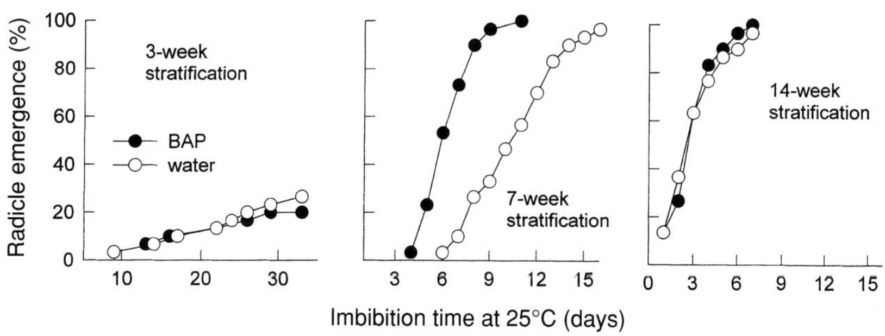

Fig. 3.4. Effect of benzylaminopurine (BAP) (50 mg l^{-1}) on dormancy release in horse chestnut seeds.

10^{-5} M GA_3) did not evoke a germination response. It was known from the literature that the content of endogenous gibberellins in horse chestnut seeds remained at the same level during the dormancy period (Schildmacher and Borris, 1967). The endogenous abscisic acid (ABA) (Fig. 3.6) does not appear to be responsible for dormancy release because of its constant low level. Ethylene also did not alleviate the horse chestnut dormancy, because its precursor, aminocyclopropane carboxylic acid (100 mg l^{-1}), and the inhibitor of its synthesis, aminoxyacetic acid

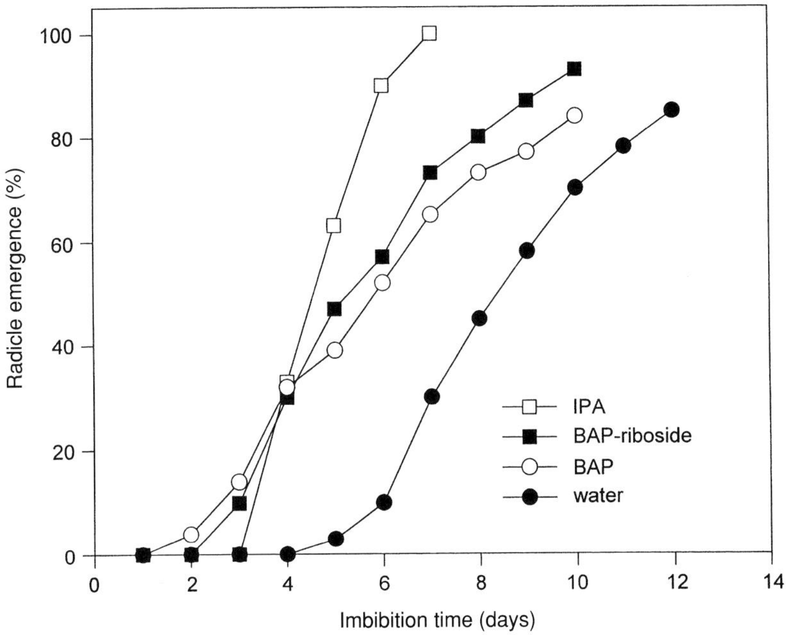

Fig. 3.5. Effect of various cytokinins (50 mg l^{-1}) on dormancy release in horse chestnut seeds after 10-week stratification. IPA, isopentenyladenine; BAP, benzylaminopurine; BAP-riboside, benzylaminopurine-riboside.

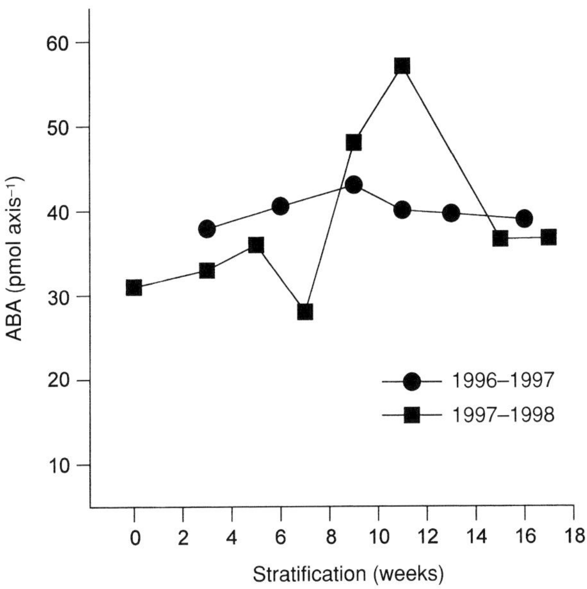

Fig. 3.6. ABA in the axial organs of horse chestnut seeds.

(100 mg l^{-1}), did not reduce the dormancy state. Therefore, in *Aesculus hippocastanum* seeds, the dormancy is relieved only by cytokinins. Neither dormancy maintenance nor germination *per se* is affected by cytokinin (Fig. 3.4); nor does it accelerate the seedling growth.

The next question was whether the processes preparing for growth initiation develop during dormancy release. The axes of dormant, less dormant and no longer dormant horse chestnut seeds were tested during chilling for the appearance of additional solutes (Table 3.3). No additional amounts of monosaccharides and potassium accumulated during the whole chilling period. No acidification of cell walls, a prerequisite for growth initiation, was observed during dormancy and dormancy release (Table 3.4). Therefore, we have no basis on which to equate the states of dormancy release and preparation for growth initiation, and may try to separate them with respect to their controlling functions.

In horse chestnut seeds, the dormancy release is a cytokinin-regulated process whereas the preparation for growth commencement is triggered by

Table 3.3. Osmotically active solutes and starch (mg $axis^{-1}$) in the axial organs of horse chestnut seeds.

Stratification (weeks)	Water content (% fresh wt)	Sucrose	Glucose	Fructose	Starch	K^+
0	65.4	1.9	0.03	0.2	3.8	—
1	65.3	2.2	0.03	0.2	3.9	—
2	65.8	2.0	0.05	0.2	4.6	—
3	65.2	2.2	0.03	0.2	4.3	—
5	64.1	2.3	0.08	0.1	4.1	0.63
6	63.2	2.2	0.04	0.2	4.0	0.58
7	63.4	2.1	0.05	0.1	4.0	0.62
9	63.4	2.4	0.06	0.1	4.2	—
11	64.7	2.7	0.07	0.1	3.8	0.54
13	64.0	2.1	0.05	0.2	4.1	0.58
15	66.1	2.1	0.05	0.1	3.0	0.66

Table 3.4. Acidification (Δ pH) of ambient solution, pH 6.3, by axial organs from horse chestnut seeds.

Water content (% fresh wt)	Stratification (weeks)	Δ pH
62.0	4	0.0
64.5	7	0.08
65.5	8	0.10
65.8	9	0.05
67.0	13	0.0

increasing water content. During dormancy and its release, the axes maintain almost the same rather high hydration level of 64–66% (see Table 3.3), which slightly exceeds the level reached by horse chestnut axes by matric forces. Such high water content is typical of recalcitrant seeds (Chin *et al.*, 1989; Tompsett and Pritchard, 1998); nevertheless, it is insufficient to initiate the germination, which usually begins at 74% water content in the axes. Any retardation of water inflow will delay the commencement of germination. We kept the seeds at the end of dormancy release in 30% polyethyleneglycol (PEG–6000), an osmoticum, and prevented further seed hydration (Fig. 3.7) and germination (Obroucheva and Antipova, 1999). Subsequently, the seeds transferred to water germinated at the same rate as the seeds kept initially in water.

Therefore, in dormant seeds, the germination *per se* as well as cell preparation before it are triggered by hydration as in quiescent or non-dormant seeds. It is the hydration that triggers the transition from dormancy release to preparation for germination. Turning back to the suggested subdivision of the transition from dormancy to germination to separate physiological events, it is clear that the appearance of a new event such as cell preparation for elongation is justified; this group of processes is not equivalent to the dormancy release.

The whole transition from dormancy to early germination (Fig. 3.8) includes dormancy release, cell preparation to elongation and growth commencement by cell elongation *per se*. Dormancy release is controlled by phytohormones (either cytokinins or, more frequently, gibberellins). It

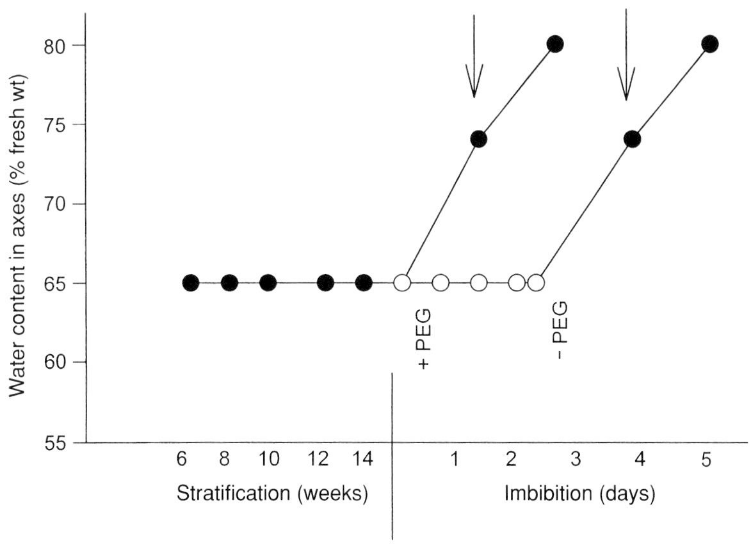

Fig. 3.7. Delayed germination of non-dormant horse chestnut seeds treated with polyethyleneglycol (PEG-6000). Arrows indicate visible germination.

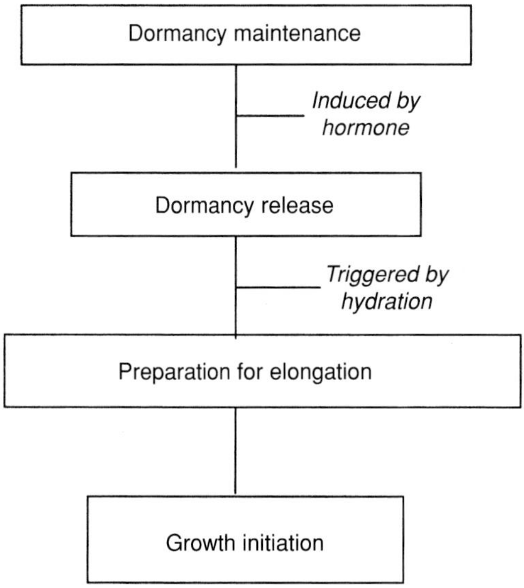

Fig. 3.8. Sequence of events during dormancy–germination transition.

occurs at a rather high water content 'automatically' achieved in the axes of dormant dry orthodox seeds mainly by matric forces during imbibition, as inferred, for example, from data on water content in dormant *Avena fatua* seeds (McIntyre and Hsiao, 1985). This hydration level allows the entire metabolic apparatus to be operative, but the rate of its operation depends on temperature. Recalcitrant dormant seeds do not desiccate at maturity and retain the water content in the axes slightly exceeding the 'automatically' achieved level.

The hormone-induced dormancy release is followed by preparation for cell elongation – the process triggered only by water and proceeding via solute accumulation and cell wall acidification, thus culminating in visible germination.

Acknowledgements

The authors are grateful to Drs A.A. Kotov and L.M. Kotova for ABA analysis. Work was supported by the Russian Foundation for Fundamental Research, grant 98-04-48630.

References

Antipova, O.V. (1997) Osmotic relations and cell wall acidification as the prerequisites of the start of elongation in the seed axial organs. In: Ellis, R.H., Black, M.,

Murdoch, A.J. and Hong, T.D. (eds) *Basic and Applied Aspects of Seed Biology.* Kluwer Academic Publishers, Dordrecht, pp. 469–478.

Chin, H.F., Krishnapillay, B. and Stanwood, P.C. (1989) Seed moisture: recalcitrant vs. orthodox seeds. In: *Seed Moisture,* Special Publication No. 14. Crop Science Society of America, Madison, Wisconsin, pp. 15–22.

McIntyre, G.I. and Hsiao, A.I. (1985) Seed dormancy in *Avena fatua.* Evidence of embryo water content as a limiting factor. *Botanical Gazette* 146, 347–352.

Obroucheva, N.V. (1997) Hydration up to threshold levels as the triggering agent of the processes preparing germination in quiescent seeds. In: Ellis, R.H., Black, M., Murdoch, A.J. and Hong, T.D. (eds) *Basic and Applied Aspects of Seed Biology.* Kluwer Academic Publishers, Dordrecht, pp. 555–562.

Obroucheva, N.V. (1999) *Seed Germination: a Guide to the Early Stages.* Backhuys Publishers, Leiden.

Obroucheva, N.V. and Antipova, O.V. (1997) Physiology of the initiation of seed germination. *Russian Journal of Plant Physiology* 44, 250–264.

Obroucheva, N.V. and Antipova, O.V. (1999) Common physiological mechanisms prepare seeds with different dormancy types for germination. *Russian Journal of Plant Physiology* 46, 363–368.

Schildmacher, H. and Borris, H. (1967) Untersuchungen über Gibberellingehalt von reifenden und keimenden Aesculussamen. *Wissenschaftliche Zeitschrift der Rostock Universität, Mathematische-naturwissenschaftliche Reihe* 16, 587–589.

Tompsett, P.B. and Pritchard, H.W. (1998) The effect of chilling and moisture status on the germination, desiccation tolerance and longevity of *Aesculus hippocastanum* seed. *Annals of Botany* 82, 249–261.

Vertucci, C.W. (1989) The effects of low water contents on physiological activities of seeds. *Physiologia Plantarum* 77, 172–176.

Sugar Metabolism in Apple Embryos

4

Stanisław Lewak[1,2], Renata Bogatek[1] and Bożenna Żarska-Maciejewska[1]

[1]University of Warsaw, Institute of Plant Experimental Biology; [2]Institute of Biochemistry and Biophysics, Polish Academy of Sciences, ul. Pawińskiego 5a, 02-106 Warszawa, Poland

Introduction

Dormancy in apple seeds

Dormancy in plants, wherever it is observed, can be defined as a temporal cessation of growth of plant meristem. All the classifications of dormancy are related to the organ and/or tissue where the inhibition of growth takes place, to the kind of factor that affects dormancy and, more recently, to the relationship between the site of action of this factor and the location of inhibited (dormant) meristem (Lang, 1987).

The mature seeds of apple are dormant. They become able to germinate and to produce normal seedlings after several weeks of moist-cold treatment (stratification). Apart from the dormancy located in the embryo (*cryogenic endodormancy* according to Lang, 1987), there is also the coat-imposed inhibition of germination (*testa paradormancy*). The embryos isolated from dormant seeds are able to germinate slowly, the germination is light-mediated (*spectral endodormancy*) and it is also affected by some hormones (Rudnicki, 1969; Smoleńska and Lewak, 1971; Lewak and Bryzek, 1974). The seedlings grown from such dormant embryos show, however, several developmental anomalies (Côme, 1970; Wyzińska and Lewak, 1978; Bogatek *et al.*, 1991). The most important are inhibition of hypocotyl and internode elongation growth, and asymmetric growth and greening of cotyledons. This morphological asymmetry is reflected by different oxygen uptake by each cotyledon of a germinating embryo (Côme and Thévenot, 1970). These symptoms of non-completely removed embryonic dormancy do not appear in germinating non-dormant embryos isolated from cold-treated (stratified) seeds. Short pre-treatment of dormant apple embryos with gaseous hydrogen cyanide eliminated all the above symptoms of dormancy (Bogatek *et al.*, 1991).

Reserve lipid mobilization

One of the most frequently postulated mechanisms for control of seed dormancy is a block in reserve mobilization (e.g. Villiers, 1972; Lewak, 1981; Drewes and Van Staden, 1991; Foley *et al.*, 1993). The block in reserve lipid mobilization in dormancy of apple seeds has been demonstrated in our laboratory (Żarska-Maciejewska and Lewak, 1976; Żarska-Maciejewska *et al.*, 1980; Żarska-Maciejewska, 1992). Another mechanism, which involves the hydrolysis of reserve proteins, has also been postulated (Żarska-Maciejewska and Lewak, 1983; Ranjan and Lewak, 1995).

The direct evidence that catabolism of triacylglycerols operates during the dormancy removal and germination of apple embryos comes from the decreased content of reserve fats (Kawecki, 1970; Żarska-Maciejewska, 1992). The gradual degradation of lipid bodies is first observed in the embryonic axis and then in the cotyledons (Dawidowicz-Grzegorzewska, 1981). Studies on the regulation of activity of certain key enzymes involved in the lipid catabolic pathway allowed us to conclude that blockage of this pathway plays an important role in the onset and maintenance of embryonic dormancy.

The first enzymes of lipid mobilization, lipases, are activated under conditions that enable the elimination of dormancy. Low temperature increases the activity of acid lipase. This enzyme is located mainly in the embryo axis, has the temperature optimum of 4°C and operates during the first 3 weeks of stratification (Smoleńska and Lewak, 1974; Żarska-Maciejewska and Lewak, 1976). Alkaline lipase is located in cotyledons, operates at the late stage of stratification and during germination and early growth. Both lipases are stimulated by light, which acts throughout the phytochrome (Lewak and Smoleńska, 1968; Smoleńska and Lewak, 1971). Also, gibberellin (GA) and jasmonate (JA) stimulate lipase activity, and abscisic acid (ABA) inhibits it (Smoleńska and Lewak, 1974; Ranjan and Lewak, 1994). It has been demonstrated that the levels of endogenous stimulators (GA, JA and cytokinins) increased and that of ABA decreased preceding the rise of lipase activity (Rudnicki, 1969; Sińska and Lewak, 1970; Borkowska and Rudnicki, 1975; Żarska-Maciejewska *et al.*, 1980; Ranjan and Lewak, 1995). The stimulatory or inhibitory effects of exogenous hormones on lipase activity were paralleled by a corresponding stimulation or inhibition of embryo germination.

Fatty acids released from storage lipids by lipases undergo β-oxidation producing acetyl-CoA. It has been proved that the capacity for β-oxidation in germinating apple embryos exceeds the potential capacity of lipases to produce fatty acids and that this is not affected by hormones (Bogatek *et al.*, 1999). It seems, therefore, that the production of acetyl-CoA is not a limiting step in lipid catabolism, which could be involved in the control of dormancy. On the other hand, the activity of isocitrate lyase, an enzyme involved in the glyoxylate cycle and gluconeogenesis, located in cotyledons of germinating apple embryos (Bogatek *et al.*, 1989), is controlled by the factors affecting removal of dormancy and germination (light, GA, JA and ABA).

All these observations allowed us to postulate that the inhibition of reserve lipid hydrolysis and further transformations of the hydrolysis products present a barrier (block) participating in embryo dormancy. The steps of the lipid catabolic pathway that are controlled by environmental (cold and light) and hormonal (GA, ABA and JA) factors affecting dormancy are indicated in Fig. 4.1.

In apple seeds the main reserves are the storage lipids and proteins (*c.* 25% and 20% of dry matter, respectively); reserve sugars are of minor importance (Wierszyłłowski, 1960). The aim of our recent studies was to demonstrate that in these seeds there also operates another block that inhibits catabolism of storage sugars.

Catabolism of Saccharides

As mentioned earlier, pre-treatment of dormant apple embryos with hydrogen cyanide improved their germination and eliminated the developmental anomalies in the seedlings. However, none of the lipid catabolism

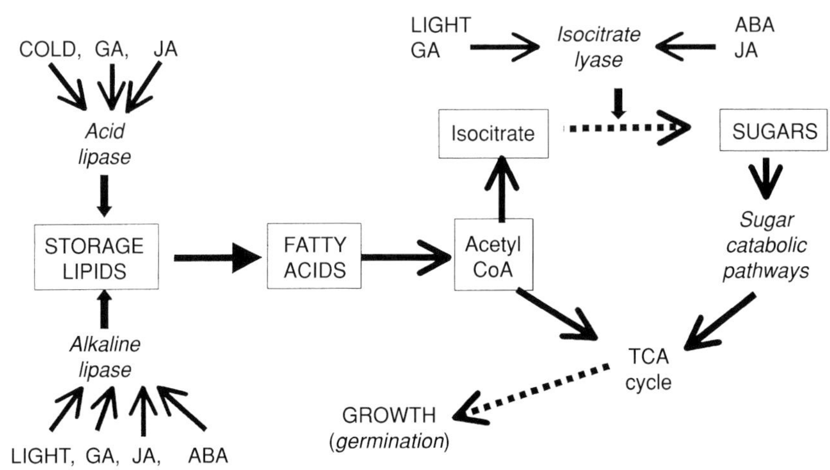

Fig. 4.1. Control of storage lipid catabolism in germinating apple embryos.

enzymes presented in Fig. 4.1 are affected by this treatment. On the other hand, it was demonstrated that the ratio of sugar catabolic pathways (C_6/C_1 ratio) changes during dormancy removal under conditions of cold stratification: the pentose phosphate oxidative pathway (PPP) dominates during the early period of stratification and later it is replaced by glycolysis. Cyanide treatment accelerated the appearance of glycolysis (Bogatek and Lewak, 1988). It has to be underlined that apple seeds contain a considerable amount of cyanogenic glycoside amygdalin, which undergoes degradation during dormancy removal, and its product, free hydrogen cyanide (HCN), is present in a relatively high concentration in seeds subjected to cold stratification (Dziewanowska *et al.*, 1979). In contrast, the embryos isolated from dormant seeds and cultured at 25°C contain no detectable free HCN (Bogatek *et al.*, 1991).

The enzymes that catalyse the key reactions of sugar catabolism are active during both cold- and light-triggered germination. Their activity is affected by cyanide, but not by the hormones that control lipid catabolism. In light-cultured dormant embryos the maximum activity of glycolytic enzymes (pyruvate kinase, ATP- and PPi-dependent phosphofructo-kinases, and 2-kinase) occurs during the first 3 days of culture; all these enzymes are stimulated by HCN pre-treatment (Bogatek and Lewak, 1991; Bogatek, 1995). The pentose phosphate path is monitored by determining the activity of glucose-6-phosphate dehydrogenase, which appeared late during embryo culture. It is slightly inhibited by cyanide (Bogatek and Lewak, 1991).

The content of monosaccharides, substrates of both pathways, constitutes only 0.2% of fresh weight in dormant apple embryos, whereas the amount of storage oligosaccharides (sucrose and raffinose) is higher by an order of magnitude (Popov, 1967; Kawecki, 1970; Bogatek *et al.*, 1999). It has been tempting to assume that: (i) the insufficient level of glycolytic substrates limits the growth of the embryo axis and that of one cotyledon, thus being responsible for dormancy symptoms; (ii) the hydrolysis of storage oligosaccharides takes place during dormancy removal thus supplying the substrates for glycolysis; and (iii) hydrolysis of oligosaccharides is controlled by HCN.

The above hypothesis was verified by determination of changes in soluble sugar contents and of oligosaccharide-hydrolysing enzyme activities in the embryonic axis and separately in both cotyledons of apple embryos (Bogatek *et al.*, 1999). The isolated, dormant embryos were cultured at 25°C in light. The pre-treatment of embryos with 1 mM HCN for 6 h markedly improved germination, eliminated the asymmetry in development of cotyledons and simultaneously resulted in marked alterations in studied sugar levels and enzyme activities.

Embryo axis

The level of sucrose, the main sugar in the axis, after a transient rise, remained constant during the culture of control embryos. HCN pre-treatment caused a sharp decrease in sucrose content; it almost disappeared in the axes of 10 day-cultured embryos. The beginning of the decrease in sucrose level was accompanied by a rise in glucose and fructose contents, which was also HCN-stimulated. Alkaline and acid invertases were operating in the axes during the first 3 days of embryo culture, but only the cytosolic alkaline invertase was stimulated by cyanide. These results, together with the data mentioned above, allow one to draw the following conclusions:

1. Inhibition of growth of the embryo axis (embryo dormancy) is, at least partly, due to the shortage in supply of monosaccharides and a low efficiency of glycolysis.
2. Cyanide participates in elimination of this block by stimulating the hydrolysis of sucrose and by enhancing the activities of certain key enzymes of glycolysis.

Cotyledons

During the culture of apple embryos, the changes in sucrose levels differed dramatically in the cotyledons. The level gradually decreased in the lower cotyledon, which was in direct contact with water. The decrease in sucrose level was accompanied by a rise in glucose content. In contrast, in the upper cotyledon (which grows slowly and remains white) the level of sucrose, after an initial drop of less than 30%, remained unchanged until the end of the experiment. The increase in glucose content was observed only late during culture. The activity of invertase was markedly decreased in the upper cotyledon as compared with the lower one. The HCN pre-treatment had no effect on the sucrose level in the lower cotyledon. On the other hand, it provoked a decrease in the level of this sugar in the upper cotyledon and accelerated the accumulation of glucose so that the pattern of sugar changes in this organ became almost identical with that in the lower, untreated, cotyledon. Moreover, HCN pre-treatment stimulated the activity of invertase. These data, together with the observations discussed earlier, have led us to the following conclusions:

1. Inhibition of growth and greening of one (upper) cotyledon in the germinating embryo (a post-effect of incompletely removed dormancy) is, at least partly, due to the shortage in supply of monosaccharides, substrates for PPP.
2. This block is eliminated by the pre-treatment with HCN, which stimulates the hydrolysis of sucrose.

The schematic presentation in Fig. 4.2 indicates the steps in sugar catabolic pathways that are controlled by HCN pre-treatment.

Metabolic Control of Dormancy

Figure 4.2 presents some metabolic events involved in the induction of germination and growth (removal of dormancy) that occur in the embryo

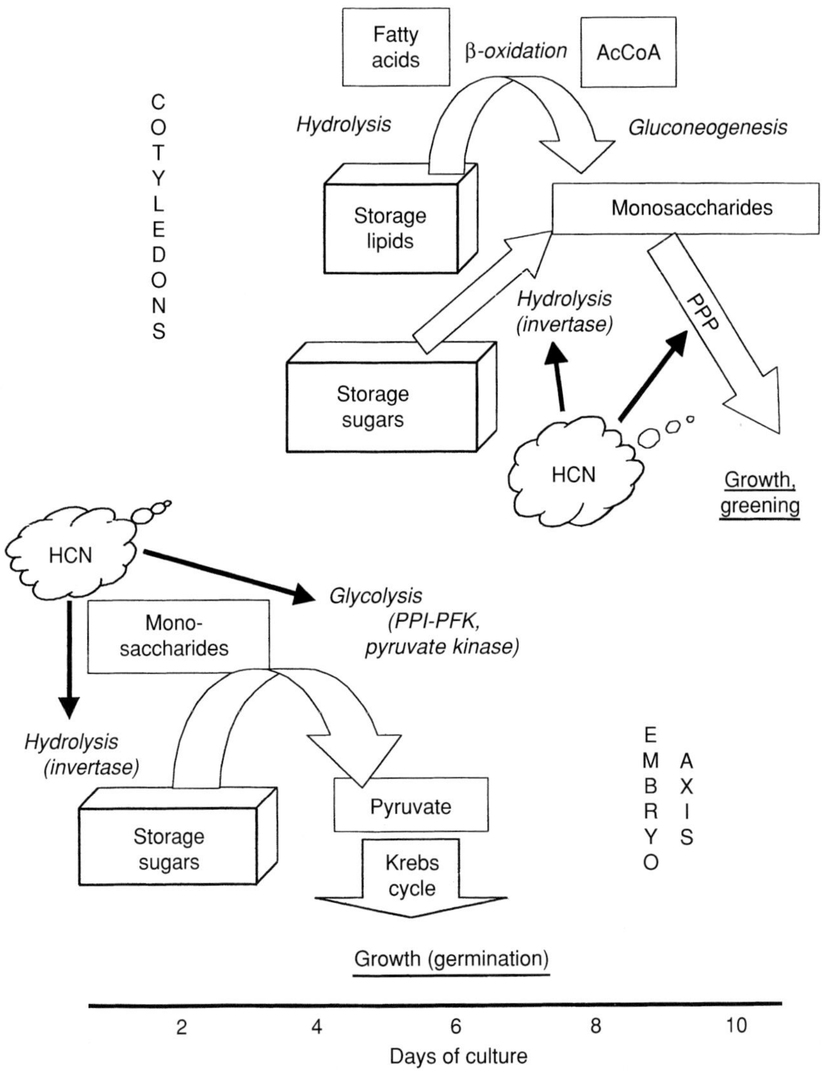

Fig. 4.2. Schematic presentation of storage lipid and saccharide mobilization in the embryonic axis and cotyledons of apple embryos pre-treated with HCN.

axis and in the cotyledons of dormant apple embryos submitted to culture in light. It shows the approximate sequence of these events: storage oligosaccharides in axes are hydrolysed and catabolized (mainly via glycolysis) just after the beginning of imbibition, whereas the mobilization of sugar reserves in cotyledons takes place when the embryo is already growing.

The embryonic axis contains very limited amounts of reserve sugars (*c.* 40 μg per axis). It seems, however, that it is sufficient for the induction of growth (germination) provided that the enzymatic system catalysing their hydrolysis and catabolism is active. This is the situation during the first days of HCN pre-treated embryos. For the further growth of the axis the import of monosaccharides and/or of the products of their catabolism from cotyledons appears to be indispensable.

The cotyledons contain 20 times more storage sucrose than the embryonic axis. In the lower cotyledon the massive hydrolysis of this sugar starts on the 4th day after the beginning of imbibition. The produced monosaccharides sustain the growth of the cotyledon, but they can also be exported to the growing axis. Stimulation of sucrose hydrolysis in the upper cotyledon by HCN pre-treatment increases the monosaccharide supply to the axis, and thus its growth. At the same time initiation of reserve lipid mobilization in cotyledons takes place (Fig. 4.1). It seems that gluconeogenesis following the production of AcCoA is an important source of sugars that sustain the growth of the cotyledons as well as that of the axis. The third source of substrates for glycolysis and PPP is photosynthesis, which starts relatively early in greening cotyledons (Ryć and Lewak, 1982).

The control of metabolic events discussed above is related to the embryo dormancy. It explains the existence and elimination of some important dormancy symptoms. It does not, however, disclose the primary mechanisms of embryonic dormancy in apple. The unanswered questions concern, for example, the role of changes in cell membrane structure and properties, induction of synthesis of specific proteins, induction of synthesis of the respective hormones and/or their activation, etc. Among the relations discussed here, the closest to the problem of the primary dormancy mechanism is the observation that the temperature optimum of acid lipase is identical to the optimum of cold-mediated dormancy removal (Smoleńska and Lewak, 1974; Żarska-Maciejewska and Lewak, 1976).

Acknowledgements

The authors are grateful to E. Głuchowska MSc for her skilful assistance. This work was partly supported by the Centre Franco-Polonais de Biotechnologie des Plantes (CNRS/KBN) and by the Joint Project offered by the French Embassy in Warsaw.

References

Bogatek, R. (1995) The possible role of fructose 2,6-bisphosphate in the cyanide-mediated removal of embryonic dormancy in apple. *Physiologia Plantarum* 94, 460–464.

Bogatek, R. and Lewak, S. (1988) Effect of cyanide and cold treatment on sugar catabolism in apple seeds during dormancy removal. *Physiologia Plantarum* 173, 406–411.

Bogatek, R. and Lewak, S. (1991) Cyanide controls enzymes involved in lipid and sugar catabolism in dormant apple embryos during culture. *Physiologia Plantarum* 183, 422–426.

Bogatek, R., Żarska-Maciejewska, B., Sińska, I. and Lewak, S. (1989) Embryonic axis controls lipid catabolism in cotyledons of apple seeds during germination. *Physiologia Plantarum* 176, 557–562.

Bogatek, R., Dziewanowska, K. and Lewak, S. (1991) Hydrogen cyanide and embryonal dormancy in apple seeds. *Physiologia Plantarum* 183, 417–421.

Bogatek, R., Côme, D., Corbineau, F., Picard, M.-A., Żarska-Maciejewska, B. and Lewak, S. (1999) Sugar metabolism as related to the cyanide-mediated elimination of dormancy in apple embryos. *Plant Physiology and Biochemistry* 37, 577–585.

Borkowska, B. and Rudnicki, R. (1975) Changes in the level of cytokinins in apple seeds during stratification. *Fruit Science Reports* 12, 1–16.

Côme, D. (1970) *Les Obstacles à la Gérmination.* Masson et Cie, Paris, pp. 54–80.

Côme, D. and Thévenot, C. (1970) La dormance des embryons de pommier (*Pirus malus* L.) en relation avec l'imbibition et l'intensité respiratoire de leurs cotyledons. *Bulletin de l'Académie et Societé Lorraines des Sciences* 9, 42–58.

Dawidowicz-Grzegorzewska, A. (1981) Anatomy, histochemistry and cytology of dormant and stratified apple embryos. III. Structural changes during the early development of seedlings in relation to embryonic dormancy. *New Phytologist* 87, 573–579.

Drewes, F.E. and Van Staden, J. (1991) Reserve mobilization during germination of *Tagetes minuta* L. *Annals of Botany* 68, 78–83.

Dziewanowska, K., Niedźwiedź, I. and Lewak, S. (1979) Hydrogen cyanide and cyanogenic compounds in seeds. II. Changes in free HCN level in apple seeds during stratification. *Physiologie Végétale* 117, 681–686.

Foley, M.E., Nichols, M.B. and Myers, S.P. (1993) Carbohydrate concentrations and interactions in after ripening-responsive dormant *Avena fatua* caryopses induced to germinate by gibberellic acid. *Seed Science Research* 3, 271–278.

Kawecki, Z. (1970) Studies on physiology of stratified apple seeds. *Roczniki Nauk Rolniczych* A 126, 2–22 (in Polish).

Lang, G.A. (1987) Dormancy: a new universal terminology. *HortScience* 22, 817–824.

Lewak, S. (1981) Regulatory pathways in removal of apple seed dormancy. *Acta Horticulture* 120, 149–159.

Lewak, S. and Bryzek, B. (1974) The influence of cytokinins on apple embryo photosensitivity and acid phosphatase activity during stratification. *Biologia Plantarum* 16, 334–340.

Lewak, S. and Smoleńska, G. (1968) Le systeme phytochrome dans la gérmination de semences de Pommier. *Physiologie Végétale* 6, 403–408.

Popov, Y.U.G. (1967) Carbohydrate metabolism in apple tree seeds during post-harvest maturing. *Fizjologya Rastenij* 14, 153–156 (in Russian).

Ranjan, R. and Lewak, S. (1994) Interaction of jasmonic acid with some plant growth regulators in the control of apple (*Malus domestica*) embryo germination. *Plant Growth Regulation* 114, 59–66.

Ranjan, R. and Lewak, S. (1995) Interaction of jasmonic and abscisic acid in the control of lipases and proteases in germinating apple embryos. *Physiologia Plantarum* 93, 421–426.

Rudnicki, R. (1969) Studies on abscisic acid in apple seeds. *Planta* 186, 63–69.

Ryć, M. and Lewak, S. (1982) Hormone interactions in the formation of the photosynthetic apparatus in dormant and stratified apple embryos. *Zeitschrift für Pflanzenphysiologie* 107, 15–24.

Sińska, I. and Lewak, S. (1970) Apple seed gibberellins. *Physiologie Végétale* 8, 661–667.

Smoleńska, G. and Lewak, S. (1971) Gibberellins and the photosensitivity of isolated embryos from non-stratified apple seeds. *Planta* 199, 144–153.

Smoleńska, G. and Lewak, S. (1974) The role of lipases in the germination of dormant apple embryos. *Planta* 116, 361–370.

Villiers, T. (1972) Seed dormancy. In: Kozlowski, T.T. (ed.) *Seed Biology*, Vol. 2. Academic Press, New York, pp. 220–276.

Wierszyłłowski, J. (1960) Anatomical structure and chemical composition of apple and pear seeds. *Roczniki Nauk Rolniczych* 81-A-1, 205–228 (in Polish).

Wyzińska, D. and Lewak, S. (1978) Morphological aspects of apple seedling early development in relation to embryonal dormancy. *Biologia Plantarum* 20, 53–60.

Żarska-Maciejewska, B. (1992) Lipolytic activity during dormancy removal in apple seeds. *Plant Physiology and Biochemistry* 130, 65–70.

Żarska-Maciejewska, B. and Lewak, S. (1976) The role of lipases in the removal of dormancy in apple seeds. *Planta* 132, 177–181.

Żarska-Maciejewska, B. and Lewak, S. (1983) The role of proteolytic enzymes in the release from dormancy of apple seeds. *Zeitschrift für Pflanzenphysiologie* 110, 409–417.

Żarska-Maciejewska, B., Sińska, I., Witkowska, E. and Lewak, S. (1980) Low temperature, gibberellin and acid lipase activity in removal of apple seed dormancy. *Physiologia Plantarum* 148, 532–535.

5

Population Responses to Temperature for Loss and Induction of Seed Dormancy and Consequences for Predictive Empirical Modelling

Alistair J. Murdoch, Landing Sonko* and Ermias Kebreab

Department of Agriculture, The University of Reading, PO Box 236, Earley Gate, Reading RG6 6AT, UK

Introduction

Within a population of seeds, individuals vary in their response thresholds to regulatory factors (Bradford, 1996). The literature on seed dormancy, viability and germination contains numerous examples where this seed-to-seed variation is used to construct predictive empirical models of the behaviour of the seed population (e.g. Roberts, 1972; Murdoch *et al.*, 1989; Bradford, 1996). This variation within homogeneous seed lots can usually be described by the normal distribution function and can therefore be quantified by the mean and standard deviation of the distribution. Examples include germination progress curves as a function of thermal time (Kebreab and Murdoch, 1999c) and loss of dormancy during dry afterripening of seeds during seed storage (e.g. Roberts, 1965; Favier and Woods, 1993). Transforming the proportions of seeds responding to probits or normal equivalent deviates will then linearize the response as a function of the dose applied. Sometimes, and especially when germination responses to varying doses of chemical or light treatments are involved, the doses are expressed on a logarithmic scale (Murdoch and Ellis, 2000).

Sometimes, however, the observed responses are clearly non-linear. This non-linearity is most obvious when a very wide range of treatment doses are applied and may sometimes be understood quite simply in terms

* Present address: Department of Agricultural Services, Agricultural Pest Management Unit, Yundum, The Gambia.

of a supra-optimal dose. Often, however, a secondary process may be interfering with the response to a primary process. An example where overlapping processes may occur is in the response to afterripening period. Here the loss of seed dormancy during afterripening of seeds in dry storage sometimes overlaps with loss of viability (Roberts, 1965). Secondly, during prechilling or imbibed storage, seeds often show an initial loss in dormancy, while prolonging the moist storage period results in a decreased response (Totterdell and Roberts, 1979; Jones *et al.*, 1997; Kebreab and Murdoch, 1999a). A further example is where the low fluence response in light-sensitive seeds may overlap with the high irradiance reaction (Ellis *et al.*, 1986; cf. Murdoch and Ellis, 2000).

This chapter aims to show that in cases where more than one process is acting on the seed population, and where the response of the individual seed is evaluated as a binary effect (i.e. it either germinates or does not), non-linear probability models provide appropriate predictive empirical models to describe the germination behaviour of the seed population. Key hypotheses of the non-linear models are:

- seed-to-seed variation with respect to each component process is normally distributed with respect to the 'dose' variable (e.g. time);
- the processes operate concurrently and independently within each individual seed.

There are interesting inferences if the processes are indeed independent. Physiologically it implies that the component processes have different physiological mechanisms. Mathematically, it implies that the net response of the seed population is given by the product of the probabilities of the component processes. That is, the observed response may be described by a *multiplicative* probability model. Where processes clearly operate sequentially and not concurrently, an additive model should be used.

This paper describes the use of multiplicative probability models to quantify seed-to-seed variation within seed populations with respect to dormancy. Three stages of the life cycle are examined:

- loss of dormancy during afterripening in dry storage;
- loss of primary dormancy and induction of secondary dormancy of non-germinating seeds in imbibed storage;
- effects of temperature on the germination of non-dormant seeds in comparison with seeds with secondary dormancy.

The development and statistical testing of multiplicative probability models has been achieved by fitting non-linear equations in GENSTAT.

Examples are drawn in this paper from the germination biology of two genera of parasitic weeds: *Orobanche* (broomrapes) and *Striga* (witchweed). Despite their very small size and the need for specialized handling in the laboratory, seeds of the parasitic plants have proved to be useful tools for examining the effects of temperature on aspects of germination biology. There are three prerequisites for their germination:

- a period of dry afterripening;
- a short warm stratification of imbibed seeds (preconditioning);
- exposure to a chemical germination stimulant.

The stimulant is produced by roots of host plants in the natural environment while in the laboratory, artificial stimulants such as GR24 are commonly used.

Loss of Dormancy During Afterripening in Dry Storage

The overlap of the two concurrent processes of loss of dormancy and loss of viability during dry storage was evident in rice seeds (Roberts, 1972) and has recently been shown in seeds of *Striga hermonthica* stored at 50 and 60°C (Sonko, 1998; Fig. 5.1). An overall model to quantify the combined

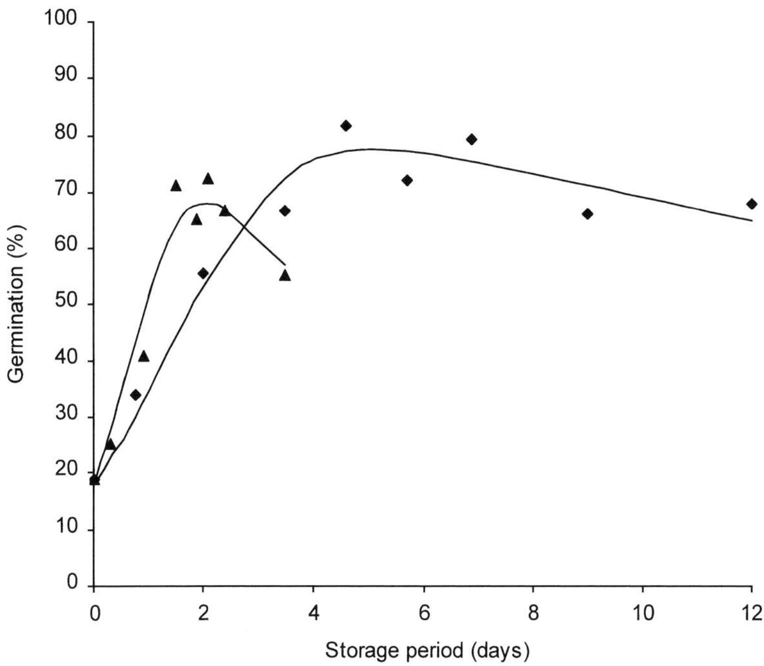

Fig. 5.1. Germination of *Striga hermonthica* seeds afterripened in dry storage for various periods at 50 (◆) and 60°C (▲). Lines were fitted according to equation (1). After storage, the seeds were preconditioned for 14 days at 30°C and then germinated for 2 days at 30°C with 1 ppm GR24. Seeds which failed to germinate were tested for viability using tetrazolium solution (Sonko, 1998).

processes in terms of the probability of germination (G) following afterripening for t days is given simply by the product of the probabilities that a seed: (i) has lost primary dormancy during afterripening; and (ii) is still viable. That is,

$$G = [\Phi^{-1} (K_d + \beta_d t)] \, [\Phi^{-1} (K_i + \beta_v t)] \quad (1)$$

where K_d is the initial proportion of non-dormant seeds in normal equivalent deviates (ned) or probits, that is the theoretical fraction of seeds without an afterripening requirement; β_d is the estimated rate of loss of dormancy during afterripening at 0°C (ned day^{-1}), K_i is the proportion of viable seeds and β_v is the rate of loss of viability (ned day^{-1}) (Sonko, 1998). The symbol Φ^{-1} represents back-transformation from ned to proportions.

Loss of Primary Dormancy and Induction of Secondary Dormancy during Imbibed Seed Storage

Imbibing seeds often show an initial sensitization to a subsequent stimulus, while prolonging the moist storage period results in a decreased response to subsequent stimulation. Such responses have been observed during imbibed storage of *Rumex* spp. during prechilling at 3–15°C (Totterdell and Roberts, 1979) and at 20°C prior to temperature shift treatments (Totterdell and Roberts, 1981). Similar effects have been shown for seeds of Sitka spruce (Jones *et al.*, 1997) and for three *Orobanche* species (Kebreab and Murdoch, 1999a). The principal processes in this latter case are arguably a loss of primary dormancy followed by induction of secondary dormancy (Totterdell and Roberts, 1979; Jones *et al.*, 1997). In each case, a multiplicative probability model has been applied to account for the non-linear responses.

Fig. 5.2. (A) Rates of loss of primary dormancy during conditioning in seeds of *Orobanche aegyptiaca* (□), *Orobanche cernua* (▲) and *Orobanche crenata* (○) as a function of temperature. Rates were fitted independently for each temperature (symbols) or lines were fitted according to the expression ($p_0 + pT$), which quantifies the rate of loss of dormancy in equation (2). The analysis was based on an experiment in which seeds were conditioned for up to 14 days on a temperature gradient plate which provided 13 temperature regimes in a constant temperature gradient between 10 and 40°C. Each species was tested separately. After conditioning, seeds were germinated at 20°C for 10 days in 9 cm Petri dishes moistened with 3 ppm GR24 (after Kebreab and Murdoch, 1999a). (B) Rates of induction of secondary dormancy in *O. aegyptiaca* (□), *O. cernua* (▲) and *O. crenata* (○) at different temperatures. Rates were fitted separately for each temperature (symbols) or constrained to an exponential model (fitted line) according to the expression ($s_m + sr^T$) which quantifies the rate of induction of secondary dormancy in equation (2). Original data are shown in Fig. 5.3 (after Kebreab and Murdoch, 1999a, with permission from Oxford University Press).

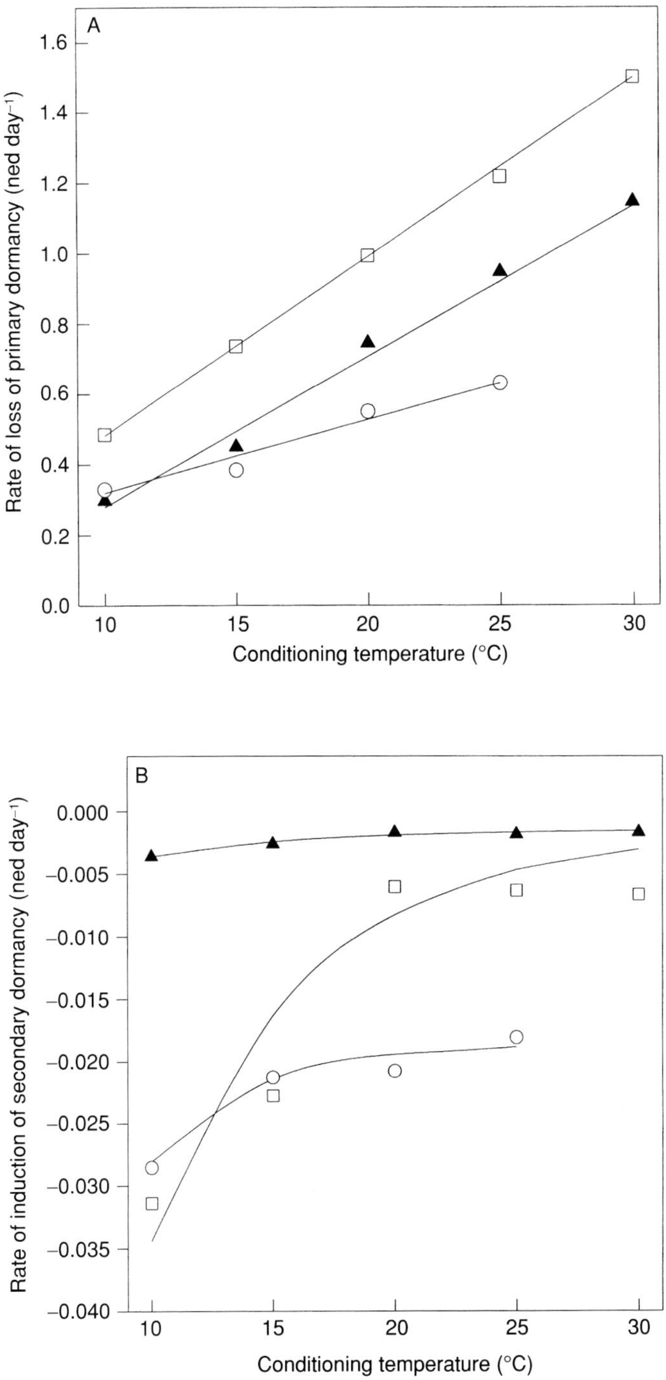
A
Rate of loss of primary dormancy (ned day^{-1})
1.6
1.4
1.2
1.0
0.8
0.6
0.4
0.2
0.0
10
15
20
25
30
Conditioning temperature (°C)
B
Rate of induction of secondary dormancy (ned day^{-1})
0.000
–0.005
–0.010
–0.015
–0.020
–0.025
–0.030
–0.035
–0.040
10
15
20
25
30
Conditioning temperature (°C)

Germination after preconditioning and stimulation was therefore the product of the probabilities that a seed: (i) has lost primary dormancy during the preconditioning; (ii) has not entered secondary dormancy; and (iii) is still viable. Kebreab and Murdoch (1999a) found that it was necessary to include loss of viability as a third concurrent process where prolonged imbibed storage treatments were applied especially at higher temperatures in three *Orobanche* species.

Using the slopes of the probit lines as indirect estimates of the rates of the first two processes within the seed population (i.e. the proportion of seeds losing primary dormancy per day, or the proportion entering secondary dormancy per day), the effects of temperature on these rates can be quantified (Fig. 5.2A and B, respectively). Note that the slopes of the probit lines, which are defined here as rates, are primarily a measure of seed-to-seed variation in the population. By definition, the standard deviation of the dormancy periods in the seed population is the reciprocal of the slope. In speaking of rates, the slopes cannot be used to infer mechanistic deductions about the rates of the processes in individual seeds. Indeed because dormancy is measured as a quantal response in the individual seed (it is either dormant or non-dormant), the concept of rate is inappropriate if applied to the individual seed.

The rate of loss of primary dormancy increased linearly with temperature over the temperature range tested (Fig. 5.2A) while the rate of induction of secondary dormancy decreased (became slower and thus the slope became less negative) and became asymptotic with increase in temperature (Fig. 5.2B). These effects are in contrast to those assumed for *Rumex crispus* and *R. obtusifolius* where the rate of loss of primary dormancy was assumed to be independent of temperature and the rate of induction of dormancy was assumed to increase with increasing temperature (Totterdell and Roberts, 1979). An overall model to account for the probability of germination (G) in the presence of 3 ppm of an artificial germination stimulant GR24 at 20°C after preconditioning (imbibed storage) for t days at T°C is as follows:

$$G = [\Phi_p^{-1}(K_d + (p_0 + pT)t)]\,[\Phi_s^{-1}(K_i + (s_m + sr^T)t)]\,[\Phi_v^{-1}(K_i + \beta_v t)]/\Phi^{-1}(K_i) \qquad (2)$$

where $\Phi^{-1}(K_d)$ is the initial proportion of non-dormant seeds, that is those without a conditioning requirement; p_0 is theoretical rate of loss of dormancy during conditioning at 0°C and p is the temperature coefficient (ned day^{-1} °C^{-1}); $\Phi^{-1}(K_i)$ is the proportion of viable seeds; s and r control the slope and direction of the response to temperature and s_m is the asymptotic rate of induction of secondary dormancy with increase of temperature; s is the range of the curve (Fig. 5.2B) between the rate of induction at 0°C and the asymptote; r defines the rate of exponential increase and β_v is the rate of loss of viability based on additional viability tests.

This model provided a good fit to the observed non-linear response to conditioning period in most cases (Fig. 5.3). In *Orobanche crenata* equation

(1) fitted well at 10°C and 15°C but relatively poorly at 20°C and 25°C where an apparent stabilization in dormancy was observed between 70 and 154 days (Fig. 5.3K–N). An example of the underlying theoretical distributions is shown in Fig. 5.4.

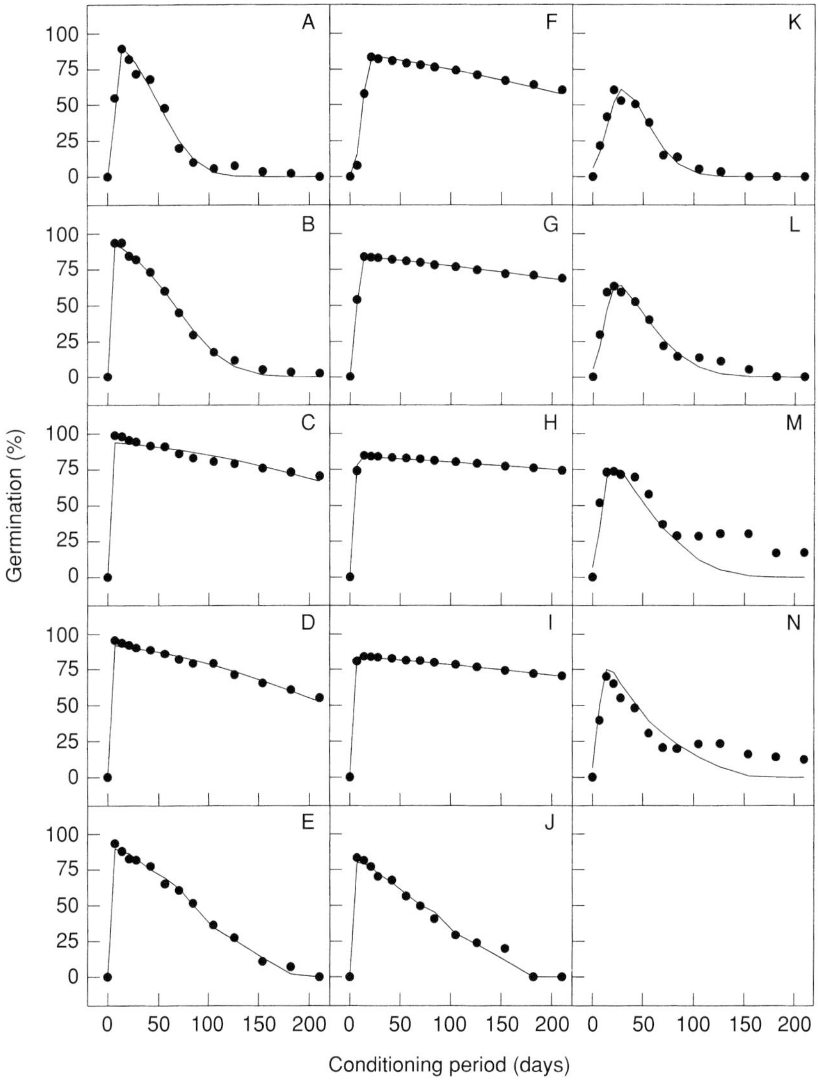

Fig. 5.3. Germination of *Orobanche aegyptiaca* (A–E), *Orobanche cernua* (F–J) and *Orobanche crenata* (K–N) seeds after conditioning at temperatures of 10 (A, F, K), 15 (B, G, L), 20 (C, H, M), 25 (D, I, N) and 30°C (E, J) for up to 30 weeks. Lines were fitted according to equation (2). Petri dishes were kept moist throughout. Seeds were germinated as described in Fig. 5.2A (after Kebreab and Murdoch, 1999a, with permission from Oxford University Press).

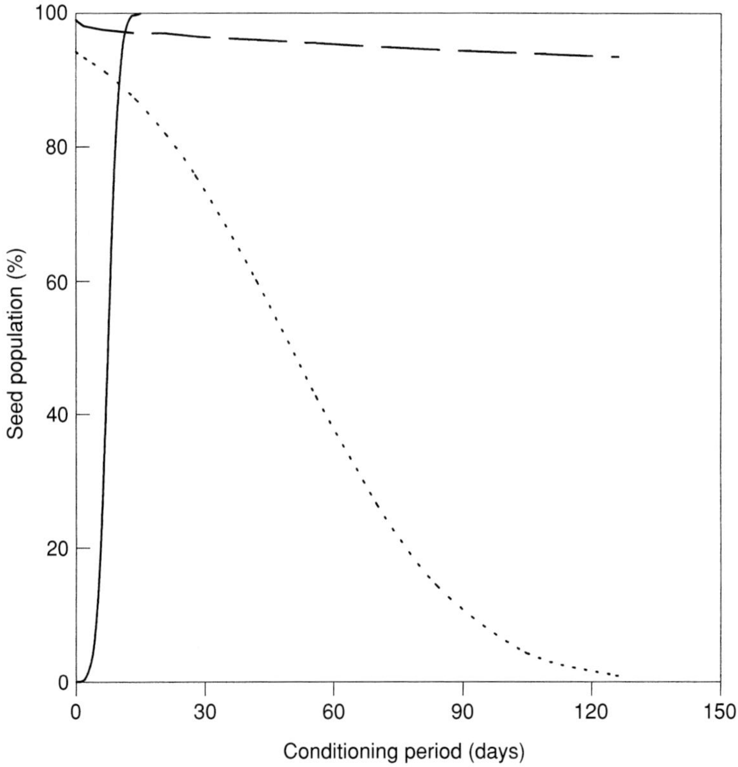

Fig. 5.4. An example of the processes of loss and induction of dormancy and loss of viability occurring in *Orobanche aegyptiaca* seeds during conditioning at 10°C according to equation (2). Sub-processes are loss of primary dormancy (solid line), induction of secondary dormancy (dotted line) and loss of viability (dashed line). The combined effect of these sub-processes on germination after conditioning is shown in Fig. 5.3A (after Kebreab and Murdoch, 1999a, with permission from Oxford University Press).

A similar model has been shown to apply to seeds of witchweed (*Striga hermonthica*; Sonko, 1998). The next stage in this research, which is currently in progress in field experiments in Ghana, is to apply the model to predict infestations of parasitic weeds for different planting dates of host crops after the onset of rain in the field.

Effects of Temperature on the Germination of partly Dormant and Non-dormant Seeds

Seeds of *Orobanche aegyptiaca* were conditioned at 20°C in water for 2 or 12 weeks. The 2-week treatment led to a complete loss of primary dormancy in most seeds while secondary dormancy was evident in some after 12 weeks of conditioning (Fig. 5.3D). The seeds were germinated on a two-dimensional temperature gradient plate which provided 169 different constant and alternating temperature regimes. Highest germination occurred along the constant temperature gradient diagonal of the temperature gradient plate and alternating temperatures never increased the germination compared with the corresponding mean temperature, irrespective of dormancy (Fig. 5.5A and C). In fact, wide amplitudes of temperature alternation decreased germination. The contour patterns including one or both temperatures above about 25°C imply that lower germination at alternating temperatures in *Orobanche* is a function of maximum temperature in particular. A multiplicative probability model is again appropriate since there appear to be two primary responses to temperature leading to seed-to-seed variation in the population. These responses appear to relate to: (i) *mean temperature,* such that the higher the mean temperature, the greater the proportion of seeds whose minimum temperature requirement for germination is satisfied; and (ii) *maximum daily temperature* such that the higher the maximum temperature, the smaller the proportion of seeds capable of germinating (Keheab and Murdoch, 1996b, 2000). Close inspection of the contour patterns where the maximum temperature effect is most evident suggests that this response to maximum temperature is modified by amplitude and thermoperiod (Fig. 5.5A and C).

An overall model to account for the probability of germination (G) in the presence of 3 ppm of the artificial germination stimulant GR24 at a range of temperatures is given by the product of: (i) the proportion of seeds whose minimum mean temperature (T_m) for germination is satisfied; and (ii) the proportion of seeds whose maximum temperature limit has not been exceeded. Mathematically this can be expressed as follows:

$$G = \Phi^{-1}(K_d + \beta_1 T_m)\ \Phi^{-1}[K_i + (s + s_1 A + s_2 A^2)\ r^{T_{max}}] \qquad (3)$$

where Φ^{-1} (K_d) is the proportion of seeds whose minimum mean temperature requirement is met at 0°C; β_1 is the slope (ned per °C increase in T_m) and Φ^{-1} (K_i) is the proportion of viable seeds; s and r control the slope and direction of the response to maximum temperature, T_{max} and s_1 and s_2 describe the effect of amplitude, A. It was found necessary to estimate the parameters independently for each thermoperiod and also for each conditioning period (Kebreab and Murdoch, 1999b).

The resulting multiplicative probability model provided a reasonable fit to the data (compare Fig. 5.5A with B, and Fig. 5.5C with D) and is the first successful attempt to apply such a model to germination at alternating temperatures. In essence the model implies that there is a temperature range for germination in the seed population and that seed-to-seed variation in this range can be modelled as two independent prerequisites for germination (Fig. 5.6). In practice, the two temperature limits may not be completely independent, but physiologically it is not surprising that

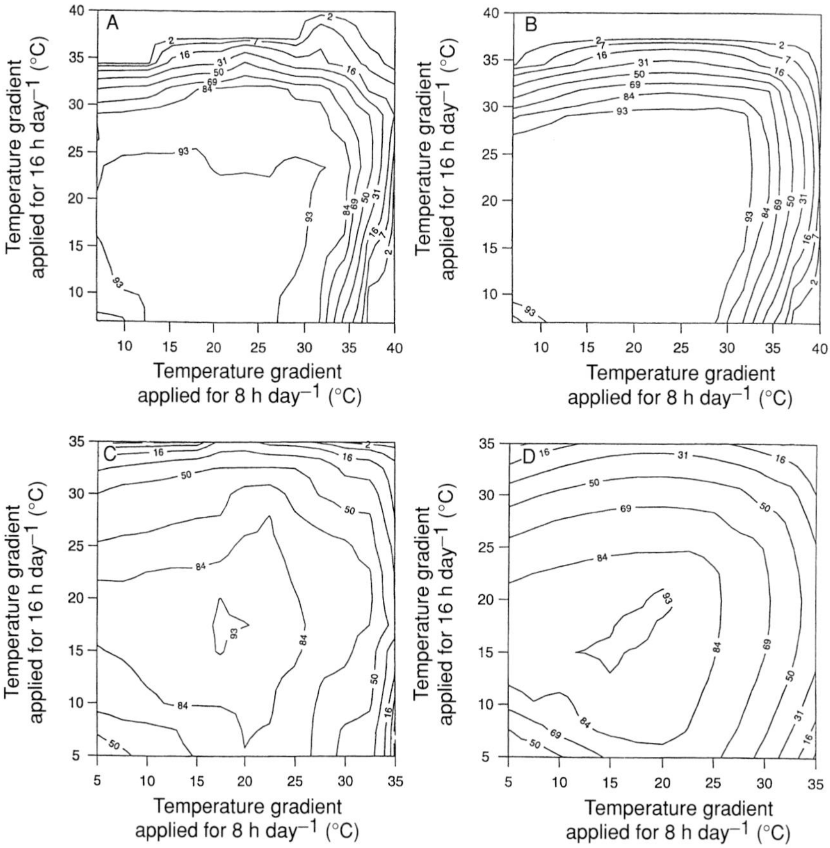

Fig. 5.5. A map of germination of *Orobanche aegyptiaca* seeds on the temperature gradient plate after being conditioned for 2 weeks (A, B) and 12 weeks (C, D) at 20°C showing the original data (A, C) and fitted values (B, D) according to equation (3). Percentage germination values are shown on the isopleths, which are drawn at 0.5 normal equivalent deviate (ned) intervals from –2 ned (2%) to 1.5 ned (93%) (after Kebreab and Murdoch, 1999b, with permission from Academic Press).

these limits not only exist, but that they vary widely and are normally distributed in the seed population. The modifying effect of thermoperiod demonstrates that the time spent at the maximum temperature each day can be an important constraint. Ecologically, the inhibitory effect of wide amplitude and high maximum temperature means that seeds will germinate better if they are buried rather than at the soil surface and also under a crop canopy.

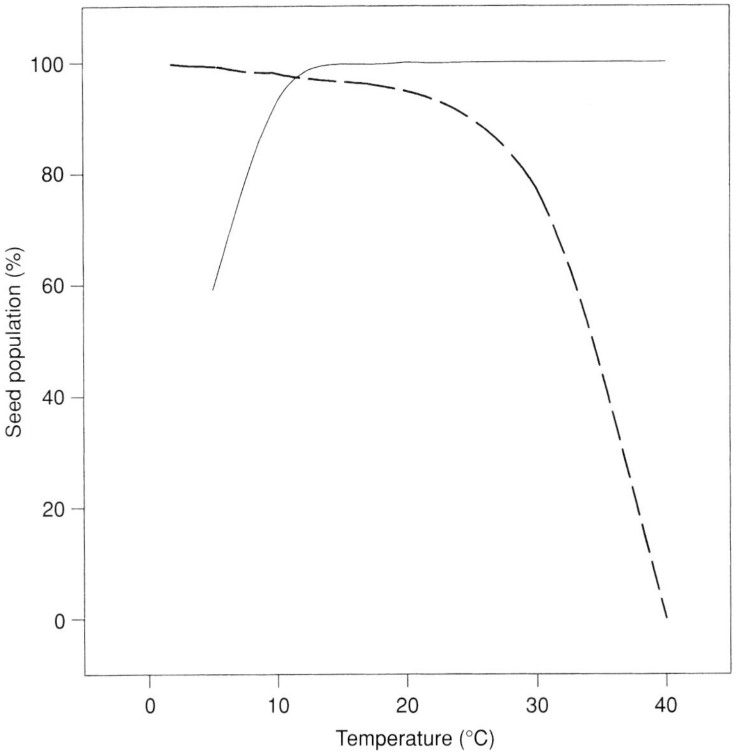

Fig. 5.6. An example of the underlying assumptions of equation (3) such that higher proportions of seeds have their minimum temperature requirement to germinate satisfied as the mean temperature increases (solid line), while a higher proportion of seeds exceed their maximum temperature limit to germinate as the maximum temperature increases (broken line). (After Kebreab and Murdoch, 1999b, with permission from Academic Press.)

Conclusions

Non-linear multiplicative probability models thus allow ecologically realistic modelling of concurrent binary responses of dormant seeds. The physiological interpretations and implications of the models must be treated with caution, until verified experimentally; for example, where mechanisms of primary and secondary dormancy are assumed to differ. The multiplicative probability population models 'may not define for us what dormancy is, but they do give a foundation on which to build' (Bradford, 1996) and hypotheses which can be tested.

References

Bradford, K. (1996) Population-based models describing seed dormancy behaviour: Implications for experimental design and interpretation. In: Lang, G.A. (ed.) *Plant Dormancy: Physiology, Biochemistry and Molecular Biology.* CAB International, Wallingford, UK, pp. 313–339.

Ellis, R.H., Hong, T.D. and Roberts, E.H. (1986) Quantal response of seed germination in *Brachiaria humidicola, Echinochloa turnerana, Eragrostis tef* and *Panicum maximum* to photon dose for the low energy reaction and the high irradiance reaction. *Journal of Experimental Botany* 37, 742–753.

Favier, J.F. and Woods, J.L. (1993) The quantification of dormancy loss in barley (*Hordeum vulgare* L.). *Seed Science and Technology* 21, 653–674.

Jones, S.K., Ellis, R.H. and Gosling, P.G. (1997) Loss and induction of conditional dormancy in seeds of Sitka spruce maintained moist at different temperatures. *Seed Science Research* 7, 351–358.

Kebreab, E. and Murdoch, A.J. (1999a) A quantitative model for loss of primary dormancy and induction of secondary in imbibed seeds of *Orobanche* spp. *Journal of Experimental Botany* 50, 211–219.

Kebreab, E. and Murdoch, A.J. (1999b) A model of the effects of a wide range of constant and alternating temperatures on seed germination of four *Orobanche* species. *Annals of Botany* 84, 549–557.

Kebreab, E. and Murdoch, A.J. (1999c) Modelling the effects of water stress and temperature on germination rate of *Orobanche aegyptiaca* seeds. *Journal of Experimental Botany* 50, 655–664.

Kebreab, E. and Murdoch, A.J. (2000) The effect of water stress on the temperature range for germination of *Orobranche aegyptiaca* seeds. *Seed Science Research* 10, 127–133.

Murdoch, A.J. and Ellis, R.H. (2000) Dormancy, viability and longevity. In: Fenner, M. (ed.) *Seeds: the Ecology of Regeneration in Plant Communities,* 2nd edn. CAB International, Wallingford, UK.

Murdoch, A.J., Roberts, E.H. and Goedert, C.O. (1989) A model for germination responses to alternating temperatures on seed of four *Orobanche* species. *Annals of Botany* 63, 97–111.

Roberts, E.H. (1965) Dormancy in rice seed. IV. Varietal responses to storage and germination temperatures. *Journal of Experimental Botany* 16, 341–349.

Roberts, E.H. (1972) Storage environment and the control of viability. In: Roberts, E.H. (ed.) *Viability of Seeds.* Chapman & Hall, London, pp. 14–58.

Sonko, L. (1998) Effects of nitrogenous compounds on the germination biology of *Striga hermonthica* and implications for integrated control. MPhil. thesis, The University of Reading, UK.

Totterdell, S. and Roberts, E.H. (1979) Effects of low temperatures on the loss of innate dormancy and the development of induced dormancy in seeds of *Rumex obtusifolius* L. and *Rumex crispus* L. *Plant, Cell and Environment* 2, 131–137.

Totterdell, S. and Roberts, E.H. (1981) Ontogenetic variation in response to temperature change in the control of seed dormancy of *Rumex obtusifolius* L. and *Rumex crispus* L. *Plant, Cell and Environment* 4, 75–80.

Pre-treatment at Controlled Seed Moisture Content as an Effective Means to Break Dormancy in Tree Seeds

6

Maria P.M. Derkx

Research Station for Nursery Stock, Rijneveld 153, PO Box 118, 2770 AC Boskoop, The Netherlands

Introduction

Seeds of many temperate trees and shrubs are dormant at maturity. The mechanisms which restrict germination vary among species. Physical (hardseededness), mechanical, chemical, morphological and physiological dormancy types have been described by Nikolaeva (1977). Combinations of these types often occur. Obviously, the pre-treatment required to overcome dormancy varies as well. Many species require a cold moist pre-treatment or prechill to overcome dormancy, other species require a warm followed by cold pre-treatment. The duration of pre-treatment required can be as short as a couple of weeks or as long as over 1 year. In order to satisfy warm and cold temperature requirements some nurseryworkers sow the freshly harvested seeds in autumn. In a cold autumn a warm temperature requirement may not be satisfied and seeds fail to respond to the subsequent winter temperatures. These seeds germinate at the earliest in the second spring after sowing. When the autumn and winter temperatures satisfy the warm and cold requirement of the seeds, seeds become ready to germinate during winter. Since seeds of many tree species can germinate at temperatures lower than 5°C, they may start to germinate in late winter or early spring. Late frosts may completely destroy all seedlings, as happens in *Fagus sylvatica*. To avoid these risks, nurseryworkers sow in spring after releasing dormancy under more or less artificial conditions. Traditionally, seeds are mixed with moist sand or sphagnum peat and exposed to temperatures outdoors during winter or during a summer–winter period. Seeds are kept at a full hydration level. This method has been often referred to as

(eds J.-D. Viémont and J. Crabbé)

stratification. Alternatively, the seeds are pre-treated in climate chambers, which have the advantage that known temperatures can be maintained. Since the temperature range for dormancy relief and germination overlap during a cold pre-treatment in many species, seeds may start to germinate during pre-treatment. If radicle emergence continues, the seedlings may easily be killed during sowing, especially when sowing machines are used. Chitted seeds may also develop into crooked seedlings. Therefore, the nurseryworker wants to sow immediately when the first radicles appear. Since weather and field conditions often do not allow sowing at this time, the seeds have to be stored at freezing temperatures of around −2°C to stop further germination of the seed lot. However, further dormancy release of deep dormant seeds within the seed lot may be stopped as well. This is a major problem in many tree species that exhibit large variation in the degree of dormancy between individual seeds within the seed lot. In *Fraxinus excelsior*, for example, it has been demonstrated that the time interval between germination of the first and the last seed within a seed lot during stratification at 4°C may be as much as 3 months (Van de Walle, 1987). Thus, some of the seeds are ready to germinate before the cold requirement of other seeds within the seed lot has been met. The deep dormant seeds will fail to germinate when the cold requirement is not further satisfied after sowing. Moreover, germination will be spread over many weeks or even over years. It is therefore necessary to find a compromising pre-treatment and sowing date such that most seeds grow into marketable plants.

Control of Seed Moisture Content

Although traditional stratification in medium can provide good seedling emergence in one year, results in another year may be erratic and low, thereby preventing efficient management at nurseries. Moreover, stratification in medium requires very intensive monitoring, especially when seeds are nearly ready to germinate. Although nursery staff have a lot of experience in stratifying tree seeds, they do not know the exact warm and cold requirement of a seed lot. The prechill demand of seeds varies according to the seed lot, within the seed lot, according to the year of harvest and to the history of seeds between maturation and start of the pre-treatment. Seed moisture content is often high when seeds are mature. Between maturation and harvest, either by picking or by collecting in nets or on the ground part of the chilling demand may be fulfilled. This variation in seed behaviour between seed lots makes traditional stratification very sensitive to errors.

The ideal pre-sowing treatment should prevent premature germination and should allow dormancy breaking of every individual seed within the seed lot, resulting in high, rapid and synchronous germination after sowing. Prolonging the pre-treatment will meet the cold requirement of both low-dormant and deep dormant seeds, but is not appropriate for nursery practice due to problems with premature germination. A simple but effective method to avoid premature germination is to reduce the amount of water during the pre-treatment. On the other hand, there should be sufficient water to allow dormancy breaking. The water requirement for dormancy release is lower than that for germination. Control of the seed moisture content is easiest when seeds are pre-treated without a medium. First attempts to reduce water during or after a partial cold pre-treatment were described some decades ago (Blundell, 1973; Danielson and Tanaka, 1978). Some years later studies started in important broadleaf species, like *Fagus sylvatica* (Suszka, 1979; Muller and Bonnet-Masimbert, 1989; Derkx and Joustra, 1997) and in conifer species from genera *Pinus, Picea, Abies* and *Pseudotsuga* (Edwards, 1986; Jones and Gosling, 1994; Poulsen, 1996; Jensen, 1997).

Control of seed moisture content in cold-requiring species

Many species, like *Acer pseudoplatanus, Acer platanoides, Acer palmatum, Fagus sylvatica, Abies nordmanniana, Picea sitchensis* and *Pseudotsuga menziesii* require a period of exposure to low temperatures, usually between 1°C and 5°C to overcome dormancy. The seed moisture content that prevents germination, but allows dormancy breaking during a cold pre-treatment varies within these species. The critical moisture content for germination can be easily determined by adjusting the seed moisture content of seed samples to several levels and subsequent storage of the seeds at a temperature of about 3°C. Premature germination can be recorded. Additionally, the level of dormancy of non-germinated seeds can be determined by transferring the seeds to conditions in which water is not limiting for germination.

An example of such a test is shown in Fig. 6.1 for two *Acer* species. Samaras (whole winged fruit, including the seed) of *A. pseudoplatanus* and *A. platanoides* were pre-treated at 3°C with different moisture contents (on a fresh weight basis). After a pre-treatment of 20 weeks germination was tested at a non-optimal temperature of 20°C. When the fruit moisture content of *A. pseudoplatanus* was 48% or higher seeds germinated during the cold pre-treatment (Fig. 6.1A). In *A. platanoides* a fruit moisture content of 42% allowed premature germination (Fig. 6.1B). Germination of *A. pseudoplatanus* was fully prevented at fruit moisture contents of 46% and lower, that of *A. platanoides* at fruit moisture contents of 40% and lower. The optimal fruit moisture content for dormancy breaking without premature germination was about 46% in *A. pseudoplatanus* and 36–40% in *A. platanoides.* In both species further reducing the fruit moisture content badly influenced dormancy breaking. It must be realized that the moisture contents are mean moisture contents of the seed lot. A seed lot consists of a population of seeds within which the moisture content varies from one seed

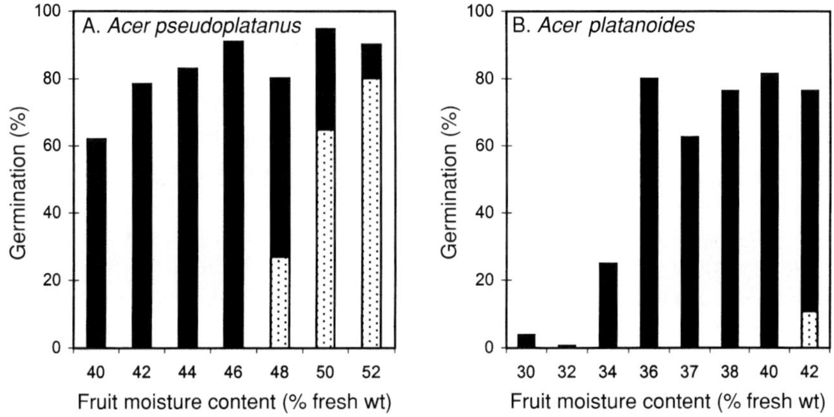

Fig. 6.1. The effect of the fruit moisture content (FW) of *Acer pseudoplatanus* (A) and *Acer platanoides* (B) on germination during mediumless pre-treatment (dotted bar) or during a subsequent germination test (black bar). Whole samaras with different moisture contents were pre-treated at 3°C for 20 weeks. Pre-treatment was in plastic bags. The samaras were then transferred to germination tests at 20°C in darkness. Germination of four replicates of 50 seeds was tested in plastic boxes on wet thick paper.

to another. This means that individual seeds may have moisture contents that are lower or higher than the optimal seed moisture content. Although both species have a full hydration level at about 63%, the optimal moisture content for dormancy breaking and at the same time avoiding premature germination is much lower and very different for the two species. Jensen (1996) suggested that the critical moisture content for germination is generally 3–8% lower than the full hydration level.

Seed or fruit moisture contents that prevent radicle protrusion but allow dormancy breaking are summarized in Tables 6.1 and 6.2 for several cold-requiring species. Results were obtained in the present study and from the literature.

Effect of the prechill duration

The effect of the prechill duration at controlled seed moisture content on dormancy release was studied in seeds of *A. platanoides*, *A. pseudoplatanus*, *Fagus sylvatica* and *Pseudotsuga menziesii* (Fig. 6.2). After several intervals of pre-treatment germination was tested at full hydration level over a range of temperatures. As long as this range becomes wider, dormancy breaking is not complete (Vegis, 1964; Karssen, 1982). Freshly harvested seeds of the three broadleaf species hardly germinated at test temperatures of 10°C and higher. Germination at 3°C was slow and could be attributed to dormancy

Table 6.1. Seed or fruit moisture contents that prevent radicle protrusion and allow good dormancy breaking in several broadleaf species that require a cold pre-treatment to overcome dormancy.

Species	Moisture content (% FW)	Reference
Acer palmatum (de-winged samara)	35	Present study
Acer platanoides (de-winged samara)	35–37	Present study
Acer platanoides (whole samara)	36–40	Present study
Acer pseudoplatanus (de-winged samara)	46–48	Present study
Acer pseudoplatanus	44–50	Suszka *et al.* (1996)
(whole samara)	44–46	Present study
Berberis thunbergii	38–42	Present study
Fagus sylvatica (nut)	30–32	Suszka *et al.* (1996)
	30	Derkx and Joustra (1997)
	30–32 (after harvest)	Muller *et al.* (1999b)
	32–34 (after dry storage)	Muller *et al.* (1999b)

release, followed by germination. Non-chilled seeds of *P. menziesii* gave reasonable to good germination at test temperatures between 20°C and 30°C, failed to germinate at 3°C and 10°C and gave only little germination at 15°C and 35°C.

In the three broadleaf species the temperature range of germination became wider during the first 16–20 weeks of pre-treatment, indicating that this duration of pre-treatment is required to get optimal dormancy breakage. Generally, nurseryworkers stratify these species for a much shorter duration in medium, depending on the start of premature germination. Obviously, the optimal duration of prechill at controlled seed moisture content will be longer than the optimal duration of a traditional treatment at full hydration level. However, due to premature germination, such an optimal duration is not possible during traditional stratification.

Seeds of *P. menziesii* highly benefited from 18 and 37 weeks of prechill at 35% moisture content in terms of maximum percentage germination over a broad range of temperatures, including low temperatures. Since conifer seeds are usually sown when soil temperatures are still low, this prechill treatment will significantly improve germination in the seedbed. After a prechill treatment of 8 weeks, germination at low temperatures was still poor. Traditionally, this species has been stratified at full hydration level for 6 weeks, which is clearly not sufficient to achieve germination at sub-optimal temperatures (Fig. 6.2D; Jones and Gosling, 1994). Beneficial

Table 6.2. Seed moisture contents that prevent radicle protrusion and allow good dormancy breaking in several conifer species that require a cold pre-treatment to overcome dormancy.

Species	Moisture content (% FW)	Reference
Abies amabilis	30	Leadem (1986)
Abies grandis	30	Edwards (1986)
Abies lasiocarpa	30	Edwards (1986)
Abies nordmanniana	33–34	Jensen (1997)
Abies procera	25–30	Tanaka and Edwards (1986)
	30–34	Poulsen (1996)
Picea abies	30	Bergsten (1989)
Picea glauca	30	Downie *et al.* (1993)
Picea mariana	30	Downie *et al.* (1993)
Picea sitchensis	25–30	Gosling and Rigg (1990)
	31	Jones and Gosling (1994)
	24–28	Poulsen (1996)
Pinus banksiana	30	Downie *et al.* (1993)
Pinus contorta	35	Jones and Gosling (1994)
Pinus strobus	30	Downie *et al.* (1993)
Pinus sylvestris	30	Bergsten (1989)
Pseudotsuga menziesii	35	Edwards (1986)
	37	Jones and Gosling (1994)
	30–34	Poulsen (1996)
	32	Muller *et al.* (1999a)
	35–37	Present study

effects of a prolonged prechill treatment have been previously reported by Jones and Gosling (1994).

The enormous widening of the temperature range of germination during a prolonged pre-sowing treatment has the important implication that temperature conditions at the time of sowing in spring are not very critical. It also means that the seeds are less sensitive to induction of secondary dormancy in the seedbed, which is a major problem in many species, including *A. platanoides* and *A. pseudoplatanus*.

The duration of prechill did not only affect the final percentage of germination, but also the rate of germination. In *A. pseudoplatanus* a prechill treatment of 20 weeks resulted in a tenfold decrease of the mean germination time (MGT, as defined by Jones and Gosling, 1994) compared with non-prechilled seeds (Fig. 6.3). A tenfold decrease in MGT was also found during the first 24 weeks of prechill of *F. sylvatica* seeds. In *P. menziesii* 18 weeks of prechill resulted in an enormous increase in speed of germination in comparison with untreated seeds. Although final percentages of germination were similar after 18 and 37 weeks of prechill, speed of germination was significantly increased after the latter treatment (Table 6.3).

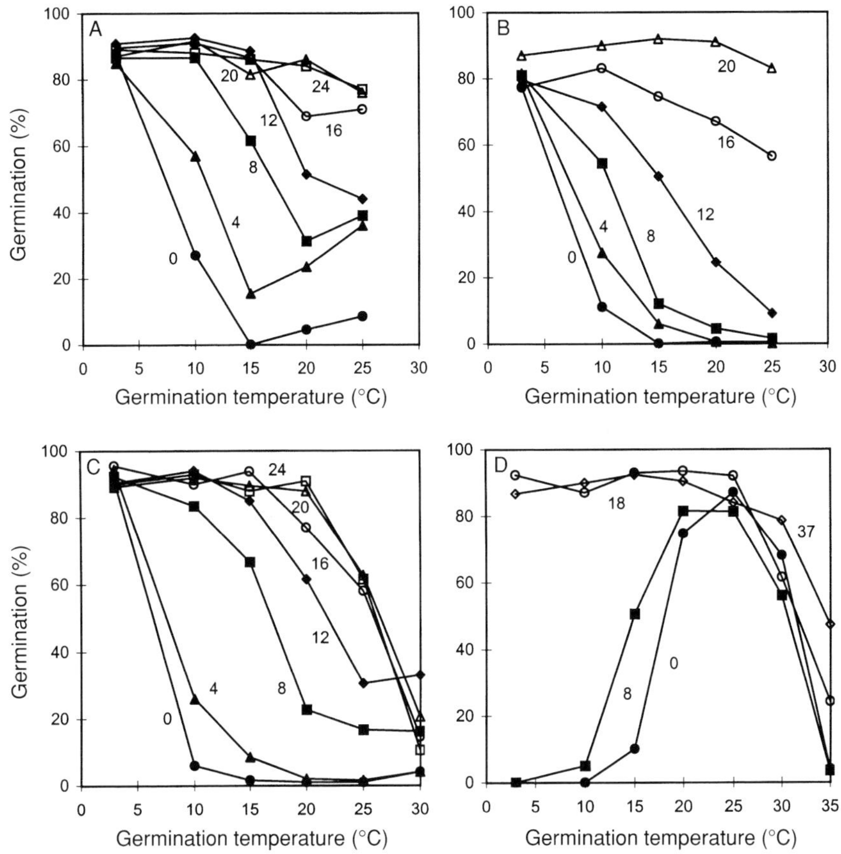

Fig. 6.2. The effect of the duration of pre-treatment on the temperature range of germination of *Acer pseudoplatanus* (A), *Acer platanoides* (B), *Fagus sylvatica* (C) and *Pseudotsuga menziesii* (D) seeds. Moisture content (FW) during pre-treatment was 46, 36, 30 and 35%, respectively. Seeds were pre-treated without medium at 3°C. After pre-treatment during 0 (●), 4 (▲), 8 (■), 12 (◆), 16 or 18 (○), 20 (△), 24 (□) and 37 (◇) weeks, germination was tested in darkness at constant temperatures. Germination tests as in Fig. 6.1.

Effects of very long prechill treatments up to 76 weeks have been studied in some conifer species (Jones and Gosling, 1994). Germination capacities of *Picea sitchensis* and *Pinus contorta* were not harmed by a prechill duration of 76 weeks, but seeds of *P. menziesii* were affected. So far, there is little information on the harm of prolonged prechill treatments in broadleaf species. In *F. sylvatica* the temperature range of germination started to narrow after about 24 weeks of pre-treatment (Fig. 6.4). It is most likely that the decline in germination can be attributed to a decrease in viability, although some induction of secondary dormancy cannot be fully

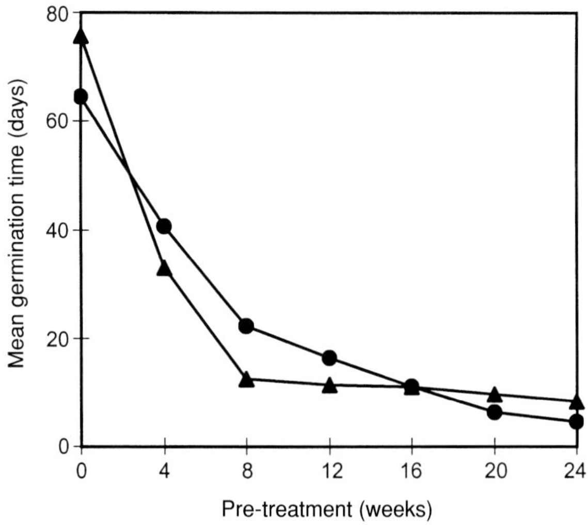

Fig. 6.3. The effect of the duration of pre-treatment on the mean germination time of *Acer pseudoplatanus* (●) and *Fagus sylvatica* (▲) seeds. Pre-treatment as described in Fig. 6.2. Germination was tested at 3°C as described in Fig. 6.1.

Table 6.3. Mean germination time (in days) at different test temperatures of *Pseudotsuga menziesii* seeds after prechill at 3°C at 35% moisture content.

Duration of prechill (weeks)	Germination temperature (°C)						
	3	10	15	20	25	30	35
0	nd	nd	nd	100.9	29.1	38.5	nd
18	74.5	17.3	6.2	4.1	3.3	3.8	5.6
37	25.1	7.0	3.0	2.4	2.2	2.2	2.5

nd = not determined.

excluded after these long pre-treatments. After about 40 weeks of prechill, germination capacity also started to decline at low temperatures. Root tips looked damaged after these long pre-treatments.

Control of seed moisture content in species requiring warm and cold

Many species first require a period of warm stratification to overcome some physical, mechanical, chemical or morphological barrier before they can respond to cold stratification. In species that require a combination of warm and cold pre-treatment in order to overcome different dormancy types, precise control of the moisture content during the warm phase is not necessary since germination is not yet possible during this phase. Control of

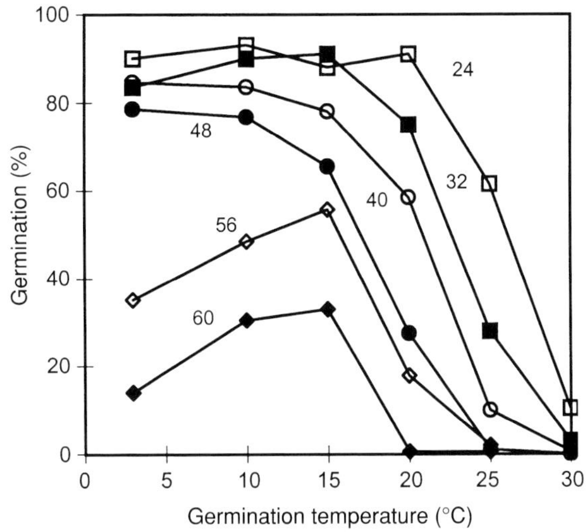

Fig. 6.4. The effect of prolonged pre-treatment on the temperature range of germination of *Fagus sylvatica* seeds. Seeds with a moisture content of 30% were pre-treated without medium at 3°C during 24 (□), 32 (■), 40 (○), 48 (●), 56 (◇) and 60 (◆) weeks. Germination was then tested in darkness over a range of constant temperatures. Germination tests as in Fig. 6.1.

other conditions, like temperature and the presence and composition of a medium proved to be successful in several species, including *Fraxinus excelsior* and *Tilia cordata.*

Fraxinus excelsior

Dormancy in this species can be attributed to at least three factors: a pericarp that is impermeable to gases, an immature embryo that needs a period of further growth to reach final size and a deeply dormant embryo (Villiers and Wareing, 1964). Recently, Finch-Savage and Clay (1997) demonstrated that the embryo is constrained by its enclosing tissues. Under natural conditions the seeds only begin to germinate in the second spring after fruit dispersal. It is generally accepted that increase in embryo length is the main indicator of the progress of the warm phase. Figure 6.5 shows that embryo growth is particularly sensitive to the conditions of the warm phase. Embryo growth was much slower in fruits pre-treated without medium compared with stratification in a medium of sphagnum peat and sand. In the medium some degradation of the pericarp was observed, whereas it remained much more intact in naked fruits. Since restriction of oxygen supply by the intact pericarp may retard embryo growth (Villiers and Wareing, 1964), differences in pericarp breakdown may explain differences in embryo growth. Under natural conditions embryo growth of fruits that were not embedded in the soil was strongly retarded in

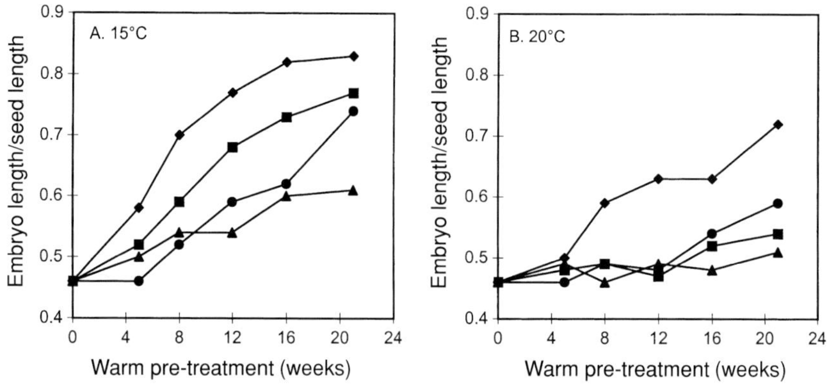

Fig. 6.5. Embryo growth of *Fraxinus excelsior* seeds during pre-treatment at 15°C (A) or 20°C (B) without medium (●), in vermiculite (▲), in sand (■) or in a sphagnum peat sand mixture (1:1)(◆). Three volume parts of medium were mixed with one volume part of samaras. Fruit moisture content during pre-treatment was 60–66%.

comparison with buried fruits (Wagner, 1996). In media of inorganic material like sand and vermiculite, embryo growth was reduced compared with stratification in a medium of sphagnum peat and sand. A lower microbial activity and a decrease of oxygen availability in inorganic media may be responsible for this.

In our study embryo growth was better and faster at 15°C than at 20°C. A temperature of 20°C has often been used in previous studies and has resulted in good germination during or after the cold phase (Van der Walle, 1987; Tylkowski, 1990; Muller, 1993). Embryo growth continued for at least 20 weeks. Full occupancy of the seed by the embryo was not essential for embryos to respond to low temperatures, as embryo growth could continue after transfer of the fruits to chilling temperatures, although at a slower rate (Fig. 6.6A). Therefore it is not necessary to leave the fruits at a high temperature for 20 weeks. Embryos did not grow at chilling temperatures without a previous initiation at high temperatures. The period of cold needed was reduced in fruits that had longer embryos at the end of the warm phase, as was previously reported by Krauss and Köhler (1985). A cold pre-treatment of at least 20 weeks can be recommended after a warm pre-treatment of 16 weeks, or alternatively 20 weeks of warm pre-treatment followed by at least 16 weeks of cold (Fig. 6.6B). Control of the fruit moisture content has been proven to be successful during the cold pre-treatment. A fruit moisture content of 45% did not completely prevent premature germination during the later stages of the cold phase. In a subsequent experiment we found no premature germination when the fruit moisture content was 42% (data not shown). In previous studies moisture contents between 52 and 60% have been recommended (Tylkowski, 1990; Muller, 1993; Suszka *et al.*, 1996).

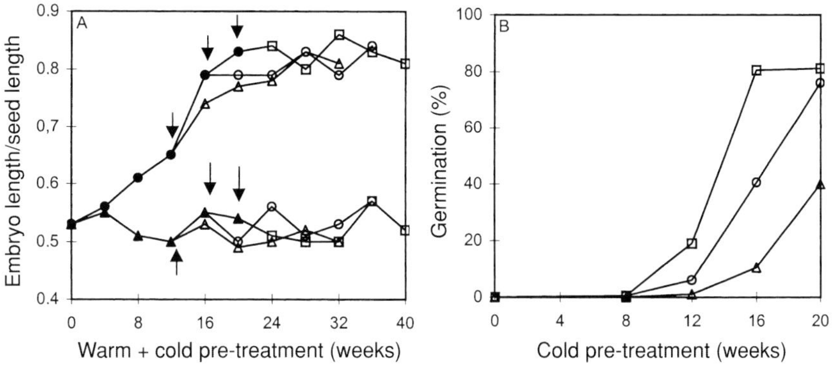

Fig. 6.6. (A) Embryo growth of *Fraxinus excelsior* seeds during pre-treatment at 15°C followed by pre-treatment at 3°C. Pre-treatment at 15°C was without medium (▲) or in a mixture of sphagnum peat and sand as in Fig. 6.5 (●). Fruit moisture content was about 60%. After 12 (△), 16 (○) or 20 (□) weeks of warm pre-treatment the medium was removed (arrows). All fruits were dried to a moisture content of 45% and further pre-treated at 3°C. (B) Germination tests at 10°C after warm pre-treatment during 12 (△), 16 (○) or 20 (□) weeks followed by the indicated durations of cold pre-treatment. Germination tests as in Fig. 6.1.

Tilia cordata

Dormancy in *Tilia cordata* can be attributed to at least three factors: impermeability of the seed coat to water, a tough pericarp and a dormant embryo (Brinkman, 1974). Warm stratification is required to overcome the constraints of the pericarp and the seed coat. Low temperatures are required to overcome the embryo dormancy. So far, the only way of determining whether the warm phase has been satisfied is to see if the seeds respond to the subsequent cold phase. Dormancy breaking during the warm phase was affected by the temperature and by the presence of a medium consisting of sphagnum peat and sand. When seeds were pre-treated in medium, dormancy release at 25°C was strongly reduced compared with that at 20°C. When the seeds were pre-treated without medium, dormancy release at 20°C and 25°C was comparable (Fig. 6.7). In Polish seed lots 25°C proved to be more effective than 20°C (Suszka *et al.*, 1996). In a previous study we found that 20°C was more effective than 15°C (unpublished results). The seed lot used in this study did not ultimately require a warm pre-treatment. The seeds became ready to germinate during a long cold pre-treatment that was not preceded by a warm pre-treatment (Fig. 6.8). However, a mediumless warm pre-treatment at 20°C reduced the cold requirement of the seeds. A warm pre-treatment of 16 weeks followed by 24 weeks of cold pre-treatment resulted in germination of all viable seeds. Further prolonging the cold pre-treatment gave some reduction in germination. After a warm pre-treatment of 24 weeks, the response to cold started to decline earlier and the maximum germination

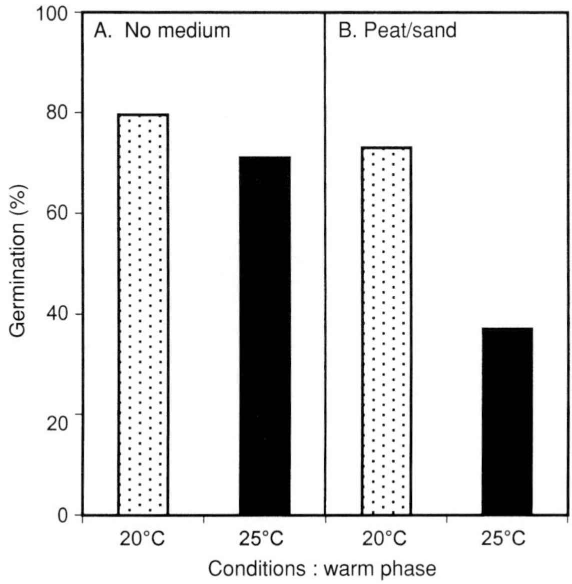

Fig. 6.7. The effect of the temperature and the presence of a mixture of sphagnum peat and sand during warm pre-treatment on dormancy breaking of *Tilia cordata* seeds. Seeds with pericarp were pre-treated at 20°C (dotted bar) or 25°C (black bar) without medium (A) or in a mixture of sphagnum peat and sand (B), as described in Fig. 6.5. The duration of warm pre-treatment was 16 weeks. Fruit moisture content was about 50%. The medium was removed and the fruits were dried to a moisture content of 40% and pre-treated at 3°C for 24 weeks. Germination was subsequently tested at 5°C/15°C (12 h/12 h) in darkness in Petri dishes on wet filter paper.

percentage achieved was lower than after 16 weeks of warm pre-treatment. Control of the fruit moisture content at 40% fully prevented germination during the cold phase. Little germination was observed when the fruit moisture content was 45% (data not shown). Table 6.4 summarizes recommended seed or fruit moisture contents of several species with combined dormancy.

Storage of Seeds after a Controlled Pre-treatment

The ability to dry and store non-dormant seeds after pre-treatment would allow distribution of seeds that can be sown immediately, without any need for further pre-treatment by nurseryworkers. Sowing can be easily delayed when field or weather conditions are bad. It also means that centralized pre-treatment becomes possible with good equipment and control by specialized personnel. Seeds can be easily transported and flow properly through sowing machines. A permanent stock of ready-to-sow seeds also

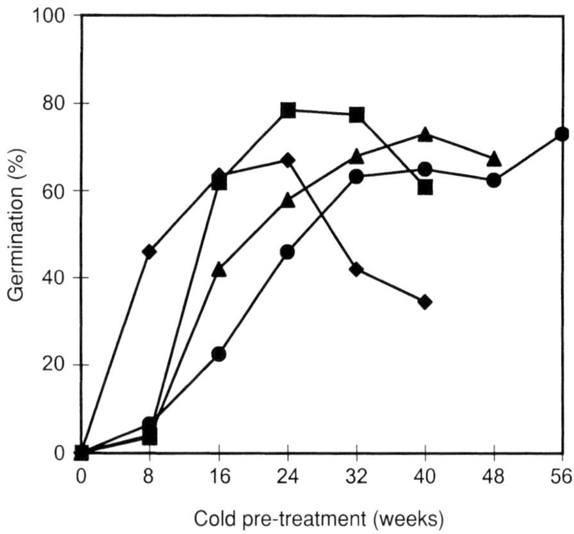

Fig. 6.8. The effect of the duration of warm and cold pre-treatment on dormancy breaking of *Tilia cordata* seeds with pericarp. Warm pre-treatment was at 20°C without medium during 0 (●), 8 (▲), 16 (■) or 24 (◆) weeks. Fruit moisture content was 50%. After the warm pre-treatment the fruits were slightly dried to a moisture content of 45% and further pre-treated at 3°C over the indicated times. Germination was tested at 15°C in darkness in Petri dishes on wet filter paper.

Table 6.4. Seed or fruit moisture contents during cold pre-treatment that prevent radicle protrusion and allow good dormancy breaking in several broadleaf species that require a warm followed by cold pre-treatment to overcome dormancy.

Species	Moisture content (% FW)	Reference
Fraxinus excelsior	55–60	Suszka *et al.* (1996)
(samara)	42–44	Present study
Tilia cordata	40–43	Present study
(seed plus pericarp)		
Prunus avium	27–29	Suszka *et al.* (1996)
(stone)		Present study
Amelanchier lamarckii	43–45	Present study
(seed plus pericarp)		
Sorbus aucuparia	43–45	Present study
(seed plus pericarp)		

means that problems with irregular cropping can be overcome. Beech, for instance, has a good mast year every 5 years.

Broadleaves

In France redrying and storage has been successfully developed for *Fagus sylvatica*, although some decline in germination capacity cannot be avoided. Beech nuts dried to a moisture content of 7–8% and stored at −7°C for periods of up to 3 years showed 60–70% germination after storage (Muller *et al.*, 1999b). Drying temperature (La Croix, 1986; Poulsen, 1993), duration of prechill (Derkx and Joustra, 1997; Muller *et al.*, 1999b) and rate of imbibition of dry seeds (Muller and Bonnet-Masimbert, 1980) are critical. When seeds were pre-treated during 8 weeks and then dried to a moisture content of 9%, no decrease in germination capacity occurred. Storage of these seeds at −2°C for 16 weeks also did not reduce germination (Fig. 6.9A). However, when less dormant seeds were dried, germination capacity declined. A further decline occurred during 16 weeks of storage (Fig. 6.9B). This phenomenon may be explained by the start of early germinative events in seeds in which dormancy breaking is complete, as a result of which the seeds lose desiccation tolerance.

Conifers

Redrying and storage have been studied in a number of conifer species (Edwards, 1986). Seeds of several species, including those of *Pinus taeda* (Barnett, 1972) and *Pinus ponderosa* (Danielson and Tanaka, 1978) lost part of the prechill effect during drying, necessitating the need for re-stratification. In *Picea sitchensis* redrying did not reverse the prechill effect on germination capacity, but the rate of germination was slowed down (Jones *et al.*, 1993; Jinks and Jones, 1996). This could be explained by the fact that the seeds need more time to imbibe water. Subsequent storage of the redried seeds reimposed dormancy (Jones, 1995). In *Pseudotsuga menziesii* the benefits of long prechill treatments were preserved during drying and storage at 6.7% moisture content for periods of up to 17 months (Muller *et al.*, 1999a).

Conclusions

Controlling the moisture content of seeds is a very effective method of preventing germination of non-dormant seeds during pre-treatment, while still allowing the remaining seeds to respond to chilling temperatures. Since no premature germination occurs, the pre-treatment can be prolonged until

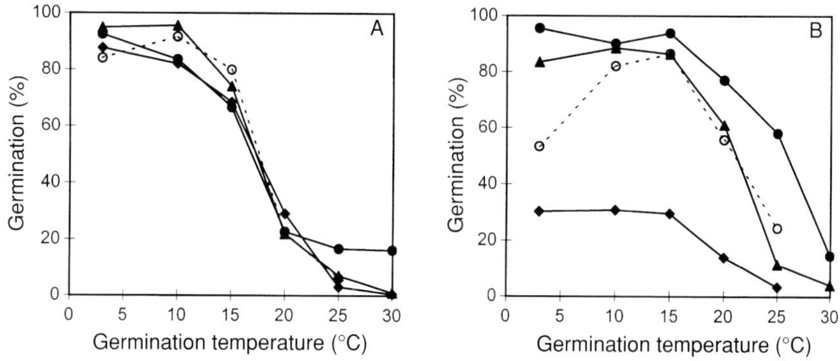

Fig. 6.9. Effects of dehydration and dry storage on germination capacity of *Fagus sylvatica* seeds, pre-treated at 3°C during 8 (A) or 16 (B) weeks. Moisture content during pre-treatment was 30%. Pre-treated seeds were dried to a moisture content of 9% and stored at −2°C. Germination was tested over a range of temperatures after pre-treatment (●), after dehydration (○) and after dry storage during 8 (▲) and 16 (◆) weeks. Germination tests as in Fig. 6.1.

even the most dormant seeds are ready to germinate. After sowing, germination is rapid and uniform, because the variability in cold requirement within the seed lot is completely overcome. Plant quality will be better because less crooked seedlings grow from unchitted seeds. As a prolonged pre-treatment gives an enormous widening in the temperature range of germination, the sowing conditions become less critical and the risk of inducing secondary dormancy in the seed bed is reduced. In many species studied so far, the duration of prechill at controlled moisture content was not very critical, thereby adding flexibility in the sowing time. Improvement of seedling emergence and more certainty on the expected number of plants give new perspectives for more efficient and environment-friendly management and continuity at nurseries.

The ability to redry and store seeds after a controlled pre-treatment would further improve flexibility in seed handling. Promising results have been obtained in some species, but further research is required to optimize drying and storage of tree seeds.

References

Barnett, J.P. (1972) Drying and storing stratified loblolly pine seeds reinduces dormancy. *Tree Planters' Notes* 23, 10–11.

Bergsten, U. (1989) Temperature tolerance of invigorated seeds of *Pinus sylvestris* L. and *Picea abies* (L.) Karst. using the TTGP test. *Forestry Supplement* 62, 107–115.

Blundell, J.B. (1973) Rootstock seed growth improved. *Gardeners Chronicle* 174, 16–19.

Brinkman, K.A. (1974) *Tilia* L. Basswood, linden. In: Schopmeyer, C.S. (ed.) *Seeds of Woody Plants in the United States,* Agriculture Handbook No. 450. US Forest Service, Department of Agriculture, Washington, DC, pp. 810–812.

Danielson, H.R. and Tanaka, Y. (1978) Drying and storing stratified ponderosa pine and Douglas-fir seeds. *Forest Science* 24, 11–16.

Derkx, M.P.M. and Joustra, M.K. (1997) Dormancy breaking and short-term storage of pre-treated *Fagus sylvatica* seeds. In: Ellis, R.H., Murdoch, A.J. and Hong, T.D. (eds) *Basic and Applied Aspects of Seed Biology.* Kluwer Academic Publishers, Dordrecht, pp. 269–278.

Downie, B., Bergsten, U., Wang, B.S.P. and Bewley, J.D. (1993) Conifer seed germination is faster after membrane tube invigoration than after prechilling or osmotic priming. *Seed Science Research* 3, 259–270.

Edwards, D.G.W. (1986) Special prechilling techniques for tree seeds. *Journal of Seed Technology* 10, 151–171.

Finch-Savage, W.E. and Clay, H.A. (1997) The influence of embryo restraint during dormancy loss and germination of *Fraxinus excelsior* seeds. In: Ellis, R.H., Murdoch, A.J. and Hong, T.D. (eds) *Basic and Applied Aspects of Seed Biology.* Kluwer Academic Publishers, Dordrecht, pp. 245–253.

Gosling, P.G. and Rigg, P. (1990) The effect of moisture content and prechill duration on the efficiency of dormancy breakage in Sitka spruce (*Picea sitchensis*) seed. *Seed Science and Technology* 18, 337–343.

Jensen, M. (1996) Breaking of tree seed dormancy at controlled moisture content. *Combined Proceedings International Plant Propagators' Society* 46, 296–304.

Jensen, M. (1997) Moisture content controls the effectiveness of dormancy breakage in *Abies nordmanniana* (Steven) Spach seeds. In: Ellis, R.H., Murdoch, A.J. and Hong, T.D. (eds) *Basic and Applied Aspects of Seed Biology.* Kluwer Academic Publishers, Dordrecht, pp. 181–190.

Jinks, R.L. and Jones, S.K. (1996) The effect of seed pre-treatment and sowing date on the nursery emergence of Sitka spruce (*Picea sitchensis* [Bong.] Carr.) seedlings. *Forestry* 69, 335–345.

Jones, S.K. (1995) Sitka spruce (*Picea sitchensis* [Bong.] Carr.) seed germination in relation to seed development, dormancy and storage. PhD thesis, University of Reading, Reading, UK.

Jones, S.K. and Gosling, P.G. (1994) 'Target moisture content' prechill overcomes the dormancy of temperate conifer seeds. *New Forests* 8, 309–321.

Jones, S.K., Gosling, P.G. and Ellis, R.H. (1993) The effects of redrying moist, prechilled, and chitted sitka spruce (*Picea sitchensis* Bong. Carr.) seeds. In: Côme, D. and Corbineau, F. (eds) *Basic and Applied Aspects of Seed Biology,* vol. 3. ASFIS, Paris, pp. 1015–1020.

Karssen, C.M. (1982) Seasonal patterns of dormancy in weed seeds. In: Khan, A.A. (ed.) *The Physiology and Biochemistry of Seed Development, Dormancy and Germination.* Elsevier Biomedical Press, Amsterdam, pp. 243–270.

Krauss, N. and Köhler, K.H. (1985) Ein Beitrag zur Kenntnis über die Stratifikation und Keimung von Eschensamen (*Fraxinus excelsior* L.). *Flora* 177, 91–105.

LaCroix, Ph. (1986) Conservation et levée de dormance des graines feuillues. *Revue Forestière Française* 3, 205–212.

Leadem, C.L. (1986) Stratification of *Abies amabilis* seeds. *Canadian Journal of Forest Research* 16, 755–760.

Muller, C. (1993) Combination of dormancy breaking and storage for tree seeds: new strategies for hardwood species. In: Edwards, D.G.W. (ed.) *Dormancy and Barriers to Germination. Proceedings of an International Symposium of IUFRO Group P2-04-00.* Pacific Forestry Centre, Victoria, BC, pp. 79–85.

Muller, C. and Bonnet-Masimbert, M. (1980) Long term storage of beechnuts. Results of large scale trials. In: Wang, B.S.P. and Pitel, J.A. (eds) *Proceedings International Symposium on Forest Tree Seed Storage.* Canadian Forestry Service, Ottawa, pp. 178–183.

Muller, C. and Bonnet-Masimbert, M. (1989) Breaking dormancy before storage: an improvement to processing of beechnuts (*Fagus sylvatica* L.). *Seed Science and Technology* 17, 15–26.

Muller, C., Falleri, E., Laroppe, E. and Bonnet-Masimbert, M. (1999a) Drying and storage of prechilled Douglas fir, *Pseudotsuga menziesii*, seeds. *Canadian Journal of Forest Research* 29, 172–177.

Muller, C., Laroppe, E. and Bonnet-Masimbert, M. (1999b) Further developments in the redrying and storage of prechilled beechnuts (*Fagus sylvatica* L.): effect of seed moisture content and prechilling duration. *Annals of Forestry Science* 56, 49–57.

Nikolaeva, M.G. (1977) Factors controlling the seed dormancy pattern. In: Khan, A.A. (ed.) *The Physiology and Biochemistry of Seed Dormancy and Germination.* Elsevier/North Holland, Amsterdam, pp. 51–74.

Poulsen, K.M. (1993) Predicting the storage life of beech nuts. *Seed Science and Technology* 21, 327–337.

Poulsen, K.M. (1996) Prolonged cold, moist pre-treatment of conifer seeds at controlled moisture content. *Seed Science and Technology* 24, 75–87.

Suszka, B. (1979) Seedling emergence of beech (*Fagus sylvatica* L.) seed pre-treated by chilling without any medium at controlled hydration levels. *Arboretum Kórnickie* 24, 111–135.

Suszka, B., Muller, M. and Bonnet-Masimbert, M. (1996) *Seeds from Forest Broadleaves from Harvest to Sowing* (translated by A. Gordon). INRA, Paris.

Tanaka, Y. and Edwards, D.G.W. (1986) An improved and more versatile method for prechilling *Abies procera* Rehd. seeds. *Seed Science and Technology* 14, 457–464.

Tylkowski, T. (1990) Mediumless stratification and dry storage of after-ripened seeds of *Fraxinus excelsior* L. *Arboretum Kórnickie* 35, 143–152.

Vegis, A. (1964) Dormancy in higher plants. *Annual Review Plant Physiology* 15, 185–224.

Villiers, T.A. and Wareing, P.F. (1964) Dormancy in fruits of *Fraxinus excelsior* L. *Journal of Experimental Botany* 14, 359–367.

Wagner, J. (1996) Changes in dormancy levels of *Fraxinus excelsior* L. embryos at different stages of morphological and physiological maturity. *Trees* 10, 177–182.

Walle, C. van de (1987) Germination uniformity of *Fraxinus excelsior* controlled by seed water content during cold treatment. *Physiologia Plantarum* 69, 645–650.

Organismic and Environmental Controls of Bud Growth in Tropical Trees*

7

Rolf Borchert

Division of Biological Sciences, Haworth Hall, University of Kansas, Lawrence, KS 66045-2106, USA

Introduction

Periodic rather than continuous growth is almost universal among trees. In tropical as in temperate trees periods of shoot growth (flushing) alternate with periods of bud rest or dormancy. Various types of dormancy have been identified in woody plants adapted to temperate climates with a relatively short growing season and large annual variations in temperature and photoperiod which strongly synchronize vegetative and reproductive tree development (Fig. 7.1A; Crabbé, 1994: Table 1; Crabbé and Barnola, 1996: Fig. 6.1). In the tropics monthly mean temperatures often vary by < 1–2°C and annual variation in photoperiod is 1 h or less (Fig. 7.2). Climatic seasonality is therefore mainly determined by duration and severity of the dry season. For example, in the semideciduous tropical forest of Guanacaste, Costa Rica, where most research to be discussed here was done, less than 5% of annual precipitation normally falls during the 5–6 month-long dry season (Fig. 7.2).

In contrast with winter cold, severe seasonal drought does not strongly synchronize periodic development of tropical trees, and bud break in vegetative or flower buds during the dry season is common (Fig. 7.1B). Periodic tree development is even less synchronized in tropical forests with a short

* Dedicated to the memory of Charles Coster (1895–1943), last Dutch Director of the Buitenzorg Botanical Garden in Java. His pioneering studies of growth periodicity, dormancy and girth growth in tropical trees, published more than 70 years ago, address many of the questions discussed in this paper.

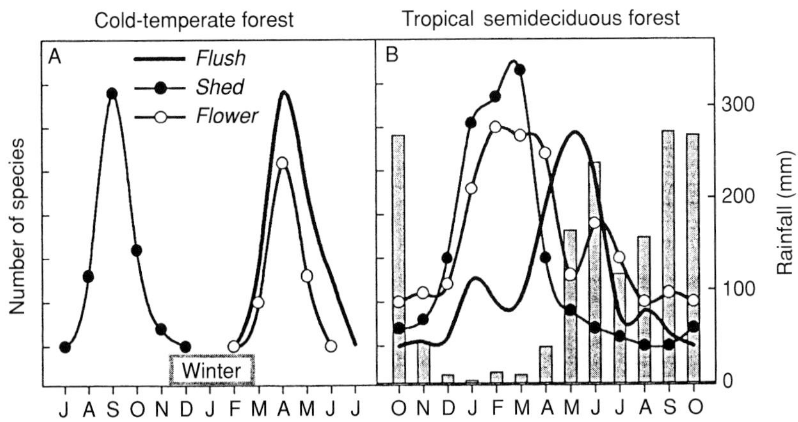

Fig. 7.1. Seasonal development (phenology) of broadleaved trees in a cold-temperate forest (A) and in a tropical semideciduous forest with a long, severe dry season (B; Guanacaste, Costa Rica). The sequence of months on the x-axis was chosen such that periods of climatic stress coincide in both graphs. Redrawn from Borchert (1998).

dry season (Borchert, 1998). The periodic arrest of shoot growth in such forests should be caused mainly by developmental constraints inherent in trees as large, complex, long-lived plants, not by environmental cues (Borchert, 1991, 1992; Crabbé and Barnola, 1996). These endogenous constraints are the consequence of size- and time-dependent changes in the functional interactions among a tree's organs and will be referred to as *correlative control*. With increasing climatic seasonality, the role of environmental changes causing the temporary arrest of bud growth become more important (e.g. control of bud dormancy of cold-temperate trees by annual variation in temperature and photoperiod). The progression from correlative to environmental control of bud growth in tropical trees will be discussed here.

Functional Tree Types in Tropical Dry Forests

In analogy to the synchronization of shoot growth periodicity by low winter temperatures (Fig. 7.1A), one might expect tropical forests with a long, severe dry season to be dominated by deciduous trees which shed leaves during the early dry season and resume growth after the first rains of the wet season (Fig. 7.2A; Fig. 7.3B,C). Instead, tropical dry forests constitute a mosaic composed of several functional tree types adapted to seasonal drought in different ways (Borchert, 1994a,b; Borchert and Rivera, unpublished). Of the functional tree types to be described here (Fig. 7.3), only *deciduous species* tolerate desiccation to very low water potentials (Fig. 7.4). Deep-rooted *evergreen* and *brevideciduous species* avoid severe water stress

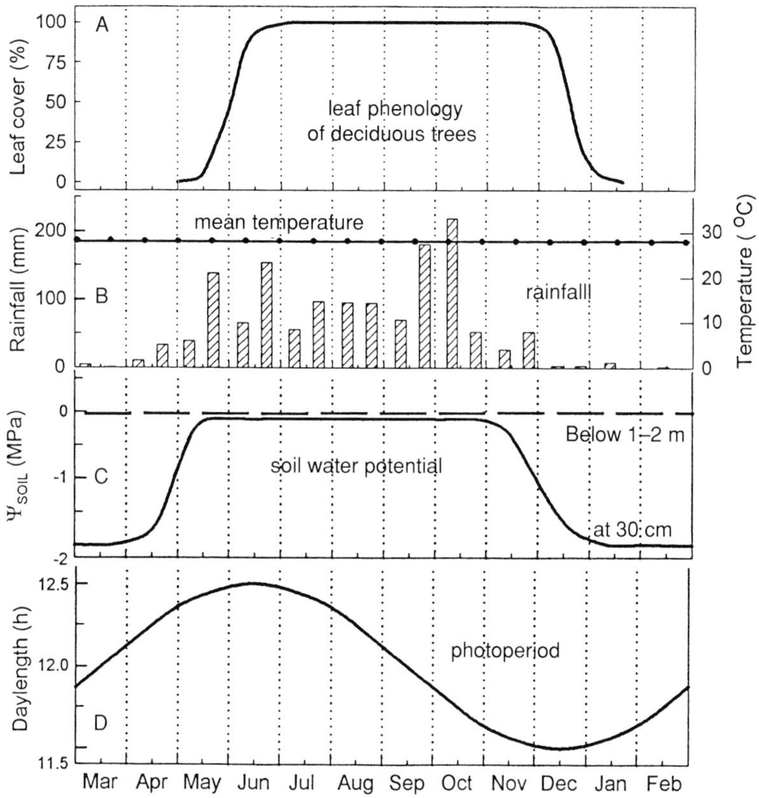

Fig. 7.2. Leaf phenology of deciduous trees (A), annual variation in rainfall and temperature (B), soil water potential (C) and photoperiod (D) in a tropical semideciduous forest in Guanacaste, Costa Rica.

during the dry season by extracting water from the moist subsoil (Fig. 7.2C), and *deciduous stem succulents* store large amounts of water in their stem tissues. Availability of stored water during climatic drought enables growth of vegetative and flower buds during the dry season and results in a large variety of species-specific patterns of bud growth (Fig. 7.3; Borchert, 1994a). Whereas the control of flower bud growth has received little attention in studies of dormancy in temperate trees, analysis of the large variety of flowering patterns (Fig. 7.3, F1–F12) is crucial to understanding the control of bud growth in tropical trees.

Deciduous species

Deciduous species have mesic leaves and often a shallow root system. As the topsoil dries out during the early dry season (Fig. 7.2, Dec), leaf and stem water potentials decline to very low values and leaves abscise (Fig. 7.4A).

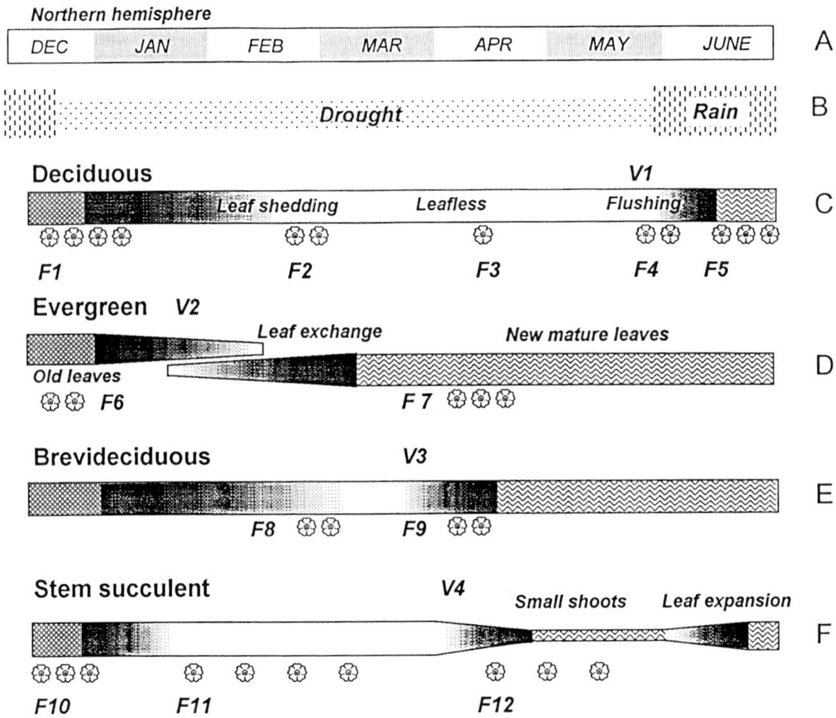

Fig. 7.3. Vegetative phenology (bars) and flowering episodes (❀) in trees of four functional types (C–F) as observed during the dry season (A, B) in a semideciduous tropical forest in Guanacaste, Costa Rica. V1–V4, bud break of vegetative buds; F1–F12, bud break of preformed flower buds or appearance of terminal inflorescences. Causes of bud break discussed in the text: rain, V1, F3, F4; declining photoperiod, F1, F6, F10; increasing photoperiod, V3, V4, F12; leaf shedding, V2, F2, F8, F11; flowering on new shoots without identifiable cause, F5, F7, F9.

Bud break of vegetative buds occurs after the first rains have resaturated the topsoil (Fig. 7.2, May–June; Fig. 7.3, V1). At moist sites, deciduous species may exchange leaves during the dry season, i.e. they become evergreen. Flowering times vary widely, both between and within species. Preformed, resting flower buds open during the dry season at various times after leaf shedding (Fig. 7.3, F2, F3) or during bud break in vegetative buds (Fig. 7.3, F4). Flowers expand in the axils of leaves on new shoots (Fig. 7.3, F5) or apical meristems are transformed into terminal inflorescences during the mid- or late wet season (Fig. 7.3, F1).

Evergreen species

Evergreen species have long-lived, coriaceous leaves and deep roots. They are confined to microsites with subsoil water reserves and maintain

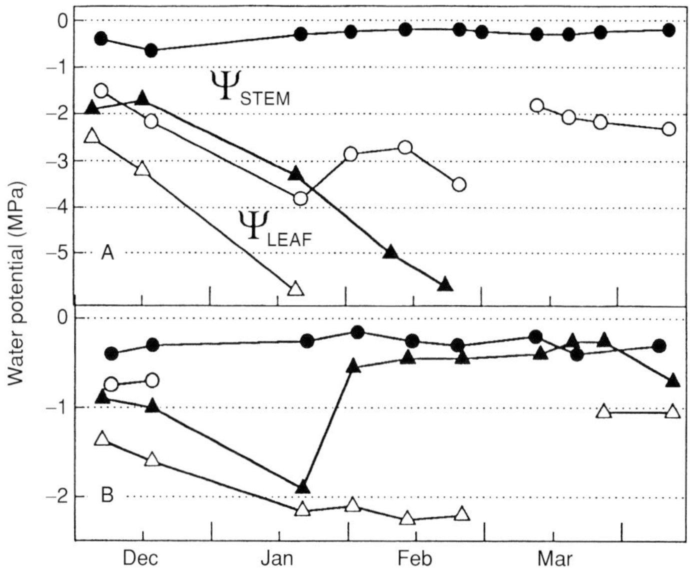

Fig. 7.4. Changes in predawn stem water potential (Ψ_{STEM}; filled symbols) and midday leaf water potential (Ψ_{LEAF}; empty symbols) associated with dry-season leaf phenology in trees of four functional groups. Gaps in Ψ_{LEAF}-values indicate that trees are leafless. A, evergreen (*Hymenea courbaril*; circles); deciduous (*Tabebuia ochracea*; triangles). B, brevideciduous (*Dalbergia retusa*; triangles); deciduous stem succulent (*Bombacopsis quinata*; circles).

relatively high stem water potentials throughout the year (Fig. 7.4A). Bud break of vegetative buds normally occurs during the early or mid-dry season shortly before or immediately after the completion of leaf shedding, i.e. trees 'exchange leaves' (Fig. 7.3D, V2; Longman and Jenik, 1988). Shoot growth often ends with the formation of a terminal inflorescence 1–2 months after flushing (Fig. 7.3D, F7).

Brevideciduous species

Brevideciduous (lat. *brevis* = short) species are similar in habit, distribution and water relations to evergreen species. However, synchronous bud break of vegetative buds generally occurs after mid-March and is preceded by a period of deciduousness which varies with the timing of leaf fall from several weeks to a few days (Figs. 7.3E, V3; Fig. 7.4B; Borchert *et al.*, unpublished). In some species preformed lateral flower buds expand when trees are leafless (Fig. 7.3E, F8), in others axillary flowers form on new shoots (Fig. 7.3E, F9).

Deciduous stem succulent species

Deciduous stem succulents have shallow root systems and mesic leaves, which are shed soon after the end of the wet season (Fig. 7.3F). Well adapted to very dry sites, succulent trees store large amounts of water in their trunks and maintain high stem water potentials throughout the dry season (Fig. 7.4B; Borchert, 1994b). In some species flushing is induced by the first heavy rains, in others bud break occurs during the late dry season, but stem elongation and leaf expansion are delayed until the first heavy rains resaturate the top soil (Fig. 7.3F, V4; Bullock and Solis-Magallanes, 1990). Trees flower at various times during the dry season (Fig. 7.3F, F10–F12).

Analysis of Bud Growth Patterns

Growth patterns discussed here were observed between 1991 and 1999 under normal climatic conditions (Figs 7.2 and 7.5) and as modified by large-scale natural experiments (Fig. 7.6: abnormal drought induced by the 1997 El Niño episode, abnormal dry-season rainfall in 1999) or irrigation (Fig. 7.6B; Borchert, 1994a,c; Borchert and Rivera, unpublished). Control mechanisms involved in bud break or growth arrest are inferred from temporal correlations between changes in environmental conditions, tree development and tree water status. Synchronous bud break of many trees in a landscape indicates an inductive environmental change such as rainfall or change in photoperiod. Correlative control is indicated by asynchronous bud break among conspecific trees in the absence of notable environmental change.

Changes in tree water status and soil water availability correlated with the initiation or arrest of bud growth were measured with a pressure chamber. Predawn stem water potential (Ψ_{STEM}) measured in defoliated twigs indicates the water status of tree branches at equilibrium with available soil water. The difference between Ψ_{STEM} and midday leaf water potential (Ψ_{LEAF}) reflects the balance between transpiration and water uptake and is an indirect measure of stomatal control (Borchert, 1994b; Borchert and Rivera, unpublished).

Correlative Control of Bud Growth

Correlative control of bud growth reflects functional interactions among the organs of a tree. Continuous shoot growth requires the provision of water, nutrients and growth regulators by other organs at rates exceeding consumption by growing shoots. As a tree increases in size and hence in the number of growing shoots and transpiring leaves, maintaining this

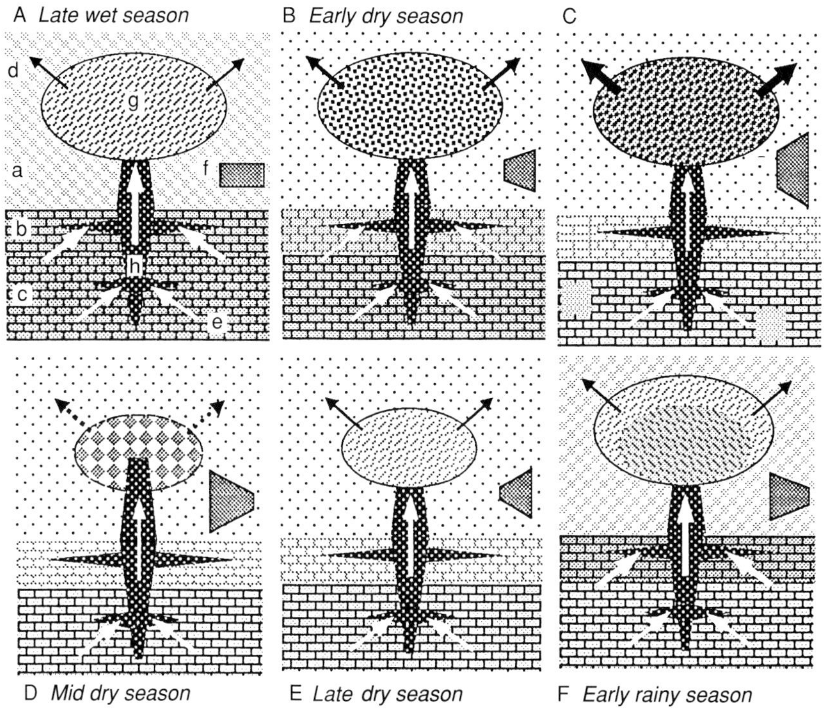

Fig. 7.5. Environmental and organismic changes resulting in leaf exchange of tropical trees during the dry season and in a second flush during the rainy season (schematic). A, Late rainy season: relative humidity (a) high; water content of topsoil (b) and subsoil (c) high; transpiration (d) low; water absorption (e) high; water stress (f) low [$\Psi_{STEM} > -1$ MPa]. B, Early dry season: relative humidity decreases; topsoil is drying out; transpiration increases; water stress increases. C, Several weeks after the last rain: topsoil is dry; water absorption is limited to deep roots (h) in moist subsoil; water stress enhances leaf senescence (g); transpiration by senescent leaves is very high; water stress becomes very high [$\Psi_{STEM} < -3$ MPa]. D, Abscission of senescent leaves: shedding of senescent leaves eliminates transpirational water loss; water absorption from subsoil continues; water stress becomes small [$\Psi_{STEM} > -0.5$ MPa]. E, Bud break and flushing: rehydration causes bud break; as leaves expand, transpiration increases; water stress increases and eventually arrests shoot growth. F, Early rainy season: relative humidity increases; top soil becomes water saturated; water absorption increases; water stress becomes small; bud break in vegetative buds initiates a second flush.

functional equilibrium becomes progressively more difficult (Borchert, 1991). In large, cold-temperate trees shoot growth therefore ceases well before the end of the growing season under climatic conditions permitting the continuing growth of tree saplings and herbs. Shoot growth may be arrested in the *absence* of adverse environmental conditions by a shortage in any one of the substances provided by other organs, i.e. by loss of the functional equilibrium among the tree's organs (Borchert, 1991; Crabbé and Barnola, 1996). Correlative control of tree growth periodicity should be carefully distinguished from *spatial controls* such as apical dominance, which determine tree architecture and will not be discussed here. As the physiological mechanisms of correlative and spatial control differ substantially, applying the same term, paradormancy, to both mechanisms appears questionable (Lang *et al.*, 1987; Crabbé, 1994).

Defoliation soon after the end of seasonal shoot growth induces bud break and flushing in temperate trees (Borchert, 1991; Crabbé and Barnola, 1996). This indicates that trees have a limited carrying capacity for leaves and attain an optimum leaf area during each growing season. Leaves are involved in the correlative feedback inhibition arresting shoot growth once this optimum has been attained. Changes in tree water balance play a central role in the correlative control of bud rest and bud break of tropical trees during the dry season, when water is the principal factor limiting tree growth (Borchert, 1991, 1992, 1994a,b).

Leaf exchange of tropical trees during the dry season

Periodic leaf exchange in evergreen trees raises several questions. First, what causes leaf exchange in trees of moist forests with moderate seasonal drought? Second, why does leaf exchange always occur during the dry season? Third, given that cell expansion is strongly inhibited by water stress, how is it possible that in semideciduous forests bud break and shoot growth of evergreen and brevideciduous trees take place during climatic drought (Fig. 7.3)? The following multi-step model of the environmental and organismic changes resulting in leaf exchange during the dry season (Fig. 7.5) provides answers to these questions.

At the end of the wet season, tree tissues and soil are nearly water-saturated and relative humidity is high (Fig. 7.5A). After the cessation of rainfall, water content of the topsoil and relative humidity of the atmosphere decline and the tree water balance becomes negative (Fig. 7.5B). The resulting water stress enhances ageing and senescence of leaves, which at that time are 8–11 months old and approaching the end of their normal life span. Reduced stomatal control of senescent leaves (Reich and Borchert, 1988) in conjunction with increased evaporative demand of the atmosphere results in high transpiration rates at a time when the topsoil has become dry and water absorption is confined to deep roots in the moist subsoil (Fig. 7.5C). The resulting increase in leaf water stress (Fig. 7.4, decline in Ψ_{LEAF}) causes abscission of senescent leaves and hence elimination of water loss by transpiration (Fig. 7.5D). Continuing absorption of subsoil water reduces tree water stress (Fig. 7.5D), and rehydration of trees (Ψ_{STEM} near 0) causes bud break in vegetative or flower buds (Fig. 7.3, V2, F2, F8; Fig. 7.5E; Borchert, 1994a,b). At the height of climatic drought, leaf shedding thus causes a

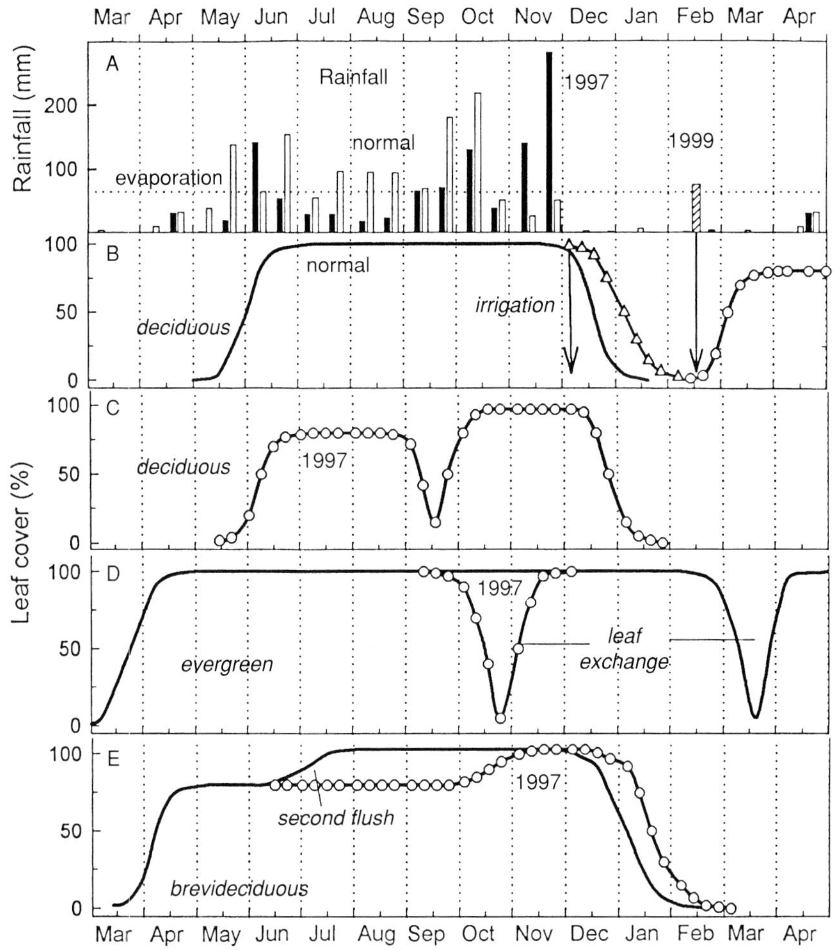

Fig. 7.6. Modification of seasonal development in a semideciduous forest of Costa Rica by abnormal drought or rain, and by irrigation. A, Normal rainfall (empty bars; mean 1991–1996), rainfall in 1997 (filled bars) and abnormal dry season rain in 1999 (crosshatched bar, Feb). B, Normal leaf phenology of deciduous trees (continuous line), delayed leaf abscission after irrigation (Dec–Jan; triangles) and flushing of leafless trees after rain during the dry season (Feb–Mar; circles). C, D, Abnormal leaf exchange caused by 1997 drought in deciduous (C) and evergreen species (D). E, Delayed second flush of brevideciduous species caused by 1997 drought.

dramatic shift in the functional relations between the root and shoot system, which results in rehydration and bud break. Shedding of senescent leaves, not environmental change, thus constitutes the indirect trigger of bud break (Borchert, 1992). The tight temporal correlation between leaf abscission and bud break of tropical evergreen and semideciduous trees has been described in several studies of tropical tree phenology, most recently for trees of a tropical savanna in northern Australia (Williams *et al.*, 1997).

Quantitative differences in the driving variables of the model may cause large variations in the timing of leaf exchange and flowering. Generally, at moist microsites rehydration after leaf shedding is faster and the interval between leaf fall and bud break is shorter than at relatively dry sites (Borchert, 1994a,b). Along an altitudinal gradient of increasing water stress during the dry season, the phenology of several species changes gradually from an evergreen to a deciduous pattern, because leaves abscise progressively earlier and bud break occurs later with increasing drought (Borchert, 1991: Fig. 10.8). The 1997 El Niño episode interrupted the normal wet season with a severe 10-week drought in June–August (Fig. 7.6A; Borchert and Rivera, unpublished). The resulting water stress caused precocious leaf abscission and an abnormal leaf exchange in deciduous and evergreen species (Fig. 7.6C,D). In contrast, the 1998/99 La Niña event brought lower than normal temperature and wind speeds during the early dry season and delayed soil drying, leaf abscission and leaf exchange by 1–2 months relative to years with severe drought.

In all trees, the periodicity of meristem activity is punctuated by the periodic replacement of leaves, and the duration of each cycle is determined primarily by leaf longevity. Thus, leaves with a life span of 11–12 months determine the annual leaf exchange of evergreen and brevideciduous trees, and some species with short-lived leaves regularly exchange leaves twice a year (Borchert, 1991: Fig. 10.8, *Erythrina*). The endogenous periodicity resulting from the finite life span of leaves can be secondarily entrained to climatic seasonality by drought-induced acceleration of leaf shedding or delay of bud break (Fig. 7.6).

Flowering after leaf shedding

In several deep-rooted deciduous species and deciduous stem succulents preformed flower buds open soon after leaf shedding and branch rehydration during the dry season (Fig. 7.3, F2, F11; Borchert, 1994a,b).

Asynchronous development within tree crowns

Many travellers to the tropics are surprised to see different stages of seasonal development displayed at the same time within the crown of large trees. The upper, leafless branches may be covered with flowers, while lower branches bear senescent leaves. In keeping with the above model (Fig. 7.5), this asynchronous development is likely to be caused by differences in water relations among parts of the crown. Leaves are often shed first in branches which experience the greatest water stress, namely those located on top of the crown or facing south (Fig. 7.7A). If soil water reserves are adequate, leaf shedding may trigger flowering, and growth of vegetative buds may start soon after flower abscission (Fig. 7.7B–D; Borchert, 1991: Fig. 10.8, *Erythrina*). As this developmental sequence progresses in the upper branches, lower branches with late leaf abscission pass through earlier stages of the sequence until the whole crown is again covered with mature leaves.

Asynchronous development may even occur on the same branch. In several deciduous species (e.g. *Gliricidia sepium, Caesalpinia eriostachys*) each axillary flower bud is released from correlative inhibition as soon as the subtending leaf has abscised (Borchert, unpublished). The correlative inhibition exerted by each senescent leaf thus extends just to its axillary bud. Accordingly, some sections of an annual shoot may bear old leaves, while axillary flowers or inflorescences are expanding on others.

Correlative arrest of shoot growth

Flushing after leaf exchange during the dry season is generally arrested within 1–2 months after bud break (Fig. 7.5D,E). Many evergreen and brevideciduous species (e.g. *Anacardium excelsum, Cassia grandis, Samanea saman*) regularly flush again during the rainy season (Figs 7.5E and 7.6E, July). In 1997, this second flush was delayed in many trees by abnormal drought until rainfall increased in September (Fig. 7.6E, Oct). These observations indicate that shoot growth is arrested by water stress resulting from an increase in leaf area and transpiration during the dry season (Fig. 7.5E). Bud growth resumes once a positive water balance has been re-established during the wet season (Fig. 7.5F). Accordingly, early leaves formed during the dry season are generally smaller than later leaves expanding during the wet season (Fig. 7.8). In some species naked terminal buds are arrested by correlative inhibition and resume growth after the first rains, in others second flushes originate from resting lateral or terminal buds of the first

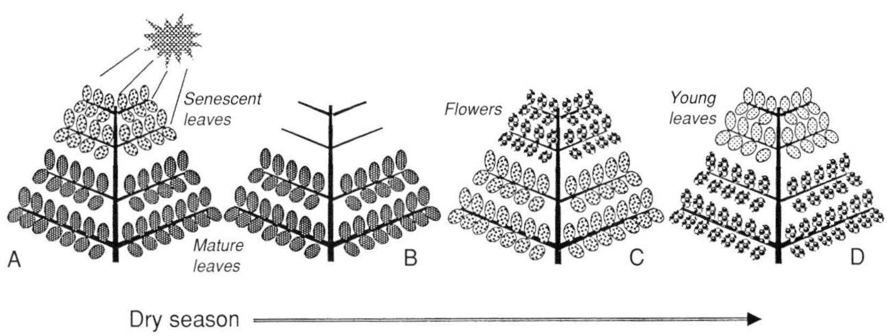

Fig. 7.7. Asynchronous dry season phenology within a tree crown (schematic). For explanation see text.

Fig. 7.8. Dry season and rainy season flushes. Left: second half of a long, continuous flush with larger leaves and axillary flower buds (*Diospyros nicaraguensis*). Right: second flush arising from an axillary bud below the terminal inflorescence on the first flush (*Godmania aesculifolia*; mature leaves cut off).

flush (Fig. 7.8). Similarly, resumption of root growth after transfer of pot-bound saplings to a larger container causes bud break in oak (Borchert, 1991). Any organismic or environmental change affecting tree water balance may thus cause the arrest or resumption of shoot growth.

Control of Bud Growth by Environmental Factors

Tropical rain forest trees tolerate neither cold nor severe drought (Borchert, 1998), but trees adapted to strongly seasonal climates have evolved tolerance of the climatic extremes of their habitat. In cold-temperate trees, buds in a state of dormancy imposed by low temperature (ecodormancy) resume growth as soon as temperature is raised to a permissive level. In autumn, complex, poorly understood interactions between declining temperature and photoperiod induce cold hardiness and endodormancy, during which buds are unable to grow even when transferred to favourable conditions (Borchert, 1991; Crabbé, 1994; Crabbé and Barnola, 1996).

Drought-imposed Ecodormancy in Deciduous Trees

During the dry season, irrigation or rainfall exceeding 20–30 mm induce rapid rehydration and expansion of preformed flower buds in deciduous trees of tropical dry forests in Central America and Mexico (Fig. 7.3, F3, F4;

Opler *et al.*, 1976; Reich and Borchert, 1982; Bullock, 1986; Borchert, 1994c). Irrigation or an exceptional 75-mm rainfall in mid-February 1999 caused rapid flushing in most leafless trees, scattered bud break on leafless twigs of leafy trees, and no bud break in trees with a full crown of old leaves (Fig. 7.6B; Borchert, 1994c; Borchert *et al.*, unpublished). Similarly, irrigation did not cause earlier than normal flushing in leaf-bearing evergreen and semideciduous trees in Panama and northern Australia (Wright and Cornejo, 1990; Myers *et al.*, 1998). Neither irrigation nor abnormal dry-season rain induced bud break in leafless stem succulent trees (e.g. *Bombacopsis, Bursera, Plumeria*; Borchert and Rivera, unpublished).

Elimination of water deficits in leafless trees by irrigation or rainfall during the dry season thus induces rapid flushing or flowering in many species. This indicates that bud rest during the dry season is generally imposed by water stress, i.e. resting buds are in a state of ecodormancy with low water potentials inhibiting bud growth. In contrast to the induction of cold hardiness in temperate trees by declining temperature and photoperiod, there is no evidence for a specific induction of drought tolerance, i.e. generally deciduous species are drought-tolerant and evergreen species are not. Irrigation or rain cannot overcome correlative inhibition of bud break by senescent leaves, which – as strong sinks for available water – are likely to inhibit rehydration of their axillary buds (Figs 7.4 and 7.5). Lack of bud break in leafless stem succulents with a high Ψ_{STEM} (Fig. 7.4B) indicates that buds are in a state of endodormancy similar to that of dormant temperate trees. In a pioneering study, Coster (1923) applied treatments known to break winter dormancy in temperate woody plants to cuttings of the stem succulent *Bombax malabaricum.* He was unable to break endodormancy at the height of the dry season, but achieved rapid bud break shortly before flushing of trees in the field.

Photoperiodic Control of Bud Growth

All instances of bud break discussed so far are preceded by environmental (rainfall, irrigation) or correlative changes (leaf abscission) which cause bud break by reducing water stress. This mechanism cannot explain the following examples of synchronous bud break not preceded by notable climatic or correlative changes. After several months of rest, lateral or terminal buds resume growth during the late rainy season and form inflorescences on leafy branches, i.e. they escape correlative bud inhibition by old leaves (Fig. 7.3, F1, F6, F10; Fig. 7.9B, Oct–Dec). In brevideciduous and stem succulent species vegetative buds expand during the late dry season, weeks before the first rains and rehydration of the top soil (Fig. 7.3, V3, V4; Fig. 7.9B, Mar–Apr; Borchert *et al.*, unpublished). In brevideciduous trees the rate of subsequent stem and leaf expansion varies widely with microsite water availability and in stem succulents shoot expansion is

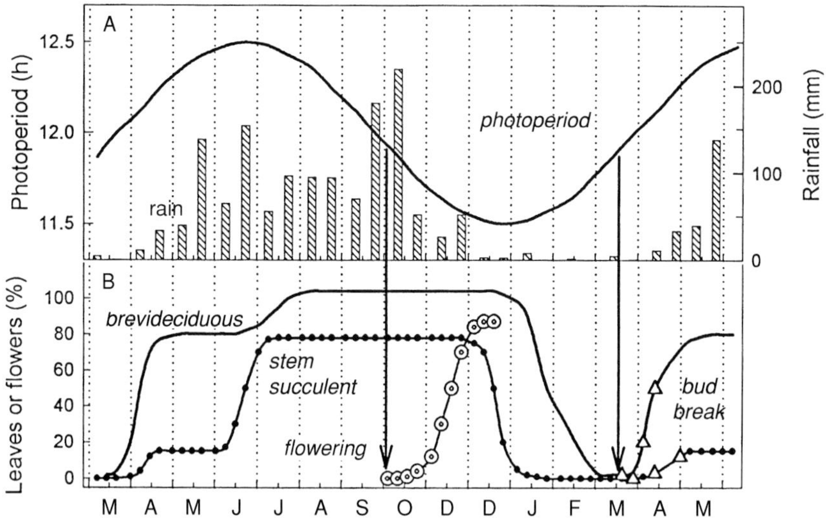

Fig. 7.9. Photoperiodic control of bud break (schematic). A, Seasonal variation in photoperiod and rainfall in Guanacaste, Costa Rica. B, Induction of flowering by declining photoperiod (Oct/Nov; dotted circles) and of bud break in brevideciduous and stem succulent species by increasing photoperiod (Mar–Apr; triangles).

delayed until after the first rains. Bud break always occurs within 1–2 weeks in all observed conspecific trees (Table 7.1). This distinct synchrony of flowering and bud break soon after the autumn and spring equinox, respectively, suggests that small changes in daylength constitute the environmental cue for bud break (Fig. 7.9; Thomas and Vince-Prue, 1997).

Experimental analysis of photoperiodic control of seasonal development is impractical in large trees. The following criteria were therefore established to screen field observations for potential photoperiodic induction of bud break. None of these criteria will prove photoperiodic control, but confirmation of several criteria in more than 20 tropical tree species provides strong indirect evidence for the control of flowering and bud break by small changes in daylength. Only a few representative observations made in 1997–1999 can be described here.

Synchrony and low inter-annual variation of bud break

Because of local variation in microsite water availability, patchiness of rainshowers, and inter-annual variation in climate, the timing of bud break induced by rehydration of twigs varies widely among conspecific trees (see section on correlative control). In contrast, bud break induced by changes in photoperiod should be synchronous throughout a landscape and occur each year at the same species-specific time.

Table 7.1. Representative trees and herbs with photoperiodic control of bud break in the semideciduous forest of Guanacaste, Costa Rica. Dates of first observation are given for short-day induced flower buds and for long-day induced bud break. N, number of observed trees. References: a, Borchert, 1996; b, Rivera and Borchert, unpublished; c, Borchert *et al.*, unpublished.

Date	Species	N	Figure	Reference
Short-day-induced flowering in herbs				
20–30 September	*Hyptis* and other herbs	> 50		b
20–30 September	*Hyparrhenia* and other grasses	> 50		b
Short-day-induced flowering in trees				
1–10 October	*Lonchocarpus minimiflorus*	> 50		b
10–20 October	*Bauhinia ungulata*	> 50		b
20–31 October	*Calicophyllum candidissimum*	> 50	7.10C	a, b
1–10 November	*Luehea speciosa*	> 30	7.10A	a, b
1–10 November	*Ochroma pyramidale*	> 20		b
20–30 November	*Cordia alliodora*	> 50	7.10B	a, b
20–30 November	*Cochlospermum vitifolium*	> 40		a, b
Long-day-induced bud break				
	Brevideciduous trees			
20–30 March	*Cassia grandis*	> 30		a, c
20–30 March	*Dalbergia retusa*	> 40	7.11A	c
20–30 March	*Guazuma ulmifolium*	> 50	7.11B	a, c
	Stem succulent trees			
10–20 March	*Plumeria rubra*	> 20	7.12	c
1–10 April	*Bursera simarouba*	> 40	7.11C	a, c
20–30 April	*Bombacopsis quinatum*	> 30		c

Between October and December, synchronous initiation of flower bud development was observed at a wide range of microsites and elevations in leafy shoots of several Costa Rican dry-forest species (Table 7.1; Rivera and Borchert, unpublished). In Mexico, more than 20 trees of four species flowered synchronously during September/October in 4 consecutive years (Bullock and Solis-Magallanes, 1990). Herbarium specimens collected over many years show flowering in such species to begin in September/October (Fig. 7.10; Borchert, 1996). Among Costa Rican trees and herbs, flower bud development induced by declining daylength started at different times between late September and December, indicating interspecific variation in critical daylength (Table 7.1).

Synchronous bud break of brevideciduous and stem succulent trees after mid-March was observed repeatedly in the semideciduous forest of Costa Rica (Fig. 7.11; Table 7.1). In a dry monsoon forest in Northern Thailand, sets of ten trees in several species flushed during the same month

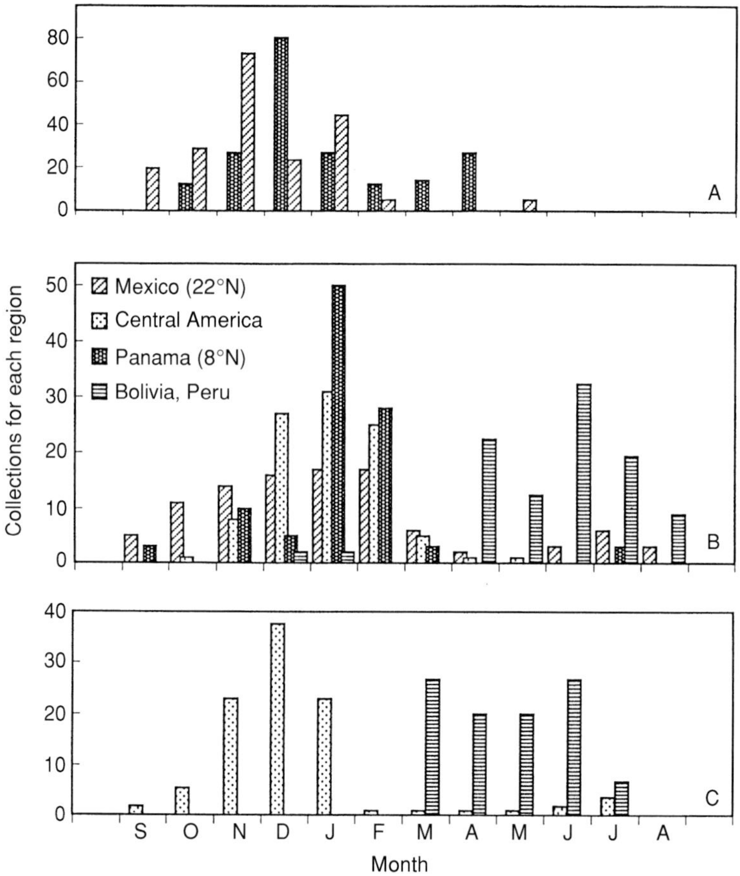

Fig. 7.10. Flowering induced by declining photoperiod in four species ranging from Mexico to South America. Flowering periodicity obtained from collection dates of herbarium specimens of the Missouri Botanical Garden, St Louis, USA (number of collections for each species in parentheses). A, *Luehea speciosa* (97); B, *Cordia alliodora* (401); C, *Calicophyllum candidissiumum* from Central America (109; left) and *Calicophyllum multiflorum* from Bolivia and Peru (15; right).

in 3 consecutive years (Borchert *et al.* unpublished). Herbarium collections of species flowering on new shoots (Fig. 7.3, F9, F12) provide long-term evidence for bud break in March/April (Fig. 7.11, right; Borchert, 1996).

Time of bud break

Short-day induced flowering and long-day induced synchronous bud break was not observed before the equinoxes. This suggests that at low latitudes, where annual variation in daylength is 1 h or less, changes in photoperiod

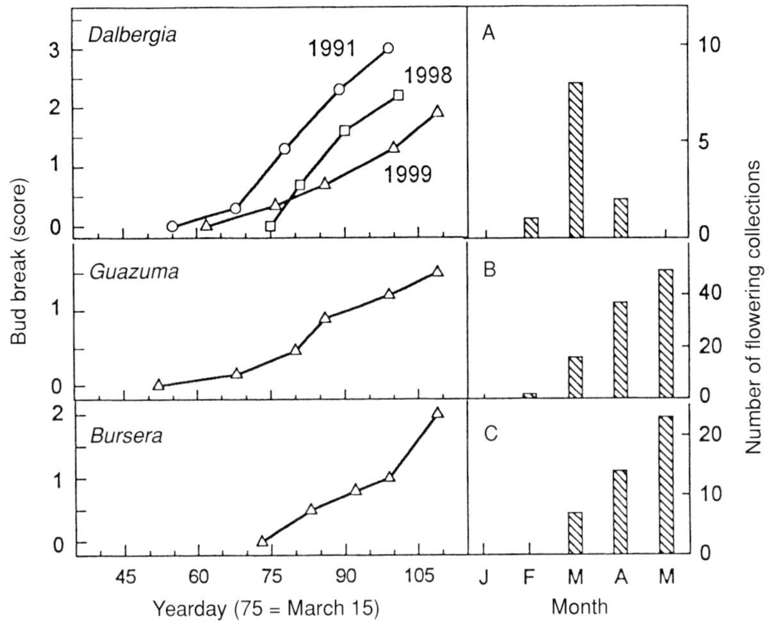

Fig. 7.11. Bud break induced by increasing photoperiod in two brevideciduous species (A, *Dalbergia retusa* (11); B, *Guazuma ulmifolia* (104) and a deciduous stem succulent (C, *Bursera simarouba* (46)). Left: bud break in various years. Right: long-term periodicity of flowering on new shoots obtained from collection dates of herbarium collections (see Fig. 7.10).

sufficiently large to trigger bud break occur only around the spring and autumn equinox, when daylength change is fastest (Fig. 7.9A).

Variation of bud break with latitude

Within wide-ranging species or genera the timing of bud break should differ by 6 months between the Northern and Southern hemispheres, as observed for several widely distributed neotropical trees (Fig. 7.10B,C; Coster, 1923; Justiniano and Fredericksen, 2000). Such species flower year-round near the equator (Borchert, 1996: Figs 3F and 7J).

Short-day induced flowering starts about 1 month earlier in subtropical Mexico (20–23°N) than in tropical Panama and northern South America (5–8°N; Fig. 7.10A,B), and flowering in subtropical Bolivia (18°S) is 6 months out of phase with Mexico (Fig. 7.10C; Justiniano and Fredericksen, 2000). These observations suggest that the critical daylength for short-day flowering is slightly longer than 12 h in ecological races of the subtropics, where rainfall ceases earlier than at low latitudes. Variation in critical daylength with latitude is well known for wide-ranging temperate tree species (Thomas and Vince-Prue, 1997).

Experimental extension of daylength

The effect of a 1.5 h daylength extension on flowering and bud break was studied in several Costa Rican herbs and tree saplings.

- *Suppression of short-day-induced flowering.* Flowering was suppressed by 1.5 h of supplementary light in several herbs and grasses flowering during the late rainy season (Table 7.1; Rivera and Borchert, unpublished).
- *Endodormancy in* Plumeria. Under natural daylength, the apical meristem of potted, watered saplings of the stem succulent *Plumeria rubra* remained dormant from September to early March, i.e. buds were in a state of endodormancy like those of *Bombax malabaricum* (Coster, 1923). Under natural daylength saplings decapitated in August had formed no shoots or very small shoots with a few leaves by December; if exposed to 1.5 h of supplementary light, plants formed shoots with large leaves (Fig. 7.12A). During a 6-week period beginning 15 January, apical cuttings stored moist in plastic bags remained dormant under natural daylength (Fig. 7.12B). If exposed to supplementary light, cuttings formed new shoots and, on the cut basal surface, callus and adventitious roots (Fig. 7.12C; Borchert and Rivera, unpublished). Thus, endodormancy of apical meristems in *Plumeria* was induced and maintained by photoperiods of <12 h, but broken by photoperiods of ~13 h.

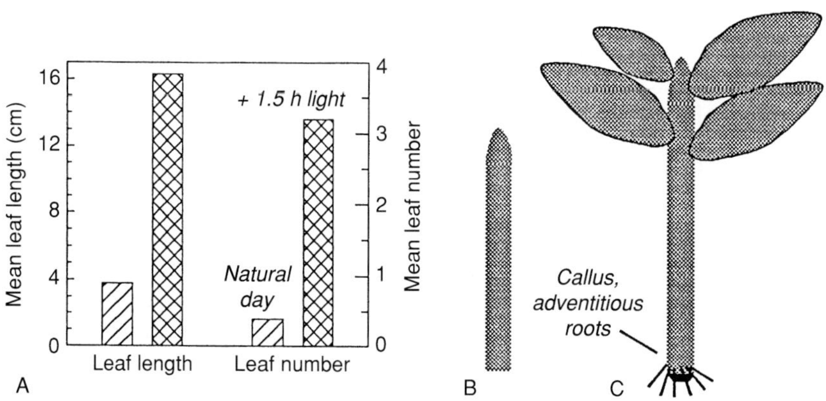

Fig. 7.12. Photoperiodic control of endodormancy in *Plumeria rubra*. Effect of 1.5 h supplementary light on leaf growth of decapitated saplings from August to December 1998 (A). Development of 20-cm-long cuttings from 15 January to 28 February, 1999 under natural daylength (B) or with 1.5 h daylight extension (C).

The combination of phenological field observations, herbarium collections and experimental results strongly supports the hypothesis that in tropical trees changes in photoperiod of < 30 min may control endodormancy and induce the transformation of vegetative into flowering meristems. Seasonal rainfall in the tropics is tightly linked with the zenithal position of the sun (Nieuwolt, 1977). Declining and increasing photoperiods thus indicate the imminent end of the rainy season and of seasonal drought, respectively (Fig. 7.9). Short-day induction of flowering is common in tropical herbs, which complete seed set before dying during seasonal drought (Ison and Hopkinson, 1985). It appears that with the exception of the woody short-day plant *Euphorbia pulcherrima* (Poinsettia), photoperiodic flower induction has not been reported for tropical or temperate trees. Stimulation of shoot and leaf growth of tropical and temperate trees by long photoperiods is well known, but the control of endodormancy by small variations in daylength has been described only once, for the African species *Hildegardia barteri* (Thomas and Vince-Prue, 1997).

Conclusions

Current concepts of types and mechanisms of bud dormancy in woody plants are based mainly on the analysis of seasonal development in cold-temperate species adapted to a climate which imposes a single pattern of vegetative and reproductive development on most broadleaved trees (Fig. 7.1A). Among tropical trees various patterns of periodic development have evolved in response to variation in the severity of seasonal drought (Fig. 7.3). Where seasonal water stress is moderate, tree growth periodicity is mainly the result of endogenous, correlative controls. The analysis of tropical trees therefore reveals characteristic aspects of correlative control, which are masked in temperate trees by the predominant role of environmental controls. Periodic replacement of leaves with a species-specific, finite life span constitutes the endogenous timer of shoot growth periodicity, and period length varies with leaf life span from 6 to 12 months. Bud break is caused mainly by changes in tree water relations resulting from leaf abscission (Fig. 7.5). Intra- and interspecific variation in the timing of bud break in vegetative or flower buds is large (Fig. 7.3).

Severe seasonal drought may entrain the endogenous growth periodicity of tropical trees to climatic seasonality by enhancing leaf abscission or delaying rehydration required for bud break (ecodormancy). Small changes in daylength may induce synchronous flowering or impose and release endodormancy in the absence of climatic change. The variety of seasonal growth patterns observed in tropical trees of different functional types (Fig. 7.3) reflects the predominance of different controls of bud growth. Spatial controls of tree architecture and correlative controls of periodic growth operate in all trees, but are supplemented in trees adapted to seasonal water stress by eco- or endodormancy induced by environmental factors.

Acknowledgements

The Program of Terrestrial Ecology of the Andrew W. Mellon Foundation supported recent field research reported here. I thank Dr Missy Holbrook, Harvard University, Dr Craig Martin, University of Kansas and Dr Guillermo Rivera, Universidad de Cordoba, for critically reading the manuscript.

References

Borchert, R. (1991) Growth periodicity and dormancy. In: Raghavendra, A.S. (ed.) *Physiology of Trees.* John Wiley & Sons, New York, pp. 221–245.

Borchert, R. (1992) Computer simulation of tree growth periodicity and climatic hydroperiodicity in tropical forests. *Biotropica* 24, 385–395.

Borchert, R. (1994a) Water storage in soil or tree stems determines phenology and distribution of tropical dry forest trees. *Ecology* 75, 1437–1449.

Borchert, R. (1994b) Water status and development of tropical trees during seasonal drought *Trees* 8, 115–125.

Borchert, R. (1994c) Induction of rehydration and bud break by irrigation or rain in deciduous trees of a tropical dry forest in Costa Rica. *Trees* 8, 198–204.

Borchert, R. (1996) Phenology and flowering periodicity of neotropical dry forest species: evidence from herbarium collections. *Journal of Tropical Ecology* 12, 65–80.

Borchert, R. (1998) Responses of tropical trees to rainfall seasonality and its long-term changes. *Climatic Change* 39, 381–393.

Bullock, S.H. (1986) Observations and an experiment on synchronous flowering. *Madroño* 33, 223–226.

Bullock, S.H. and Solis-Magallanes, J.A. (1990) Phenology of canopy trees of a tropical deciduous forest in Mexico. *Biotropica* 22, 22–35.

Coster, C. (1923) Lauberneuerung und andere periodische Lebensprozesse in dem trockenen Monsungebiet Ost-Javas. *Annales du Jardin Botanique Buitenzorg* 33, 117–189.

Crabbé, J. (1994) Dormancy. In: *Encyclopedia of Agricultural Sciences,* vol 1. Academic Press, San Diego, pp. 597–611.

Crabbé, J. and Barnola, P. (1996) A new conceptual approach to bud dormancy in woody plants. In: Lang, G.A. (ed.) *Plant Dormancy: Physiology, Biochemistry and Molecular Biology.* CAB International, Wallingford, UK, pp. 83–113.

Ison, R.L. and Hopkinson, J.M. (1985) Pasture legumes and grasses in warm climate regions. In: Halevy, A.H. (ed.) *CRC Handbook of Flowering.* CRC Press, Boca Raton, Florida, pp. 203–251.

Justiniano, M.J. and Fredericksen, T.S. (2000) Phenology of tree species in Bolivian dry forests. *Biotropica* (in press).

Lang, G.A., Early, J.D., Martin, G.C. and Darnell, R.L. (1987) Endo-, para-, and eco-dormancy: physiological terminology and classification for dormancy research. *HortScience* 22, 371–377.

Longman, K.A. and Jenik, J. (1988) *Tropical Forest and its Environment.* Longman, London.

Myers, B.A., Williams, R.J., Fordyce, I., Duff, G.A. and Eamus, D. (1998) Does irrigation affect leaf phenology in deciduous and evergreen trees of the savannas of Northern Australia? *Australian Journal of Ecology* 23, 329–339.

Nieuwolt, S. (1977) *Tropical Climatology.* John Wiley & Sons, London.

Opler, P.A., Frankie, G.W. and Baker, H.G. (1976) Rainfall as a factor in the release, timing and synchronization of anthesis by tropical trees and shrubs. *Journal of Biogeography* 3, 321–326.

Reich, P.B. and Borchert, R. (1982) Phenology and ecophysiology of the tropical tree, *Tabebuia neochrysantha* (Bignoniaceae). *Ecology* 64, 294–299.

Reich, P.B. and Borchert, R. (1988) Changes with leaf age in stomatal function and water status of several tropical tree species. *Biotropica* 20, 60–69.

Thomas, B. and Vince-Prue, D. (1997) *Photoperiodism in Plants.* Academic Press, San Diego.

Williams, R.J., Myers, B.A., Muller, W.J., Duff, G.A. and Eamus, D. (1997) Leaf phenology of woody species in a Northern Australian tropical savanna. *Ecology* 78, 2542–2558.

Wright, S.J. and Cornejo, F.H. (1990) Seasonal drought and leaf fall in a tropical forest. *Ecology* 71, 1165–1175.

8

Water Relations in Winter: Effect on Bud Break of Walnut Tree

Thierry Améglio,[1] Agnés Guilliot,[2] André Lacointe,[1] Jean Louis Julien,[2] Georges Alves,[2] Vincent Valentin[2] and Gilles Pétel[2]

[1]INRA, UA Bioclimatologie-PIAF, Domaine de Crouelle, F-63039 Clermont-Ferrand cedex 2, France; [2]UA Bioclimatologie-PIAF, Université Blaise Pascal, 24 Avenue des Landais, F-63177 Aubière cedex, France

Introduction

Bud break and shoot growth activity in woody plants is complex, consisting of several intra- and extrameristematic subprocesses where environmental factors, hormone, nutrients and water availability interact. Normal formation of leafy shoot results from the coordination of these subprocesses (Crabbé and Barnola, 1996), but the essential factor for growth is water. Nevertheless, to this day very few studies have been devoted to water availability to buds in a tree, during the winter period (Young and Houser, 1980; Astegiano *et al.*, 1988; Cottignies, 1990).

Investigation into the water status of the xylem sap of trees during the winter period (when trees do not bear leaves) shows that embolism (air blockage) can occur as a consequence of the frost–thaw alternation (Cochard and Tyree, 1990; Just and Sauter, 1991; Améglio *et al.*, 1995; Pockman and Sperry, 1997). When the sap freezes, previously dissolved gases escape due to their very low solubility in ice (Sperry and Sullivan, 1992). The resulting micro-bubbles expand when temperatures increase and eventually initiate embolism. The resulting embolized conduit no longer contributes to water transport. There is growing evidence that freezing-induced embolism can limit the growth, survival and geographic distribution of plant species (Tyree and Cochard, 1996). In a study by these authors, for some tree species (e.g. oak), embolism was total in spring and cambial growth started before bud break in order to build new active vessels. For other tree species (e.g. grapevine, beech) embolism can be resorbed (Sperry *et al.*, 1987, 1988a). The classical explanation (Pickard, 1989; Tyree and Yang, 1992; Yang and Tyree, 1992) involves an osmotic pressure difference between the xylem sap compartment and a neighbouring compartment (sap pressurization: Améglio *et al.*, 1995).

Because xylem is the only long-distance transport pathway available for water and nutrients in winter, we investigated the xylem water status during winter and the mechanism involved in its changes, in order to determine whether water is available for bud break.

Materials and Methods

Measurements were made on excised 1-year-old twigs of walnut (*Juglans regia* L. cv. 'Franquette' scions on wild walnut rootstocks) at the INRA PIAF station near Clermont-Ferrand, in south central France, from 12-year-old walnut in the orchard and 3-year-old walnut in a container. Embolism rate, sugar fluxes and pH were measured in respectively 20, 30 and 12 cm-long stem fragments of the 1-year-old twigs from the same tree.

Water status and embolism

Loss of hydraulic conductivity (embolism) was measured by Sperry's method (Sperry *et al.* 1988b). Branches were cut off in order to detect whether exudation occurred or not. When no exudation took place, the xylem sap was considered to be under tension. It was then possible to measure this tension on the removed part of the branch with a pressure chamber (Scholander *et al.*, 1965). When exudation took place, the exudate was collected and its osmotic pressure determined with a Roebling 13DR Automatic osmometer (Messtechnik, D-1000 Berlin, Germany). This osmotic pressure value obtained was considered as an approximation (in fact a higher value) of the pressure of the xylem sap of the intact branch.

In both cases, exudation or not, sap was extracted under moderate vacuum (Bollard, 1953) and its osmolarity was measured (osmometer). Using a spectrophotometer at 340 nm, sucrose, glucose and fructose contents were determined after enzymatic assays (Boehringer, 1984).

Sugar efflux

In order to investigate the effect of temperature on efflux of sugars in xylem, the shoot sections were sealed at their ends with Parafilm®, enclosed in a plastic bag to prevent desiccation, and conditioned for 48 h at either 1°C or 15°C (± 1°C) before measuring sugar efflux rates in special controlled refrigerators (Liebherr 6501). The tissues outside the cambium had been removed from both ends in order to avoid contamination with phloem constituents (Sauter, 1982). Each segment was rinsed (perfused) with 3 ml of a saline solution (KCl 1 mM, NaCl 1 mM, $CaCl_2$ 0.2 mM with a Mes/Tris buffer 0.5 mM adjusted to pH 6 with HCl) under pressure (0.1 MPa). Then, the last 0.5 ml of solution was collected and the osmolarity was measured with an osmometer, to check the rinsing efficiency (sap osmolarity ≈ 10 mosmol l^{-1} = osmolarity of the saline rinsing solution). The efflux rate was quantified as the enrichment of vessels in sugars, within 1 h, in a saline solution, which was substituted for the original xylem sap. After 1 h, each segment was retrieved and the sap was extracted and its osmolarity and sucrose, glucose and fructose contents measured as above.

Sugar influx

For influx rate measurements, stem segments were prepared as for efflux rate measurements. Each segment was rinsed (perfused) with 3 ml of the previous saline solution and the last 0.5 ml was complemented with 10 mM ^{14}C-sucrose (70 kBq ml^{-1}) and 200 mM ^{3}H-mannitol (400 kBq ml^{-1}) and kept for half an hour in the stem segment before parenchyma ^{14}C-sucrose content determination. We used the non-permeant ^{3}H-mannitol as a control for non-specific sugar adsorption on walls. ^{14}C-sucrose uptake was calculated with the following equation: ^{14}C absorbed = total $^{14}C - ([^{14}C]_{solution}/[^{3}H]_{solution}) \times$ adsorbed ^{3}H.

Sap pH and plasmalemma H^+-ATPase

The effect of fusicoccin (FC) and carbonyl cyanide *m*-chlorophenylhydrazone (CCCP), two effectors of the plasmalemma H^+-ATPase, were investigated using the perfusion technique (with the same saline solution). The perfused solution pH was measured continuously with a microelectrode (Mini Combo PH750, World Precision Instruments, USA) before and after the addition of 10 μM FC (ATPase activator). The same perfusion technique was used to test the effect of the uncoupler CCCP (10 μM) on the pH of the sap.

Results and Discussion

Hydraulic conductivity and sap pressure during winter

During winter, stems exhibited an alternation of high and low losses of conductivity (Fig. 8.1). It seems from our results on the effects of the tension–pressure alternation on the hydraulic conductivity, that embolism

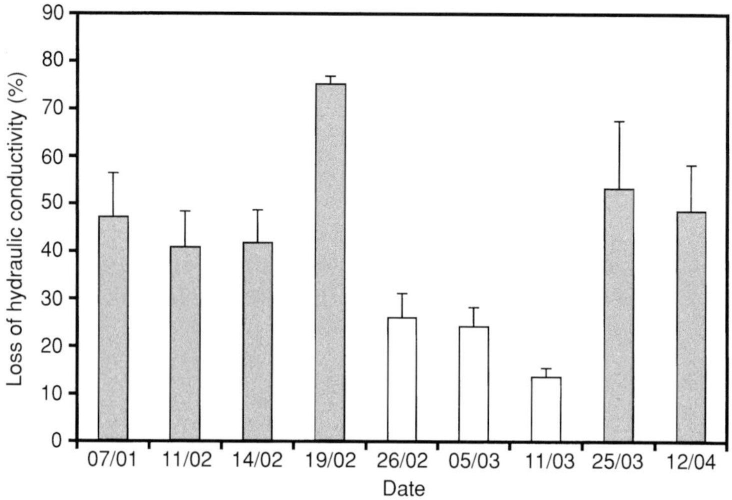

Fig. 8.1. Time course of loss of hydraulic conductivity in a twig. Alternation of high and low loss of conductivity was measured on walnut tree stems during winter. We showed that the xylem sap was under pressure when the loss of hydraulic conductivity was recovered (white bar) and under tension when hydraulic conductivity decreased (black bar). The bars represent standard errors.

is reduced in periods where sap was under pressure. At the end of winter, near bud break, hydraulic conductivity had totally recovered but at this time, root pressure could be observed which explained the vessel refilling (data not shown). These results suggested that pressure (winter and/or spring pressure) has an important effect on hydraulic conductivity.

Under natural conditions, when over a 24 h period the minimal air temperature (°C) was slightly negative followed by a slightly positive value, the sap tended to be under pressure. For temperature always above zero, the sap was under tension (Fig. 8.2). We found the same result (tension) for temperature always below zero (not shown in Fig. 8.2). In a cold greenhouse, where freezing temperatures were avoided (1°C < air temperature < 10°C), walnut trees showed a constant degradation in their water status, and the sap was under tension (Améglio and Cruiziat, 1992). Furthermore, when walnut trees previously kept in the greenhouse were transferred outside, with temperature alternating between slightly negative and slightly positive values, the water status turned positive very soon, showing strong temperature dependence. These results provided evidence that alternation between negative and positive temperatures was a major factor for sap pressure increase and *in fine* for winter embolism repair.

Sap sugar contents and osmolarity increased during winter in 1-year-old twigs (Fig. 8.3a) depending on air temperature (Fig. 8.3b). From the

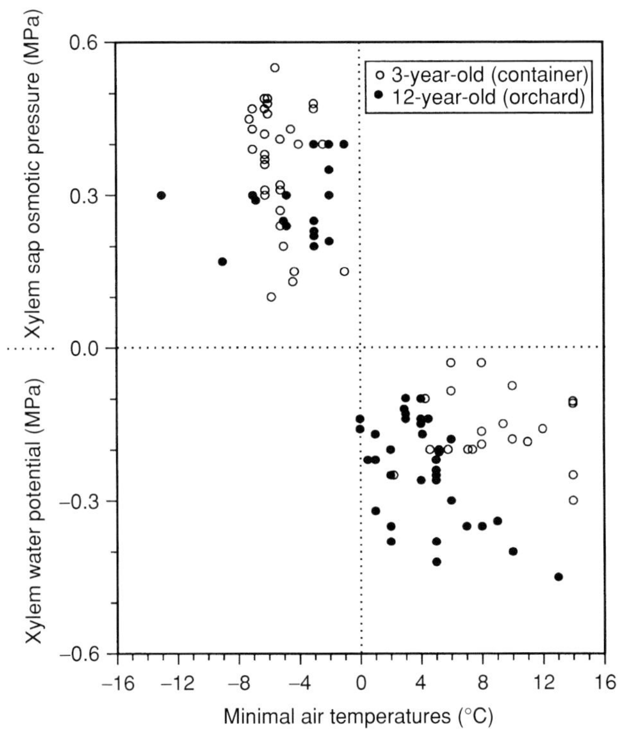

Fig. 8.2. Winter variations of xylem sap water status versus minimal air temperatures. When minimal air temperature remained above zero, the sap was under tension. We obtained the same results when daily temperature remained permanently below zero (data not shown). When minimal air temperature was slightly negative followed by slightly positive values in the same day (conditions for all dots in this figure when temperature was below zero), the sap tended to be under pressure. Closed circles represent 12-year-old walnut in an orchard. Open circles represent 3-year-old walnut in a container.

data in Fig. 8.3, we can conclude that sugars can move from parenchyma cells to xylem sap (sugar efflux) to generate pressure in the xylem allowing embolism repair when temperature was low.

Sauter (1982) described such a sugar efflux in poplar. This author also described the opposite sugar movement (from xylem sap to vessel-associated cells (VACs)) influx (Sauter *et al.*, 1973; Sauter, 1980). Influx is considered as active transport. We proposed that this sugar reabsorption by VACs at the beginning of spring allows bud growth. As presented in Fig. 8.4 and reported by several workers (Sauter, 1981; Fromard, 1990), VACs are the key point for sugar fluxes.

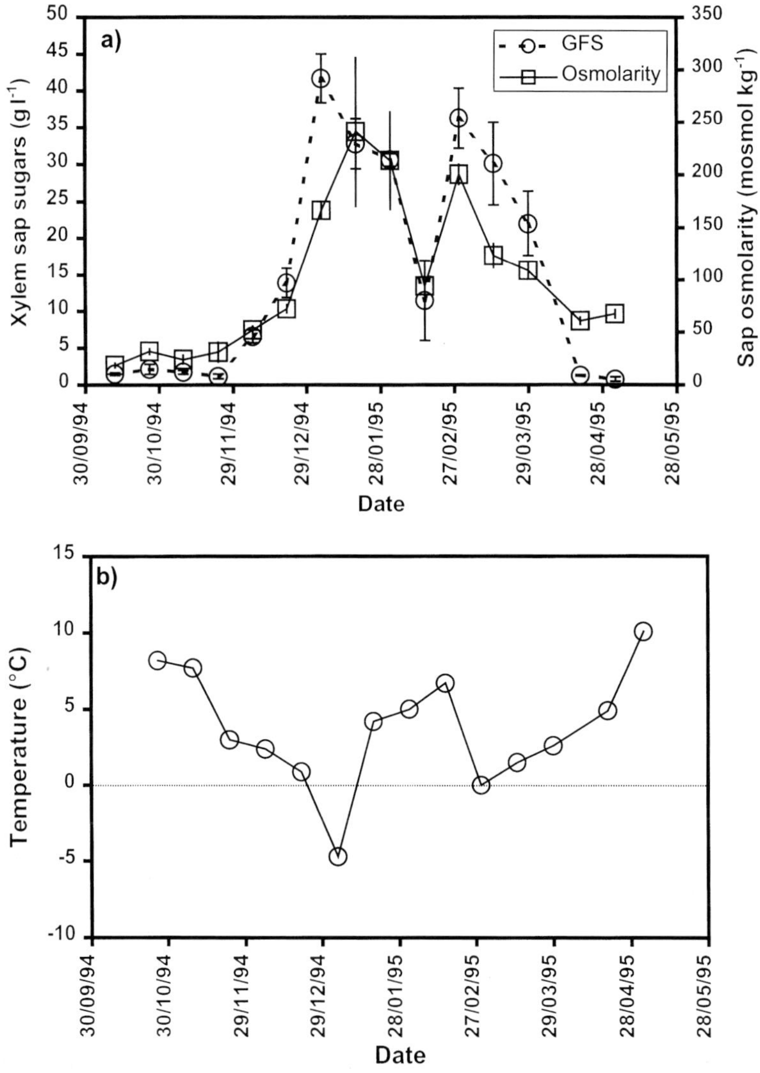

Fig. 8.3. (a) Time course of soluble sugar content and osmolarity of the xylem sap between leaf fall and bud break. Sap sugar contents and sap osmolarity increased during winter with the same dynamic. Sap osmolarity can be explained by efflux of sugars into the xylem. (b) Time course of minimal air temperature for the same day that sap was extracted. We can observe a relationship between low minimal air temperature and high osmolarity and soluble sugar content.

Estimation of sucrose flux between VACs and xylem sap

At a temperature of 1°C, efflux was high and influx was low (Fig. 8.5B), whatever the winter period tested. At 15°C, the influx rate did not increase

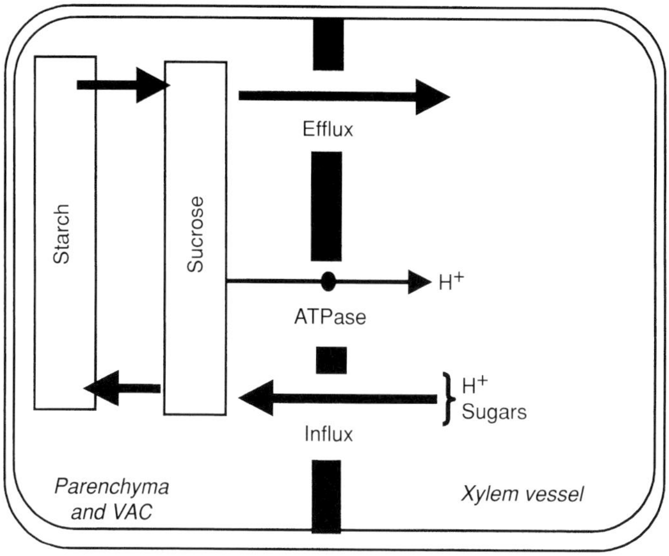

Fig. 8.4. Schematic view of sugar fluxes between parenchyma cells and xylem sap. Vessel-associated cells are a control point for sugar fluxes. In the literature, efflux is considered to be a simple diffusive flux and influx is considered to be an ATPase-dependent active transport.

but efflux decreased significantly and remained lower than influx (Fig. 8.5A). From these data, we could conclude that efflux, not influx, was temperature dependent. The high value of sucrose efflux rate after chilling temperature can be explained, as shown by Sauter (1980) on *Salix*, as the result of starch hydrolysis at 1°C, in the cells surrounding the xylem vessels. At 15°C starch hydrolysis should be lower, leading to a lower efflux. In March, influx rate was more than twice as high as in December or January. Sugar influx is generally considered to occur via H^+/sugar symports (Sauter, 1982; Fromard, 1990) depending on the pH gradient generated by a plasma membrane H^+-ATPase. In an attempt to understand the influx increase between January and March, we decided to measure the pH gradient to estimate the ATPase activity in March and January.

Plasmalemma H^+-ATPase involvement

A perfusion technique allowing the use of ATPase effectors (FC; CCCP) was carried out to study the enzyme activity (Fromard *et al.*, 1995). In walnut trees, during winter, the uncoupler CCCP reveals a low pH gradient, between the vessels of the xylem and the VAC. The H^+-ATPase activator, FC, had no effect on the pH of the perfusion solution suggesting that the enzyme is slightly active. Just before bud break, a high pH gradient was

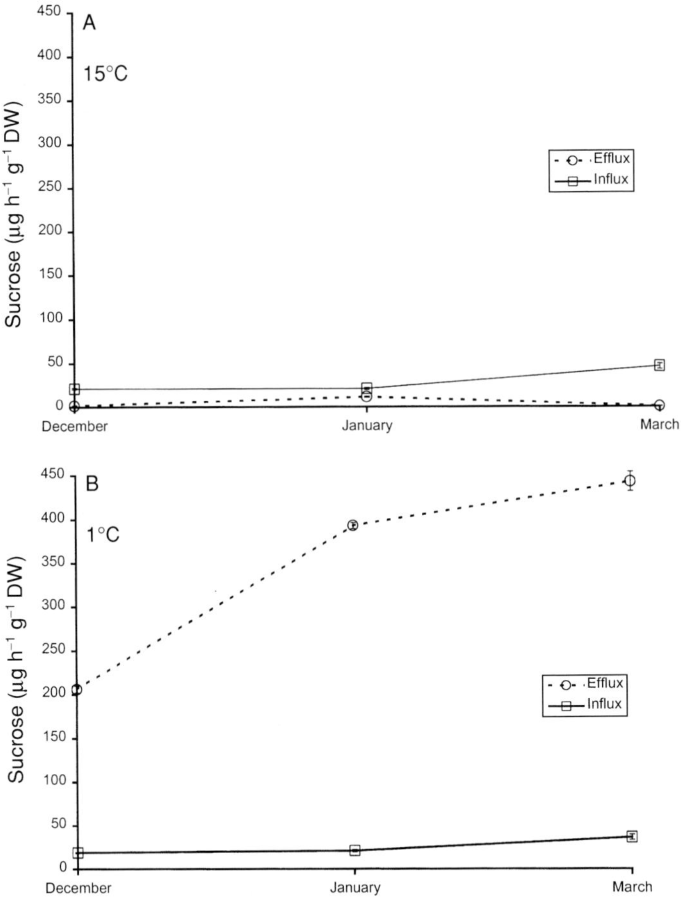

Fig. 8.5. Measurements of sugar fluxes (influx and efflux) for three dates (December, January and March) and for two temperatures (15°C and 1°C).

measured and an acidification of the perfusion solution was observed in the presence of FC (Fig. 8.6). For cold-deprived trees, kept in a greenhouse during winter at 15°C, no significant difference in the pH gradient in winter and spring was observed (data not shown). Moreover, no significant difference of the FC effect was observed between both periods. These results confirm the involvement of the H^+-ATPase in the control of the sugar influx and its role in key physiological mechanisms such as bud break and xylem refilling.

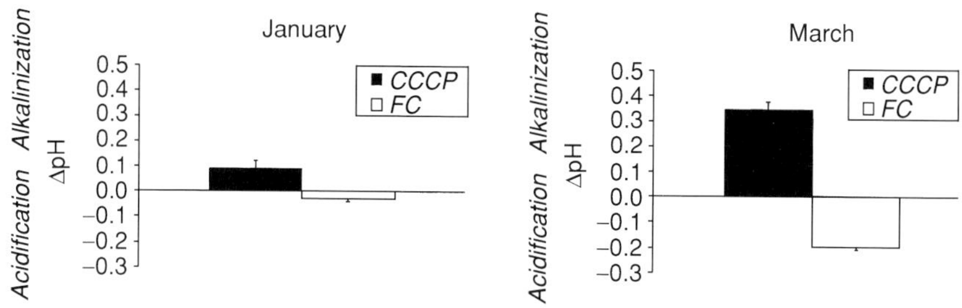

Fig. 8.6. Effect of fusicoccin (FC: white bar) and carbonyl cyanide *m*-chlorophenylhydrazone (CCCP: black bar) on the pH of the solution perfused through 1-year-old segments in winter and in early spring (mean values ± standard errors ($n > 8$)).

Conclusions

In walnut, embolism increased during the winter period, as a consequence of the frost–thaw alternation, but for this tree, embolism was totally repaired before bud break by winter stem pressurization. These variations of the water status are temperature dependent (Améglio and Cruiziat, 1992) and due to water and sugar fluxes between the different compartments of the wood. In parenchyma cells, the hydrolysis of the starch increases and starch is converted into soluble sugars, mainly sucrose. These soluble sugars are transported into the xylem vessels. Sucrose efflux increased with low temperature and this efflux increased osmotic pressure of the sap. In these conditions, by osmosis, water can then be driven into xylem vessels to repair embolism. For walnut, the VACs controlled the variation of the sugar content of the vascular sap. Efflux rate was high during winter (depending on temperature), whereas influx was low during dormancy regardless of temperature. Influx increased shortly before bud break, depending on the pH gradient generated by a plasma membrane H^+-ATPase. At this time, the vascular sap was a sugar source for growing tissues.

We are presently starting some studies based on the same hypothesis in cold-deprived conditions (greenhouse with temperature > 15°C) to understand the 'erratic' bud breaking pattern. The first results show no sugars in the sap of cold-deprived trees (15°C), but high soluble sugar contents in parenchyma cells (Améglio and Lacointe, unpublished data). This seemingly paradoxical result suggests that the sugars released into the xylem lumen (efflux) are immediately retrieved (influx), possibly due to an increased maintenance respiratory activity, resulting in a 'futile' cycle between influx and efflux. Our hypothesis, which we are currently testing, is that this strong winter activity consumes carbohydrate reserves that will be lacking for bud break and shoot outgrowth.

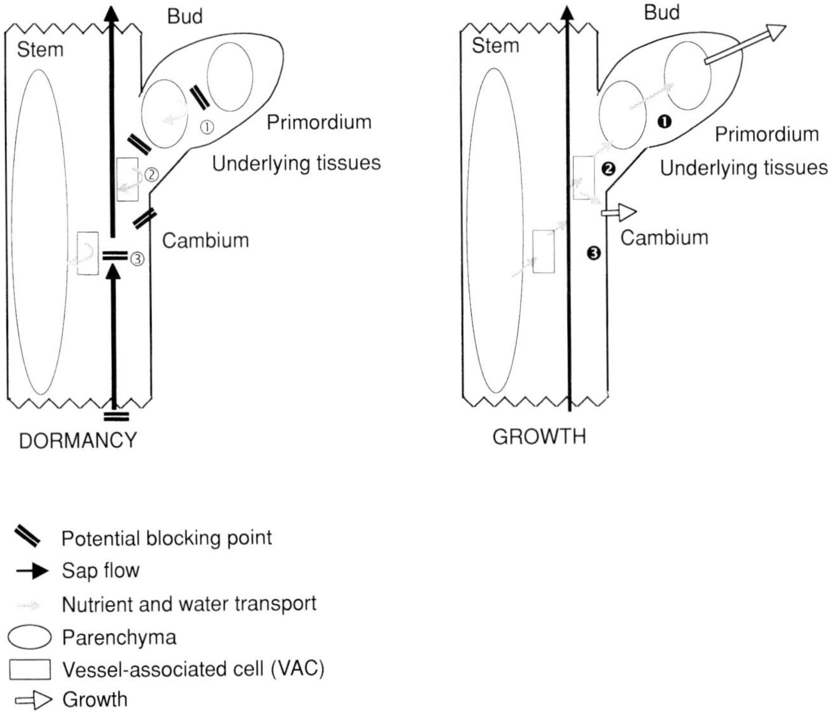

Fig. 8.7. Schematic view of putative blocking points for bud break and growth control. Vessel-associated cells (VACs) represent an important point for embolism repair and sugar influx. Before bud break, vascular sap is a sugar source for growing tissues.

In any case, the functioning of VACs seems to be a very important point for bud break and growth control in walnut (Fig. 8.7).

Acknowledgements

This work was supported in part by Action Incitative sur Programme INRA 'Biologie du Développement'. We would like to thank Christian Bodet, Marc Vandame and Stéphane Ploquin, who contributed to its progress.

References

Améglio, T. and Cruiziat, P. (1992) Tension/pressure alternation in walnut xylem sap during winter: the role of winter temperature. *Comptes Rendus de l'Académie des Sciences, Paris* (série III) 315, 429–435.

Améglio, T., Cruiziat, P. and Béraud S. (1995) Tension/pressure alternation in walnut xylem sap during winter: effect on hydraulic conductivity of twigs. *Comptes Rendus de l'Académie des Sciences, Paris* (Life sciences) 318, 351–357.

Astegiano, E.D., Maestri, M. and Estevao, M.M. (1988) Water stress and dormancy release in flower buds of *Coffea arabica* L.: water movement into the buds. *Journal of Horticultural Science* 63 (3), 529–533.

Boehringer, S.A. (1984) *Methods of Enzymatic Food Analysis using Single Reagents.* Boehringer Mannheim GmbH, Mannheim, Germany.

Bollard, E.G. (1953) The use of tracheal sap in the study of apple-tree nutrition. *Journal of Experimental Botany* 4, 363–368.

Cochard, H. and Tyree, M.T. (1990) Xylem dysfunction in *Quercus*: vessel sizes, tyloses, cavitation and seasonal changes in embolism. *Tree Physiology* 6 (4), 393–407.

Cottignies, A. (1990) Potentiel osmotique et potentiel hydrique du bourgeon terminal de Frêne, au cours du cycle annuel. *Comptes Rendus de l'Académie des Sciences, Paris* (série III) 310 (5), 211–216.

Crabbé, J. and Barnola P. (1996) A new conceptual approach to bud dormancy in woody plants. In: Lang, G.A. (ed.) *Plant Dormancy.* CAB International, Wallingford, UK, pp. 83–113.

Fromard, L. (1990) Etude du transport des glucides entre les cellules associées aux vaisseaux (CAV) et les vaisseaux chez *Acer platanoides* et *Robinia pseudoacacia.* Incidence des facteurs saisonniers et climatiques. Thèse (Dr. Univ.), Biologie et Physiologie Végétales. Faculté des Sciences, Université de Poitiers, France.

Fromard, L., Babin, V., Fleurat-Lessard, P., Fromont, J.-C., Serrano, R. and Bonnemain, J.-L. (1995) Control of vascular sap pH by the vessel-associated cells in woody species. *Plant Physiology* 108, 913–918.

Just, J. and Sauter, J.J. (1991) Changes in hydraulic conductivity upon freezing of the xylem of *Populus* × *Canadensis* Moench 'robusta'. *Trees* 5, 117–121.

Pickard, W.F. (1989) How might a tracheary element which is embolized by day be healed by night? *Journal of Theoretical Biology* 141, 259–279.

Pockman, W.T. and Sperry, J.S. (1997) Freezing-induced xylem cavitation and the northern limit of *Larrea tridentata. Oecologia* 109, 19–27.

Sauter, J.J. (1980) Seasonal variation of sucrose content in the xylem sap of *Salix. Zeitschift für Pflanzenphysiologie* 98, 377–391.

Sauter, J.J. (1981) Sucrose uptake in the xylem of *Populus. Zeitschrift für Pflanzenphysiologie* 103, 165–168.

Sauter, J.J. (1982) Efflux and absorption of sugars in the xylem. I. Seasonal changes in sucrose efflux in *Salix. Zeitschrift für Pflanzenphysiologie* 106, 325–336.

Sauter, J.J., Iten, W. and Zimmermann, M.H. (1973) Studies on the release of sugar into the vessels of sugar maple (*Acer saccharum*). *Canadian Journal of Botany* 51, 1–8.

Scholander, P.F., Hammel, H.T., Bradstreet, E.D. and Hemmingsen, E.A. (1965) Sap pressure in vascular plants. *Science* 148 (3668), 339–346.

Sperry, J.S. and Sullivan, J.E.M. (1992) Xylem embolism in response to freeze–thaw cycles and water stress in ring-porous, diffuse-porous, and conifer species. *Plant Physiology* 100, 603–613.

Sperry, J.S., Holbrook, M., Zimmermann, M.H. and Tyree, M.T. (1987) Spring filling of xylem vessels in wild grapevine. *Plant Physiology* 83, 414–417.

Sperry, J.S., Donnely, J.R. and Tyree, M.T. (1988a) Seasonal occurrence of xylem embolism in sugar maple (*Acer saccharum*). *American Journal of Botany* 75, 1212–1218.

Sperry, J.S., Donnely, J.R. and Tyree, M.T. (1988b) A method for measuring hydraulic conductivity and embolism in xylem. *Plant, Cell and Environment* 11, 35–40.

Tyree, M.T. and Cochard, H. (1996) Winter and summer embolism in oaks: impacts on water relations. *Annales des Sciences Forestières* 53, 173–180.

Tyree, M.T. and Yang, S.C. (1992) Hydraulic conductivity recovery versus water pressure in xylem of *Acer saccharum. Plant Physiology* 100, 669–676.

Yang, S.C. and Tyree, M.T. (1992) A theoretical model of hydraulic conductivity recovery from embolism with comparison to experimental data on *Acer saccharum. Plant, Cell and Environment* 15 (6), 633–643.

Young, E. and Houser, J. (1980) Influence of Siberian C rootstock on peach bloom delay, water potential and pollen meiosis. *Journal of the American Society for Horticultural Science* 105 (2), 242–245.

Morphological Structure and Water Status in Tulip Bulbs during their Transition from Dormancy to Active Growth: Visualization by Magnetic Resonance Imaging

9

Rina Kamenetsky,[1] Hanita Zemah,[1] Annette van der Toorn,[2] Henk Van As[2] and Peter Bendel[3]

[1]ARO, Department of Ornamental Horticulture, The Volcani Center, PO Box 6 Bet Dagan, 50-250 Israel; [2]Department of Biomolecular Sciences, WNMRC, Wageningen, The Netherlands; [3]Department of Chemical Services, MR Center, Weizmann Institute of Science, Rehovot, Israel

Introduction

Evolution of geophytes in climatic areas with marked seasonal changes has led to their adaptation to the periods of high and low temperatures and/or drought. One of these adaptations is a state of rest (dormancy), in which bulbs do not exhibit any visible external growth (Le Nard and De Hertogh, 1993). In many bulbs, the processes of organogenesis inside the bulb during the rest period, growth and flowering, are temperature dependent (Hartsema, 1961). In tulip, low temperatures simultaneously induce stem elongation of the current year and formation of the lateral buds for the next year (Le Nard and Cohat, 1968). Many of the morphological and biochemical changes are known as initial responses to cold induction in flower bulbs (Hartsema, 1961; Le Nard *et al.*, 1988; Le Nard and De Hertogh, 1993). However, none of the known parameters correlates significantly with the completion of chilling (Hobson and Davies, 1977; Banasik *et al.*, 1980; De Hertogh *et al.*, 1983; Walch and Van Hasslt, 1991; Lambrechts *et al.*, 1994; Rebers *et al.*, 1995). In most geophytes, the initial response to cold treatment occurs within the storage organ. Therefore, research into these processes requires destructive measures, which prohibits the continuous developmental analysis of individual plants, and of post-planting response studies. Magnetic resonance imaging (MRI) allows us to visualize morphological structures and to measure the integration and coordination of processes at the organ and whole plant level non-destructively. MRI spatially resolved images can be obtained from the same bulb as it develops during storage under various environmental conditions and/or as a result of its physiological evolution.

In general, the MRI parameters of water provide information on processes occurring at the cellular level: the distribution of water over different cell compartments, the permeability of the membranes between water compartments, and the concentrations of dissolved sugars. MRI is a powerful tool for measuring certain biophysical properties of the water, such as proton density, the diffusion constant and the relaxation times T_1 (longitudinal or spin–lattice relaxation) and T_2 (transverse or spin–spin relaxation) (Chudek and Hunter, 1997).

Recently, MRI has been used successfully in studying features of biological systems, such as water distribution in fruits, frost damage in buds, the effect of cold treatment on water status in plant tissues, and water flow in plant root and stem vascular systems (Ratkovic *et al.*, 1982; Ratcliffe, 1994; Chudek and Hunter, 1997; Clark *et al.*, 1997). It was suggested that in the fruit trees, low temperatures affect the transition of water associated with the large molecules (so-called bound water) to free water (Faust *et al.*, 1991; Liu *et al.*, 1993; Millard *et al.*, 1993; Sugiura *et al.*, 1995). In geophytes, MRI measurements have been used to estimate water status and to monitor the free water content in tulip (Iwaya-Inoue *et al.*, 1996; Okubo *et al.*, 1997) and *Allium* bulbs (Yamazaki *et al.*, 1995; Zemah *et al.*, 2000). It was suggested that physiological changes occurring during bulb storage may be reflected by the water physical state and distribution.

In our study, a MRI survey was performed using two different storage protocols to follow changes in the morphology and water status of various components of intact tulip bulbs. Special attention was given to the determination of relaxation times and diffusion parameters in different tissue types.

Materials and Methods

Plant material and temperature treatments

Bulbs of *Tulipa gesneriana* L., cv. 'Apeldoorn', 10–11 cm in circumference were obtained from the Bulb Research Centre in Lisse, The Netherlands. One hundred mature bulbs were harvested in June 1998, and subsequently sorted, cleaned and dried prior to storage at 20°C for 2 months. On 6 September, after the first MRI measurements, 50 bulbs were randomly selected and placed at 4°C for 4 months (chilled bulbs). Fifty additional bulbs were kept at 20°C (non-chilled bulbs). Sixteen intact bulbs per treatment were sampled, tagged and studied by MRI at Wageningen NMR Centre (Wageningen Agricultural University, The Netherlands). MRI experiments were performed five times at 0, 2, 4, 8 and 12 weeks of storage.

After storage, 12 bulbs of each treatment were planted in 15-cm diameter pots (one bulb per pot). The pots were placed in the Phytotron in The Volcani Center, Bet Dagan, Israel, 20/12°C day/night, respectively, under a 10-h photoperiod. Plant growth analysis was performed from planting to the completion of flowering.

Intrabulb development and bulb anatomy: microscopic observations

During storage, morphological and developmental analyses of bulbs were performed twice a month. Organ initiation and development were studied with light stereoscopy (Stemi 2000-C, Zeiss, Germany) and scanning electron microscopy (SEM) (JEOL Jsm-5410LV, Japan), at low vacuum.

MRI experiments

Single parameter images representing the proton density, relaxation time T_2 and diffusion coefficient were obtained by use of multi-echo imaging at a magnetic field strength of 0.47 T (Donker and Van As, 1996; Donker *et al.*, 1997). A custom-engineered coil/gradient set was used, containing a transmitter/receiver coil with a diameter of 50 mm and gradients in three directions with strengths up to 500 mT m^{-1} (Doty Scientific Inc., Colombia, South Carolina, USA). The cylindrical probe bore had a diameter of 45 mm and was accessible from both ends. The tulip bulbs closely fitted the bore and were inserted into the magnet on top of a sample holder. T_2 relaxation data were obtained using the multi-echo imaging sequence described by Edzes *et al.* (1998): 48 echoes, first detected echo at 7.1 ms, later inter-echo delays (IED) 4.7 ms, time-to-repeat (TR) 1.0 or 1.5 s, field of view (FOV) 45 mm, 128×128 data points, slice-thickness 3 mm, two averages.

The apparent diffusion coefficient (ADC) was obtained by using the turbo-spin-echo sequence as described by Scheenen *et al.* (1998): TR 1.5 s, first detected echo at 29.3 ms, later inter-echo delays 4.6 ms, $\delta = 4$ ms, $\Delta = 20$ ms, six diffusion gradient steps from 68 to 410 mT m^{-1}, b_{max} 3.61×10^9 s m^{-2}, slice-thickness 3 mm, FOV 45 mm, 128×128 points, four averages. In this sequence 16 encoding steps were included in a single multi-echo scan, thereby reducing the acquisition time needed for a 128×128 matrix to only eight multi-echo scans. Image acquisition was repeated six times, with increasing diffusion weighting along the *y*-direction (parallel to the central axis of the bulb).

In order to obtain three-dimensionally resolved data of the planted bulbs, a turbo spin-echo sequence was used with two phase-encoding gradients and the usual read gradient (FOV $60 \times 60 \times 45$ mm^3, TR 1 s, first detected echo at 6 ms, later inter-echo delays 5 ms, $128 \times 128 \times 128$ points, four averages, total acquisition time about 1 h).

Data Processing

The series of 48 echoes from the multi-echo experiment were Fourier transformed and the absolute value images were fitted on a pixel-by-pixel basis to a mono-exponential decay function. The fit resulted in two images. One image displayed the extrapolated amplitude of the decay curve at time 0 and the other image displayed the T_2 relaxation time. The first image represents tissue density times water content per pixel (Donker *et al.*, 1997). The other image displayed the T_2 relaxation time and was used to manually delineate the whole tulip bulb (except for the regions where the T_2 relaxation was deteriorated because the region was outside the sensitive volume of the coil, usually the upper part of the bulb), the bud and the basal plate. The scale region was determined by subtracting the bud and basal plate area from the area delineating the total bulb. The mean R_2 $(1/T_2)$ value[1] of each organ was calculated by determining the average R_2 of the pixels in the selected region of the image. Only pixels with positive relaxation rate values were used. The images of proton density were segmented with the identical procedure. However, the signal intensities in the images for each bulb could differ as a result of differences in the hardware adjustments. Therefore, the signal intensity originating from the whole bulb was used to normalize the image intensities.

The data for the ADC were obtained similarly. Diffusion-weighted data were Fourier transformed and the real parts of the phase-corrected images were used to fit them to a mono-exponential decay on a pixel-by-pixel basis. This resulted in an amplitude and ADC images for the diffusion-weighted data. Mean values of ADC were calculated separately for each organ. Since the mean values for each organ do not reveal the heterogeneity of the tissue within the organ, the numbers of pixels which had parameter values within a specified range were displayed as histograms. The histograms were produced for each bulb organ within both treatment groups and were

[1] Fast relaxation corresponds to low values of T_2 (the relaxation time) or high values of R_2 (the relaxation rate). The use of either T or R values to describe relaxation is completely equivalent, and is a matter of semantics or convenience. While the use of the T value is more popular, the use of the R value is sometimes more appropriate, because when several mechanisms contribute to the relaxation, the R_2 values from each mechanism can simply be added to obtain the overall relaxation rate. The same applies to spin–lattice relaxation (T_1 or R_1).

used to verify the trends and changes observed for mean parameter values. Student *t*-tests were used for statistical analysis of these mean values within each treatment group and between treatment groups at each time point.

Results

Bulb anatomy and development: destructive and non-destructive approaches

In August, before the storage period, bulbs were dissected and checked for their developmental stage. 'Stage G', indicating initiation of the gynoecium, was found in all dissected bulbs. During subsequent bulb development, we detected by MRI changes in size and shape of the storage scales, leaf and floral scape elongation, and the enlargement of the central daughter bulb. MRI permits visualization of the basal plate, which consists of heterogeneous components, the four storage scales, and the developing monocarpic shoot with several leaf primordia and the developing flower (Fig. 9.1).

The daughter bulb started to develop during the storage period. At this time, the scales were composed of relatively small cells of 25–28 μm in size. A gel-like substance filled up the inter-scale space, as is visible on Fig. 9.2A. The scales of the mother bulb contained large cells about 125 μm in diameter (Fig. 9.2B). The starch granules of different sizes were heavily packed in the internal cell spaces. At the end of the storage period, in

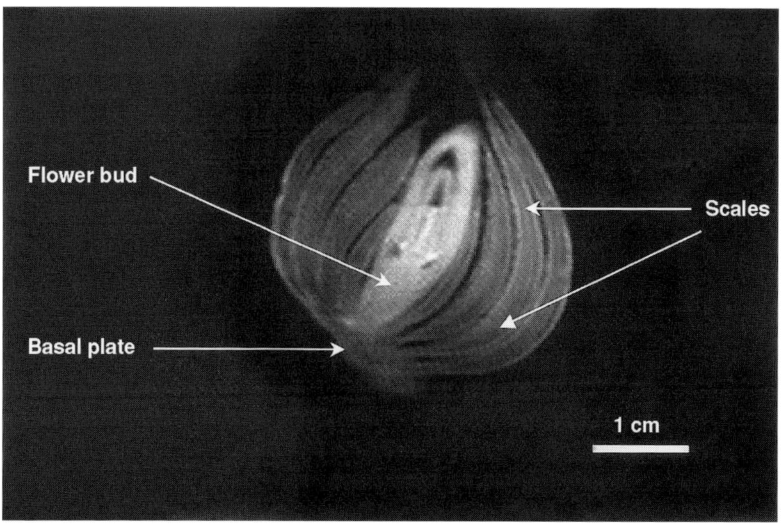

Fig. 9.1. Three-dimensionally resolved MR image of non-chilled tulip bulb before planting, December 1998. Four scales and the basal plate can be distinguished. The central bud consists of the leaf primordia and the developing flower.

December, degradation of storage material was visible within the scales. Granule destruction was initiated in both lateral parts, however, the degree of destruction was higher in the inner part of the scale. Starch degradation was observed in the scales of both chilled and non-chilled bulbs.

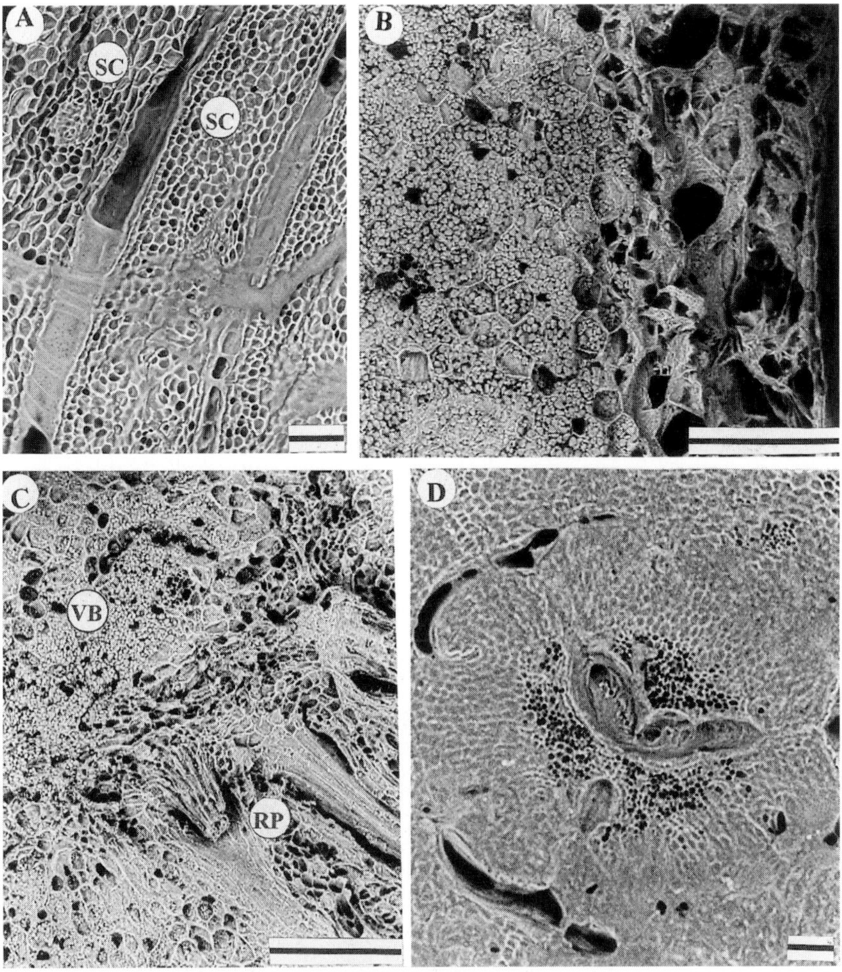

Fig. 9.2. Scanning electron microphotographs of tulip organs: (A) transversal dissection of the developing bud within the tulip bulb, December 1998. A gel-like substance is visible between the scales (SC), bar = 100 μm. (B) transversal dissection of the storage scale of the chilled tulip bulb, December 1998. A large number of the starch granules is visible. Starch destruction is visible in the lateral parts of the scale, bar = 500 μm. (C) longitudinal dissection of the basal plate of the chilled tulip bulb, December 1998. Development of root primordia (RP) and vascular bundles (VB) are visible, bar = 500 μm. (D) transversal dissection of the central bud inside a tulip bulb, December 1998. Different tissue types consisting of small cells and a large amount of air space are visible, bar = 100 μm.

Different tissues were visible in the longitudinal section of the basal plate: large storage cells, small compressed cells, vascular bundles and large granules of the storage material. In November–December, root primordia elongated and were clearly visible in the basal plate (Fig. 9.2C). During storage, the progressive development and elongation of the central bud was observed by bulb dissections and MRI observations. The MRI images were used to determine bud elongation. The bud length increased linearly during the first 8 weeks of storage at 2.0 mm week^{-1} for the chilled bulbs, but only 1.6 mm week^{-1} for the non-chilled bulbs. After the first 8 weeks, elongation became slower, and after 12 weeks, the bud length of the chilled bulbs was 26.1 ± 0.5 mm and that of the non-chilled bulbs 23.5 ± 0.6 mm. The central bud contains leaf primordia, floral scape and the flower.

In December, a transversal section of leaf primordia showed typical leaf structure, including epidermis, parenchyma and vascular bundles. Cells were relatively small (*c.* 20 μm in diameter) and free from storage granules. The developing flower consists of different tissue types, small compact cells (*c.* 10 μm) and large amount of intracellular air (Fig. 9.2D). Two weeks after planting, the chilled bulbs sprouted and produced 15–18 cm-long leaves. The floral scape was visible at the end of January, and reached 45 cm in length in February. Flowering commenced at the end of February and continued through March. By comparison, non-chilled bulbs sprouted at the end of February and produced very short (2–3 cm) leaves. The short floral scape initials reached only 2 cm in length and aborted.

MRI single parameter images during storage

Changes in the proton density and the T_2 relaxation time during storage are shown for the chilled and non-chilled bulbs in Fig. 9.3. In the proton density images (Fig. 9.3, rows 1 and 2), bright and darker regions were distinguished within the scales at the beginning of storage, indicating different water content in these regions. After 8 weeks of storage, these differences disappeared in the chilled bulbs, but were still visible in the non-chilled bulbs. During storage, the basal plates of both chilled and non-chilled bulbs showed decreases in proton density. The buds showed continuous changes in signal intensities, related to the development of the leaf primordia and flowers.

The mean proton densities obtained for the basal plate and central bud reflected the findings obtained from visual inspection of the amplitude images (Fig. 9.4A and B). At the same time, visible differences between the amplitude images of the scales of the chilled and of the non-chilled bulbs were not reflected in the mean amplitude intensities (Fig. 9.4C). In cases of tissue heterogeneity, mean intensity values can be misleading, and intensity histograms may clarify the picture (Fig. 9.5). The histograms represent the number of pixels (*y*-axis) with certain amplitude (*x*-axis) in the scale region

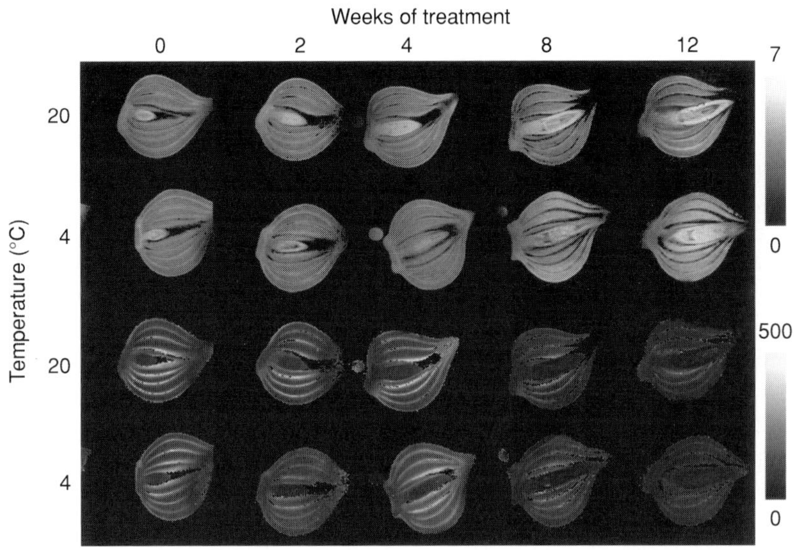

Fig. 9.3. MR images of proton density (rows 1 and 2) and T_2 relaxation time (rows 3 and 4) of the chilled (4°C) and non-chilled (20°C) tulip bulbs during storage for 12 weeks.

of the images. The mean proton density obtained from the histograms in Fig. 9.5 resulted in similar values, but the distributions of amplitudes for the chilled and non-chilled bulbs after 12 weeks of storage were clearly different. The chilled bulbs showed two maxima of proton density: a high-amplitude region within the scales (higher than before storage, represented by the bright stripes within the scales in Fig. 9.3), and a low-amplitude region (lower than before storage, represented by the darker areas between the scales, Fig. 9.3). The non-chilled bulbs showed only an enlarged range of amplitudes.

In the T_2 relaxation images (Fig. 9.3, rows 3 and 4), changes were observed in all organs of both chilled and non-chilled bulbs. In chilled bulbs, they became visible after 4 weeks of storage, indicating faster T_2 relaxation. From visual analysis, it appears that this process was faster in the chilled bulbs, than in the non-chilled ones. The mean R_2 $(1/T_2)$ relaxation rates are presented in Fig. 9.6. During storage, the relaxation rates of the basal plate and the bud become faster, and this process developed more rapidly in the chilled bulbs. Significant differences between chilled and non-chilled bulbs were observed after 4 and 8 weeks of storage for the bud and basal plate, respectively. At the same time, in the scales of both chilled and non-chilled bulbs, relaxation rates increased only after 12 weeks of storage.

ADC images are presented in Fig 9.7. All organs of the non-chilled bulbs showed an increase in the apparent diffusion of water, which was not

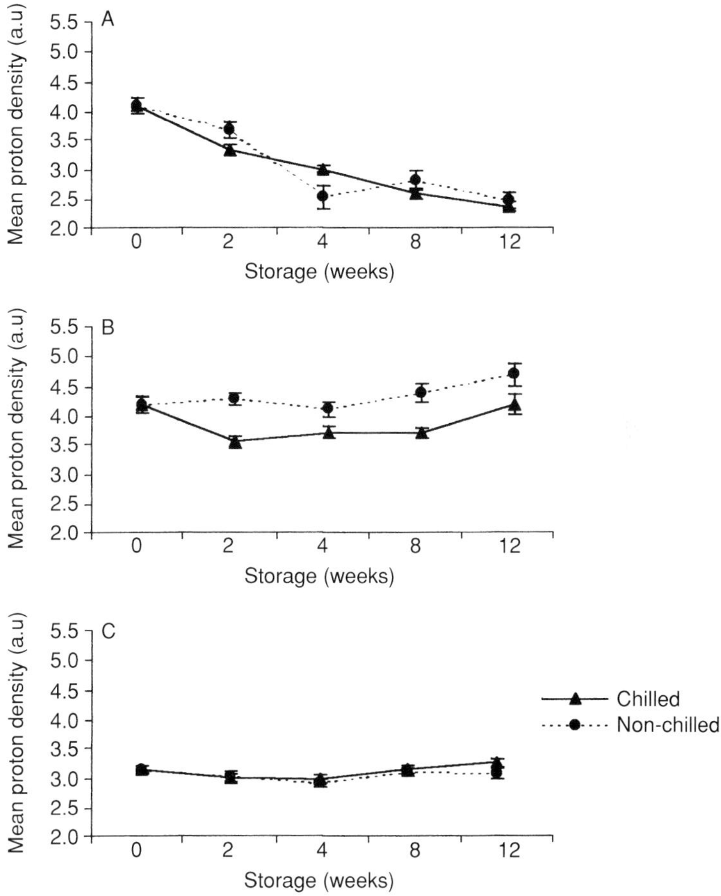

Fig. 9.4. Mean proton density of the basal plate (A), bud (B) and storage scales (C) of the chilled and non-chilled tulip bulbs during 12 weeks of storage.

readily observed in those of the chilled ones. The mean ADCs obtained for the separate organs also showed this trend (Fig. 9.8). In the basal plate and the scales, the mean ADCs of the chilled bulbs did not change markedly, while they increased significantly for the non-chilled bulbs.

Discussion

Active developmental processes occurring within tulip bulbs during postharvest storage include cell division and elongation, development of newly formed organs, and degradation of the storage materials (Rees, 1992; De Hertogh and Le Nard, 1993). In the present study, dissection and MRI observations of stored tulip bulbs showed similar processes of intrabulb

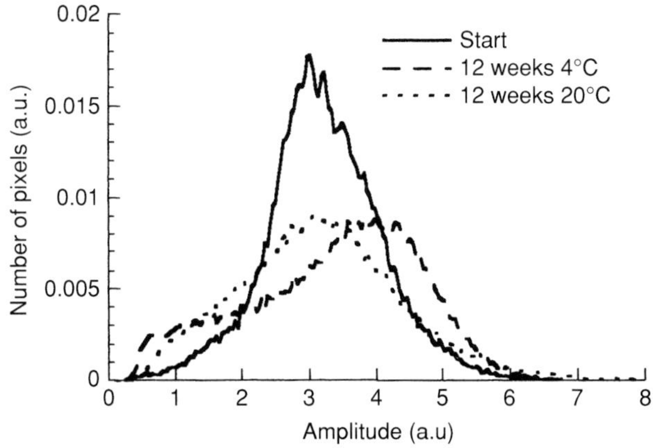

Fig. 9.5. Histograms of the distribution of proton density in the scales of the chilled and non-chilled tulip bulb. The *y*-axis represents the number of pixels (in arbitrary units) found in a certain amplitude range (the x-axis) within the scale. The contribution of all the bulbs within a treatment group is represented.

development in both chilled and non-chilled bulbs (Fig. 9.3). However, after planting, scape elongation and flowering occurred only in the chilled bulbs. Various parameters have been investigated as markers for the fulfilment of the low temperature requirement in tulips: starch and amylase (Gorin and Heidema, 1985; Lambrechts *et al.*, 1994), protein and carbohydrate content and respiration (Higuchi and Sisa, 1967; Kanneworff and Van der Plas, 1994), membrane lipids (Walch and Van Hasslt, 1991) and endogenous hormones (Rakhimbaev *et al.*, 1978; Aung and De Hertogh, 1979; Rebers *et al.*, 1995). However, none of them can explain the influence of the cold period on further bulb development and flowering.

It has been suggested that low temperatures might affect the transition of water, from association with large molecules, to free water (Faust *et al.*, 1991; Okubo *et al.*, 1997). In the present study, MRI examinations of the various organs of the chilled and non-chilled bulbs during storage indicated that changes in local water status occurred as a result of the physiological evolution of the bulb and were affected by storage temperature.

Proton density is a powerful parameter to monitor tissue dehydration and rehydration, especially in combination with information on morphological structure of the tissue. Because in most living tissues the main contribution to the proton density comes from water, the images usually represent the water density. Absolute tissue water densities could not be determined from our results, but relative water densities within the organ could be calculated with respect to the total water content of the bulb. At the beginning of the storage period, brighter and darker regions could be distinguished in the images of the bulb scales (Fig. 9.3); the lighter regions in the lateral parts of the scale may indicate highly hydrated tissues. Later,

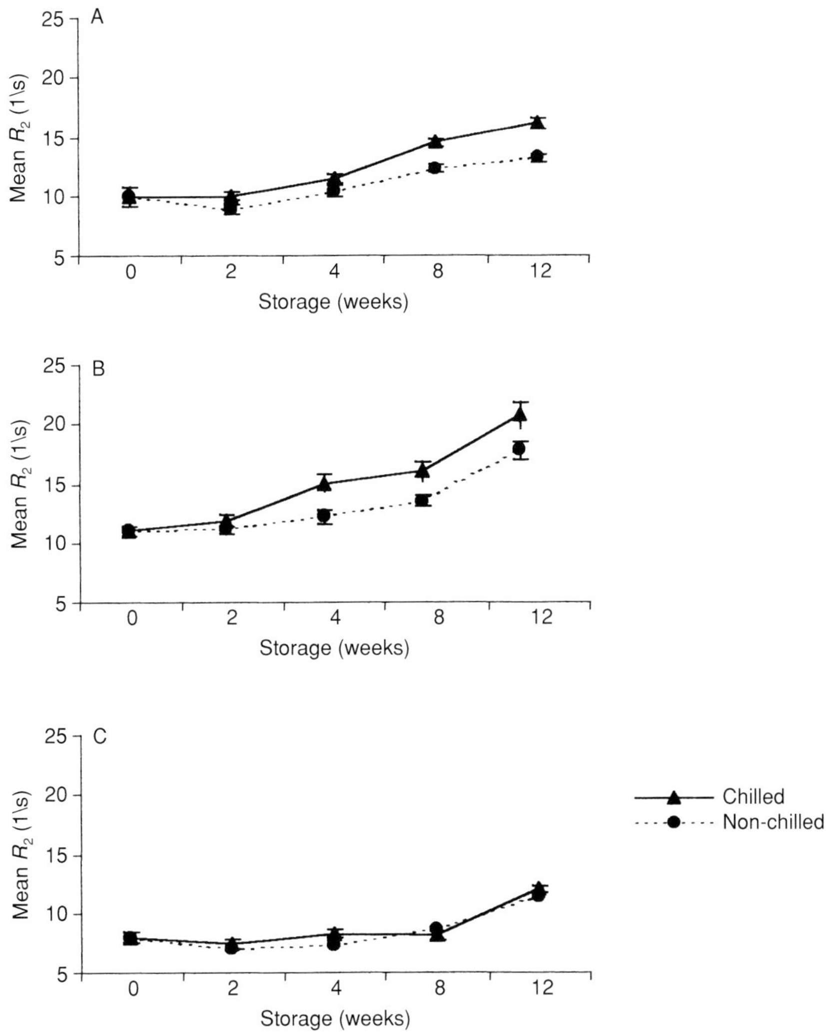

Fig. 9.6. Mean values of R_2 ($1/T_2$) of the basal plate (A), bud (B) and storage scales (C) of the chilled and non-chilled tulip bulbs during 12 weeks of storage.

lighter regions in the scales of the chilled bulbs were diminished. At the same time, inner parts of the scales of the chilled bulbs showed higher proton density, indicating a relative increase in the amount of water in these regions. It appears that redistribution of water occurred in the scales of the chilled bulbs. Then, the storage tissue was provided with extra water at the expense of the lateral parts of the scale, which were well hydrated. This effect was less pronounced in the non-chilled bulbs. It is known that many geophytes contain in their storage organs non-structural polysaccharides

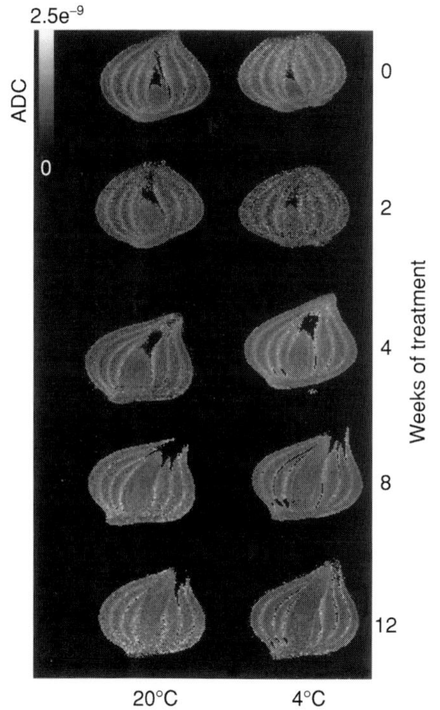

Fig. 9.7. MR images of the apparent diffusion coefficient (ADC) of the chilled and non-chilled tulip bulbs during storage for 12 weeks.

(e.g. glucomannan), which probably affect cell–water relations (Meier and Reid, 1982) and the susceptibility of the bulb scale to dehydration (Matsuo and Mizuno, 1974). Low-temperature storage has been reported to enhance degradation of glucomannans in *Lilium longiflorum* scales (Miller and Langhans, 1990). A gel-like substance was visible between the scales of the tulip (Fig. 9.2A). Further studies of the biochemical nature of polysaccharides and their role in water status are needed to explain the mechanism of water redistribution within the bulb scales. During storage, the relative amount of water increased in the floral bud and decreased in the basal plate (Figs 9.3 and 9.4), indicating that the growth of the bud may be facilitated by water transport from the basal plate.

NMR relaxation rates are related to molecular size and mobility. Rapidly rotating small molecules, such as water molecules in the bulk liquid, relax slowly (longer T_2, smaller R_2), whereas water molecules of restricted mobility or associated with macromolecules relax quickly (shorter T_2, larger R_2) (Fullerton and Camerun, 1988). In general, our data demonstrated faster T_2 relaxation in the chilled bulbs than in the non-chilled ones. This difference was more pronounced in the basal plate and the floral bud (Fig. 9.3). Published data on cold treatment and its effects on MRI relaxation time are contradictory. Faster T_1 and T_2 relaxation times have been observed in tulip bulbs (Okubo *et al.*, 1997) and for *Allium*

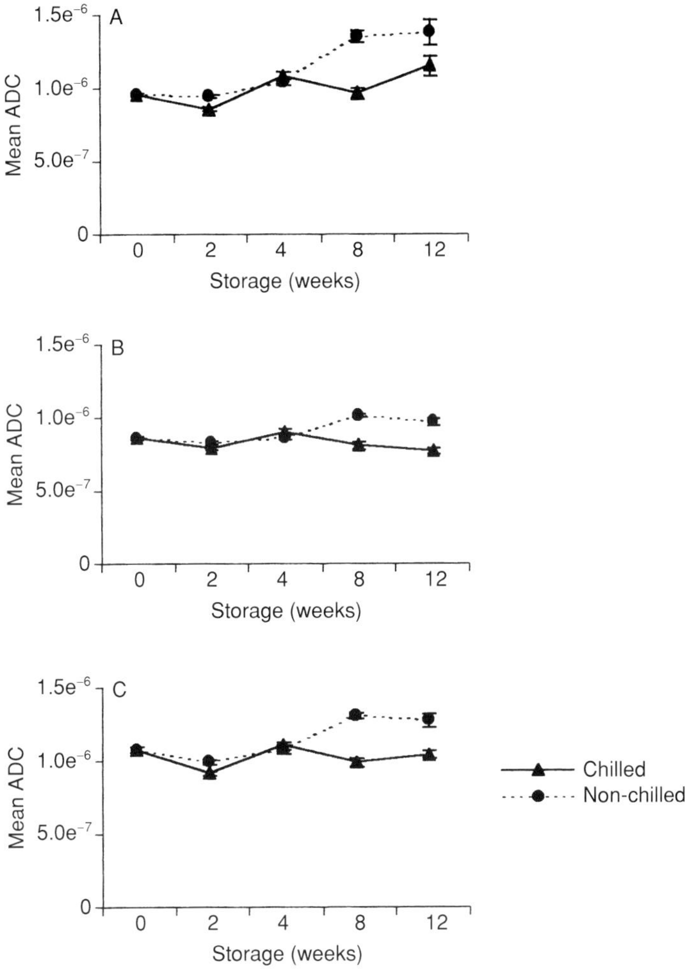

Fig. 9.8. Mean ADCs of water in the basal plate (A), bud (B) and storage scales (C) of the chilled and non-chilled tulip bulbs during 12 weeks of storage.

aflatunense and *Allium wakegi* (Yamazaki *et al.*, 1995; Zemah *et al.*, 1999). Similarly, Chen *et al.* (1978) observed significantly faster T_1 and T_2 relaxation in winter wheat crowns after 7 weeks of acclimation, and attributed this effect to changes in membrane permeability, increase of the water viscosity or changes in the relative sizes of the different water fractions. On the other hand, Iwaya-Inoue and coworkers (1996) reported slower T_1 relaxation in the scales of chilled bulbs than in those of non-chilled ones. Rowland *et al.* (1992) showed that T_2 relaxation in blueberry buds became sufficiently slow for detection only after chilling. It has been suggested that slower T_1 and T_2 relaxation indicate a transition of water from a bound to a free state in certain tissues of fruit tree buds, when dormancy is broken (Faust *et al.*, 1991).

All the above contradictions could be a result of technical variations among the MRI measurements, since the experiments were performed at differing magnetic field strengths (high and low), which might influence the water relaxation properties (Donker *et al.*, 1997). It is also possible that results were affected by the choice of echo times used for T_2 measurements, *vis-à-vis* the multi-exponential character of the transverse relaxation. Thus, the fast relaxation components may be completely missed, if the time of the first echo is too long. Consequently, the influence of chilling on the slow T_2 components may appear to be different from their effect if the complete T_2 decay curve is taken into account. The relaxation of the actual bound water molecules is usually too rapid to be detected by conventional MRI techniques. Therefore, only the relatively free water pools were usually measured, the relaxation properties of which are affected by exchange with the bound water pool. Further development of MRI protocols and examination of the techniques to allow the measurement of structured water (e.g. magnetic transfer) are essential for the resolution of these questions.

It was suggested that faster T_2 relaxation times might be affected by an increase in sap viscosity (Chen *et al.*, 1978) or a large susceptibility gradient of the tissue, caused by the presence of intracellular air (Donker and Van As, 1996). Lower T_2 values may also indicate relatively low molecular mobility of water (Usha *et al.*, 1994). In addition to the relaxation times, the data on water diffusion (ADC) provided another means to consider the mobility of water. In the non-chilled bulbs the mean ADC clearly increased in all organs, indicating increased water mobility (Figs 9.7 and 9.8). At the same time, in the chilled bulbs the ADC changes were not considerable. It is known that chilling enhances the mobilization of starch, fructans and sucrose in the scales and flower stalk of tulip bulbs (Rees, 1992; Lambrechts *et al.*, 1994). At the end of storage, the soluble sugar content in chilled bulbs was almost twice that in non-chilled ones (Lambrechts *et al.*, 1994). Hence, the sap viscosity of chilled bulbs is much higher than that of non-chilled ones, and may decrease the water mobility in the tissue. The fact that the higher water mobility in non-chilled bulbs was not accompanied by slower T_2 relaxation indicates that multiple mechanisms are present.

High heterogeneity of the bulb tissues may affect MRI measurements. For example, the basal plate is a compressed stem and contains root primordia, transport tissues and starch-containing cells (Fig. 9.2C). In the images these structures cannot be separated and, therefore, the relative contributions of the various water pools within these tissues to the T_2 relaxation are unknown. In our experiments, significant differences in mean T_2 relaxation of the basal plates between chilled and non-chilled bulbs was evident after 4 weeks of storage. Similarly, in MRI examination of the bulbs of *Allium aflatunense*, the comparison of the effects of the two storage treatments on the T_2 rate revealed significantly faster T_2 relaxation times in the basal plate of chilled bulbs than in the non-chilled ones (Zemah *et al.*, 1999).

In the central bud, leaf primordia and flower parts are presented (Fig. 9.2D). Initially, most of the bud tissue contained only small cells with thin walls. The bud developed continuously during the storage period in both chilled and non-chilled bulbs (Fig. 9.3). Significant differences between the groups in bud water status were evident after 8 weeks of storage, when the chilled bulbs had a significantly higher proton density and faster T_2 relaxation than the non-chilled ones. Scale tissue appears to be rather homogeneous (Fig. 9.2B). However, the large amount of starch granules and other reserves, and their metabolism, may influence the water properties of the tissue, leading to decreased water mobility. Visible differences in water status between chilled and non-chilled bulbs were observed in the images after 8 weeks of storage (Fig. 9.3). In the chilled bulbs, the water appears to have been redistributed within the scales. This apparently did not occur in the non-chilled bulbs.

Further development of MRI protocols and examination of the techniques that allow measurement of structured water, in combination with microscopic observations, are essential for the study of the water status of bulbs during their transition from dormancy to active growth.

Acknowledgements

We are grateful to the Wageningen NMR Centre (Wageningen Agricultural University, The Netherlands), which has helped to cover research expenses and has provided professional assistance for this work. Thanks are also due to Dr H. Franssen of the Bulb Research Centre in Lisse, The Netherlands, for supplying the tulip bulbs for the project and to Professor A.A. De Hertogh of North Carolina State University, USA, for the critical reading of this manuscript.

References

Aung, L.H. and De Hertogh, A.A. (1979) Temperature regulation of growth and endogenous abscisic acid-like content of *Tulipa gesneriana* L. *Plant Physiology* 63, 1111–1116.

Banasik, L., Rudnicki, R.M. and Saniewski, M. (1980) The physiology of hyacinth bulbs (*Hyacinthus orientalis* L.), XIII. The distribution of amylase and acid phosphatase activities and starch grain in hyacinth bulbs. *Acta Physiologiae Plantarum* 2, 145–156.

Chen, P.M., Gusta, L.V. and Stout, D.G. (1978) Changes in membrane permeability of winter wheat cells following freeze–thaw injury as determined by nuclear magnetic resonance. *Plant Physiology* 61, 878–882.

Chudek, J.A. and Hunter, G. (1997) Magnetic resonance imaging of plants. *Progress in Nuclear Magnetic Resonance Spectroscopy* 31, 44–62.

Clark, C.J., Hockings, P.D., Joyce, D.C. and Mazucco, R.A. (1997) Application of magnetic resonance imaging to pre- and post-harvest studies of fruits and vegetables. *Postharvest Biology and Technology* 11, 1–21.

De Hertogh, A.A. and Le Nard, M. (1993) Physiological and biochemical aspects of flower bulbs. In: De Hertogh, A.A and Le Nard, M. (eds) *The Physiology of Flower Bulbs.* Elsevier, Amsterdam, pp. 53–69.

De Hertogh, A.A., Aung, L.H. and Benschop, M. (1983) The tulip: botany, usage, growth and development. *Horticultural Reviews* 5, 45–125.

Donker, H.C.W. and Van As, H. (1996) NMR imaging of white button mushroom (*Agaricus bisporis*) at various magnetic fields. *Magnetic Resonance Imaging* 14, 1205–1215.

Donker, H.C.W., Van As, H., Snijder, H.J. and Edez, H.T. (1997) Quantitative H-NMR of water in white button mushrooms (*Agaricus bisporus*). *Magnetic Resonance Imaging* 15, 113–121.

Edzes, H.T., Van Dusschoten, D. and Van As, H. (1998) Quantitative T_2 imaging of plant tissues by means of multi-echo MRI microscopy. *Magnetic Resonance Imaging* 16, 185–196.

Faust, M., Liu, D., Millard, M.M. and Stutte, G.W. (1991). Bound versus free water in dormant apple buds – a theory for endodormancy. *HortScience* 26, 887–890.

Fullerton, G.D. and Camerun, I.L. (1988) Relaxation in biological tissue. In: Wherly, F.W., Shaw, D. and Kneeland, J.B. (eds) *Biomedical Magnetic Resonance Imaging. Principles, Methodology and Applications.* VCH Publishers, New York, pp.115–151.

Gorin, N. and Heidema, F.T. (1985) Starch content of freeze-dried anthers and alpha-amylase activity of their extracts as criteria that dried-stored bulbs (*Tulipa gesneriana* L.) 'Apeldoorn' have been exposed to 5°C. *Scientia Horticulturae* 26, 183–189.

Hartsema, A.M. (1961) Influence of temperatures on flower formation and flowering of bulbous and tuberous plants. In: Ruhland, W. (ed.) *Handbuch der Pflanzenphysiologie,* vol. 16. Springer Verlag, Berlin, pp. 123–167.

Higuchi, H. and Sisa, M. (1967) Serological analysis on the change of protein in scaly leaf tulip bulb caused by low temperature treatment. *Journal of Japanese Society of Horticultural Science* 36, 55–60.

Hobson, G.E. and Davies, J.N. (1977) Mitochondrial activity and carbohydrate levels in tulip bulbs in relation to cold treatment. *Journal of Experimental Botany* 28, 559–568.

Iwaya-Inoue, M., Okubo, H., Matsuo, E., Motooka, K., Ishida, N., Koizumi, M. and Kano, H. (1996) Characterizing chilling responses for tulip bulbs by ^{1}H-NMR imaging in relation to metabolic activity. *Cryo-Letters* 17, 241–248.

Kanneworff, W.A. and Van der Plas, L.H.W. (1994) Respiration of bulb scale fragments of tulip after storage at 5°C. *Plant Science* 104, 31–38.

Lambrechts, H., Rook, F. and Kolloffel, C. (1994) Carbohydrate status of tulip bulbs during cold-induced flower stalk elongation and flowering. *Plant Physiology* 104, 515–520.

Le Nard, M. and Cohat, J. (1968) Influence des temperatures de conservation des bulbes sur l'elongation, la florason et la bulbification de la Tilipe (*Tulipa gesnerina* L.). *Annales de l'Amelioration des Plantes* 18, 181–215.

Le Nard, M. and De Hertogh, A.A. (1993) *Tulipa.* In: De Hertogh, A.A. and Le Nard, M. (eds) *The Physiology of Flower Bulbs.* Elsevier, Amsterdam, pp. 617–682.

Le Nard, M., Fiala, V., Querou, Y. and Jolivet, E. (1988) L'argininge, marqueur moleculaire de l'etat physiologique du bulbede *Tulipa gesneriana* L. *Comptes Rendus des Seances de l'Academie de Science de France* (Serie III) 307, 557–562.

Liu, D., Faust, M., Millard, M.M., Line, M.J. and Stutte, G.W. (1993) States of water in summer-dormant apple buds determined by proton magnetic resonance imaging. *Journal of the American Society of Horticultural Science* 118, 632–637.

Matsuo, T. and Mizuno, T. (1974) Changes in the amounts of two kinds of reserve glucose-containing polysaccharides during germination of the Easter lily bulb. *Plant Cell Physiology* 15, 555–558.

Meier, H. and Reid, J.S.G. (1982) Reserve polysaccharides other than starch in higher plants. In: Loewus, F.A. and Tanner, W. (eds) *The Encyclopedia of Plant Physiology,* vol. 13A. Springer Verlag, Berlin, pp. 418–471.

Millard, M.M., Liu, D., Line, M.J. and Faust, M. (1993) Method for imaging the states of water by nuclear magnetic resonance in low-water-containing apple bud and stem tissues. *Journal of the American Society of Horticultural Science* 118, 628–631.

Miller, W.B. and Langhans, R.W. (1990) Low temperature alters carbohydrate metabolism in Easter lily bulbs. *HortScience* 25, 463–465.

Okubo, H., Iwaya-Inoue, M., Motooka, K., Nobuaki, I., Kano, H. and Koizumi, M. (1997) Monitoring the cold requirement in tulip bulbs by ^{1}H-NMR imaging. *Acta Horticulturae* 430, 411–417.

Rakhimbaev, I.R., Syrtanova, G.A. and Solomina, V.F. (1978) The effect of cold treatment on the level of biological activity of endogenous growth regulators of tulip bulbs. *Plant Physiology (USSR)* 25 (2), 249–253 (in Russian).

Ratcliffe, R.G. (1994) In vivo NMR studies of higher plants and algae. *Advances in Botanical Research* 20, 43–123.

Ratkovic, S., Bacic, G., Radenovic, C. and Vucinic, Z. (1982) Water in plants: a review of some recent NMR studies concerning the state and transport of water in leaf, root and seed. *Studia Biophysica* 91, 9–18.

Rebers, M., Vermeer, E., Knegt, E., Shelton, C. J. and Van der Plas, H.W. (1995) Gibberellin levels and cold-induced floral stalk elongation in tulip. *Physiologia Plantarum* 94, 687–691.

Rees, A.R. (1992) *Ornamental Bulbs, Corms and Tubers.* CAB International, Wallingford, UK.

Rowland, L.J., Millard, M. and Lin, M.J. (1992) Magnetic resonance imaging in flower buds of blueberry. *HortScience* 27, 339–431.

Scheenen, T.W.J., Van Dusschoten, D., De Jager, P.A. and Van As, H. (1998) Fast spatially resolved displacement imaging in (bio) systems. In: Blumler, P., Blumich, B., Botto, R. and Fukushima, E. (eds) *Spatially Resolved Magnetic Resonance.* Wiley-VCH, New York, pp. 481–486.

Sugiura, T., Yoshida, M., Magoshi, J. and Ono, S. (1995) Changes in water status of peach flower buds during endodormancy and ecodormancy measured by different scanning colorimetry and nuclear magnetic resonance spectroscopy. *Journal of the American Society of Horticultural Science* 120, 134–138.

Usha, K., Gambhir, P.N., Sharma, H.C., Goswami, A.M., and Singh, B. (1994) Relationship of molecular mobility of water with floral malformation in mango as assessed by nuclear magnetic resonance. *Scientia Horticulturae* 59, 291–295.

Walch, K. and Van Hasslt, P. (1991) The influence of low temperatures on the membrane lipid composition and flowering capacity of Tulip bulbs. *Acta Horticulturae* 298, 345–353.

Yamazaki, H., Ishida, N., Katsura, N., Kano, H., Nishijima, T. and Koshioka, M. (1995) Changes in carbohydrate composition and water status during bulb development of *Allium wakegi* Araki. *Bulletin of National Research Institute of Vegetables, Ornamental Plants and Tea, Japan* A 10–11, 1–11.

Zemah, H., Bendel, P., Rabinowitch, H. and Kamenetsky, R. (1999) Visualization of morphological structure and water status during storage of *Allium aflatunense* bulbs by NMR imaging. *Plant Science* 147, 65–73.

10

Seed Dormancy as One of the Survival Strategies in Annual Plant Species Occurring in Deserts

Yitzchak Gutterman

Jacob Blaustein Institute for Desert Research and Department of Life Sciences, Ben Gurion University of the Negev, Sede Boker Campus 84990, Israel

Introduction

Baskin and Baskin (1998) summarized the knowledge about seed dormancy of 1452 plant species along the rainfall and temperature gradient from rain forests to vegetation in areas with less annual rainfall and decreasing temperatures, such as areas in hot and cold deserts. They found a reduction in plant species with non-dormant seeds from 61% in tropical rain forests to about 18% in hot deserts, and only 1% in cold deserts. The percentage of plant species with dormant seeds increased from 39% in tropical rain forests to 84% in hot deserts, and up to 99% of the plant species in cold deserts. Among all the vegetation summarized, the majority of the plant species with dormant seeds had seeds with physiological dormancy. Fewer plant species had seeds with physical dormancy (Fig. 10.1A and B). Among the vegetation from more arid areas such as tropical deciduous forests, savannas and hot and cold deserts, morphological, morphophysiological dormancy, and physiological together with physical dormancy are very rare (Baskin and Baskin, 1998).

Many of the annual plant species of the Negev within the different groups of seed dispersal tactics differ in their dormancy strategies (Gutterman, 1993, 1994a, 1997a, 1998a,b). The Negev Desert of Israel is a part of the 6000 km belt of the north part of the Sahara and Arabian Deserts which receive an average annual rainfall of 50–150 mm in winter (Evenari, 1981; Shmida *et al.*, 1986). Unpredictable, very small amounts of rain and rain distribution, as well as high seed consumption, typify arid zones and deserts. Therefore, annual plant species have developed

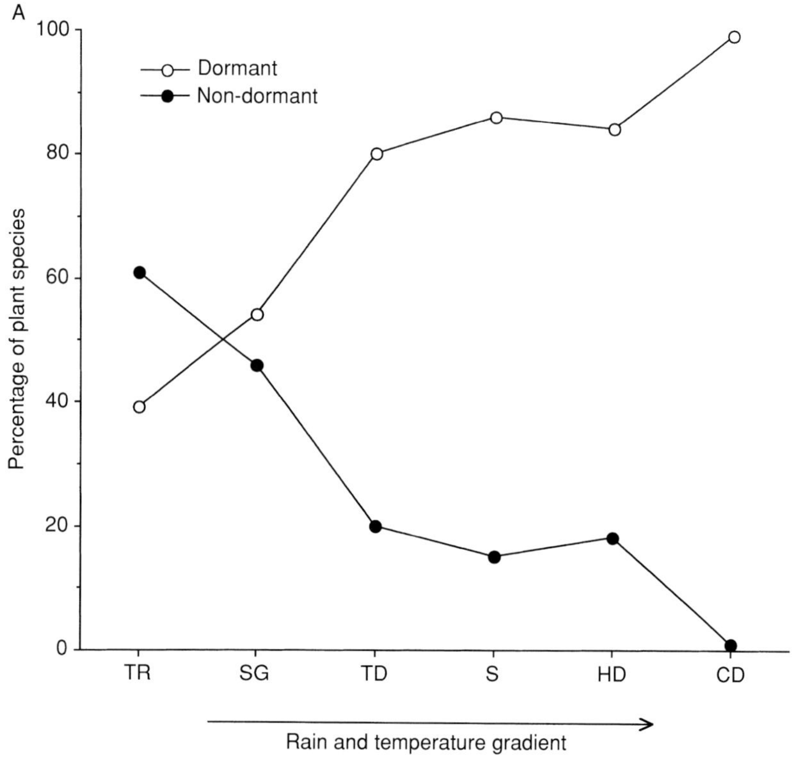

Fig. 10.1. (A) Percentages of plant species with dormant and non-dormant seeds in the different vegetation along the temperature and rain gradient from the tropical rain forests (TR), semi-evergreen (SG), tropical deciduous (TD), savanna (S) to hot deserts (HD) and to cold deserts (CD) (from Baskin and Baskin, 1998).

strategies of seed dispersal and dormancy of long-living seeds located in protected seed banks. Protected aerial seed banks of lignified plants and inflorescences, and large soil seed banks, with seeds that are dormant for short, or even very long periods, are important for the survival of these plants. This is particularly important for those occurring in the more extreme deserts (Evenari, 1981; Evenari *et al.*, 1982; Gutterman, 1993; Baskin and Baskin, 1998; Van Rheede van Oudtshoorn and van Rooyen, 1999).

Many of the plant species occurring in deserts have topochoric dispersal tactics (dispersal on the spot) (Zohary, 1937, 1962) for the long-living seeds of their seed banks. These plant species have developed: (i) protection; or (ii) escape, strategies of seed dispersal to avoid massive seed predation. In each of the two extreme strategies of seed 'escape' or 'protection', there are mechanisms which influence the germination plasticity or intermittent germination (Salisbury, 1942; Harper, 1977; Cavers and

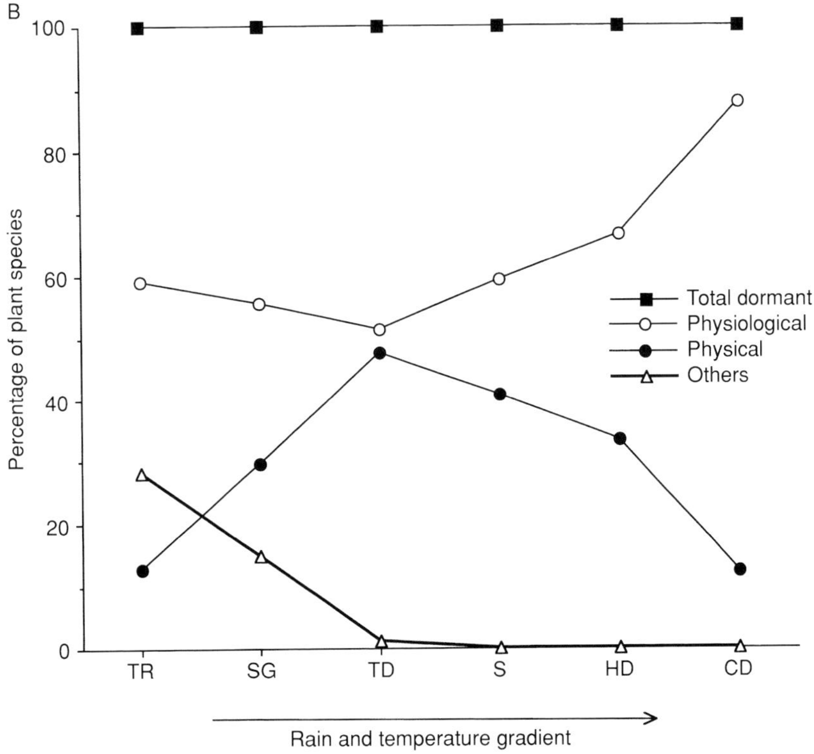

Fig. 10.1. (B) The distribution of species of the vegetation, as in (A), with dormant seeds according to their type of dormancy (from Baskin and Baskin, 1998).

Steele, 1984). Only small portions of the seeds in the species seed banks germinate after one rain event. Seed germination is influenced by daylength and seed position on the mother plant during seed maturation, as well as by post-maturation temperatures and environmental factors during the time of wetting (Gutterman, 1993, 1994a).

Seed Dispersal, Dormancy and Long-living Seeds of the Seed Banks

Aerial seed banks, synaptospermy and serotiny

Synaptospermy and seed dormancy

The topochoric tactic of seed dispersal includes topochoric synaptospermic seed dispersal (synaptospermy: two or more seeds or one-seeded fruit are joined to form a compound dispersal unit) of multi-seeded dispersal units (Zohary, 1937, 1962). Such dispersal units contain seeds with different degrees of dormancy according to their position in the dispersal unit. These dispersal units remain in the same habitat near the dry mother plants for long periods and the diaspores are protected by the lignified structures of the dispersal units.

PTERANTHUS DICHOTOMUS In *Pteranthus dichotomus* Forssk. (*Caryophyllaceae*) the seven one-seeded fruit (pseudocarps) of a three-order dispersal unit differ in their dormancy percentages according to their position in the dispersal unit and according to the germination conditions (Figs 10.2 and 10.3).

AEGILOPS GENICULATA Dormancy percentages in *Aegilops geniculata* (= *A. ovata*) (*Poaceae*) caryopses also differ according to the position of the caryopses in the one to four spikelets of the spike, which is the dispersal unit. In each spikelet there are one or two caryopses (Table 10.1) (Datta *et al.*, 1970, 1972; Evenari *et al.*, 1982; Gutterman, 1993, 1998a). A similar position effect has been found in *Aegilops kotschyi* Boiss (Wurzburger and Koller, 1976).

Plant species with protection strategies of serotiny seed dispersal

This group includes plant species of which the protected aerial seed bank may remain for many years in lignified, dry plants and inflorescences or capsules. In these plant species, the seeds have a late occurrence of 'delayed dehiscence' (serotiny) and portions of the aerial seed bank are released from the fruits when the plants and fruits are wetted by rain. Some of the seeds are then dispersed by the rain (hygriescens) (Gutterman, 1990;

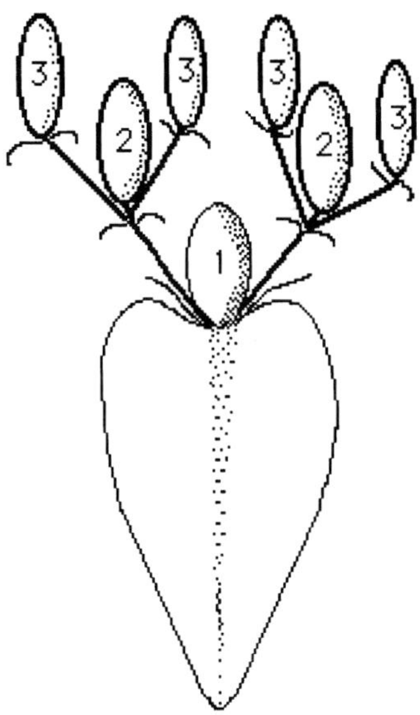

Fig. 10.2. Schematic drawing of *Pteranthus dichotomus* winged thorny inflorescence as a dispersal unit showing the arrangement of the seven pseudocarp dispersal units containing one-seeded fruit each (pseudocarps), in orders 1, 2 and 3 (adapted from Evenari *et al.*, 1982).

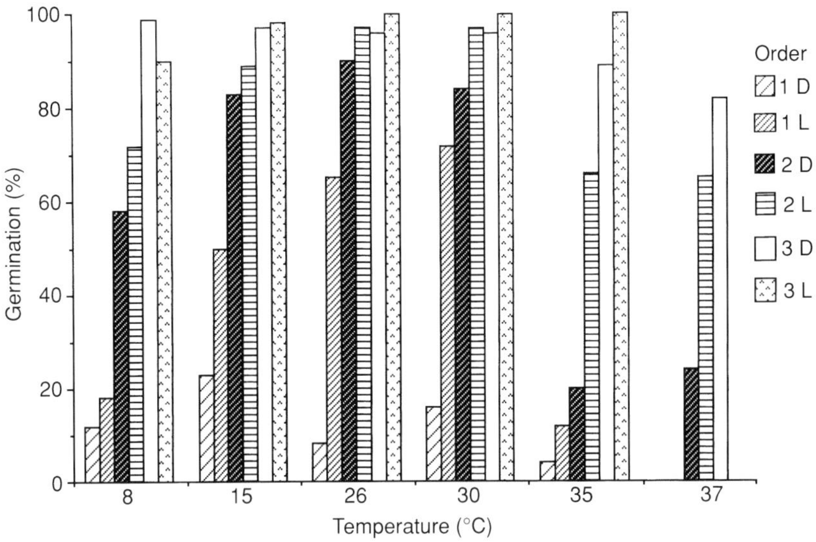

Fig. 10.3. Germination (%) of one-seeded fruit (pseudocarps) of *Pteranthus dichotomus* from the three orders of the dispersal units, 1, 2 and 3, in temperatures from 8 to 37°C in light (L) and dark (D) (see Fig. 10.2) (adapted from Evenari *et al.*, 1982).

Table 10.1. Position effect on average weight (mg) and germination (% ± SE) after 24 h in light at 15°C on *Aegilops ovata* caryopses harvested from plants originating from a_1, b_2 and c caryopses and grown under 18 h LDs at day/night temperatures of 15/10°C (from Datta *et al.*, 1972).

Three-spikelet (A, B and C) dispersal unit and position of caryopses (a_1–c)	Order of caryopses from which mother plant developed	Order of caryopses collected from mother plant	Average weight of caryopses (mg) 15/10°C	Germination (%) 15/10°C
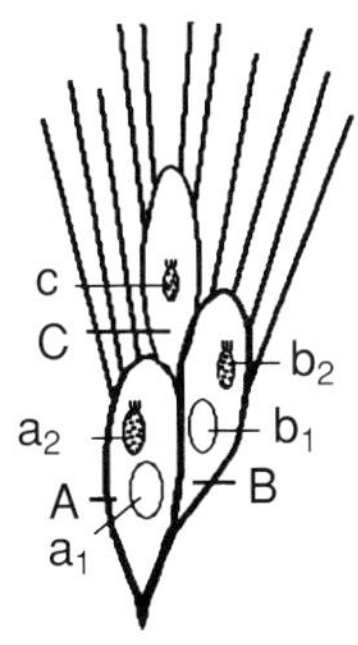		a_1	20.6 ± 0.7	84.4
	a_1	b_2	9.5 ± 0.8	10.0
		c	6.1 ± 0.6	8.5
		a_1	22.9 ± 0.5	55.0
	b_2	b_2	9.2 ± 0.2	2.3
		c	3.8 ± 0.4	0
		a_1	27.3 ± 1.9	21.2
	c	b_2	13.0 ± 0.3	0
		c	4.2 ± 0.7	0

Lamont *et al.*, 1991). Other seeds may also be dispersed further by wind after dispersal by rain (Gutterman and Ginott, 1994).

In some of these plant species all the seeds are non-dormant and may germinate a few hours after dispersal and wetting. Examples are *Blepharis* spp. (*Acanthaceae*) (Gutterman *et al.*, 1967; Gutterman, 1972) and *Anastatica hierochuntica* L. (*Brassicaceae*) (Gutterman, 1990, 1993). The rain-dispersed seeds of some other plant species are dormant for a few months after seed maturation (afterripening), as in *Plantago coronopus* L. subsp. *commutata* (Guss.) Pilger (*Plantaginaceae*) (Gutterman *et al.*, 1998). In other species, such as *Mesembryanthemum nodiflorum* L. (*Aizoaceae*), all the seeds remain dormant for several years. Then portions of the seeds germinate according to their position in their capsules during maturation (Gutterman, 1980/81, 1994b) (Fig. 10.4).

Some of the plant species of which only a portion of the rain-dispersed seeds are dispersed at one rain event have topochoric myxospermic tactics, i.e. seeds which exude mucilage from their testa or pericarpal layers (Zohary, 1937, 1962). These seeds, such as those of *Blepharis* spp., *Carrichtera annua* (L.) DC., *Anastatica hierochuntica* and *Plantago coronopus*, adhere to the wet soil surface by their mucilage. This may delay massive

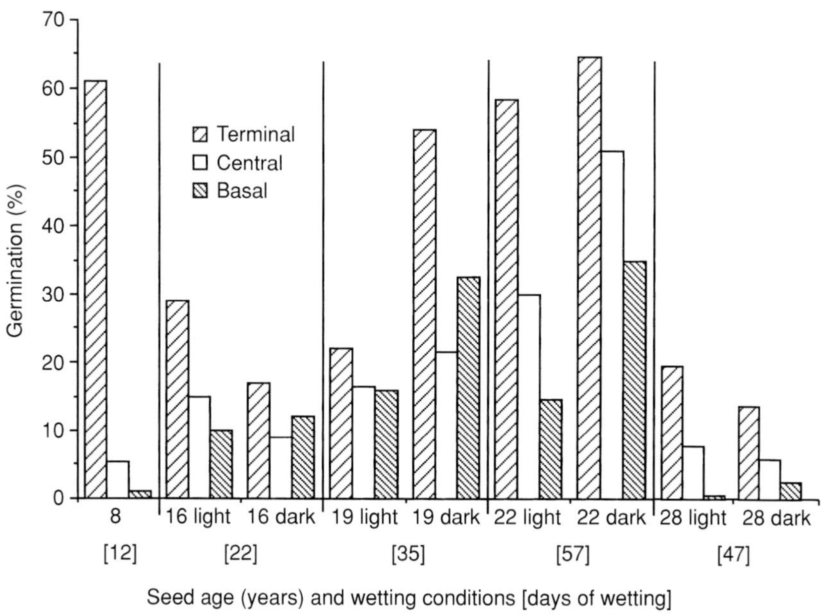

Fig. 10.4. Germination (%) of *Mesembryanthemum nodiflorum* seeds (4 × 50) according to their position in the capsule: terminals, centrals and basals, after 8, 16, 19, 22 and 28 years of storage. [..] = the number of days from wetting until they reached their germination plateaux (adapted from Gutterman, 1980/81, 1990, 1992, 1994a, in preparation).

seed collection by ants. Many of these seeds escape from ant collection because of their fast germination. Ants do not collect germinating seedlings (Gutterman and Shem-Tov, 1996, 1997a,b). The mucilaginous seeds may also become hydrochorous seeds and are dispersed by floating on runoff water when runoff develops during a more intense rainfall of more than 4 mm h^{-1}. This tactic may cause the seeds to accumulate along runnels or in depressions together with the runoff water. The seeds will later germinate in a favourable microhabitat (Friedman *et al.*, 1981; Evenari *et al.*, 1982; Huang and Gutterman, 1999).

BLEPHARIS SPP. The seeds of *Blepharis* spp. are not dormant and the dispersal mechanism is the only regulator which spreads the germination of the aerial seed bank over time and space. These seeds may germinate within a few hours of dispersal by an efficient rain. They germinate in a wide range of temperatures from 8 to 40°C, in light or dark (Gutterman *et al.*, 1967; Gutterman, 1972, 1990, 1993, 1995).

ANASTATICA HIEROCHUNTICA *Anastatica hierochuntica* seeds are not dormant and germinate within a few hours after dispersal by an efficient rain, in a wide range of temperatures, in light or dark (Friedman *et al.*, 1981), as found in *Blepharis* spp.

CARRICHTERA ANNUA The seeds of *Carrichtera annua* are also dispersed by rain and some of them are dormant. This dormancy and germination plasticity is influenced by the daylength during the last 8 days of seed maturation, when the seeds are still on the mother plants (Gutterman, 1993).

PLANTAGO CORONOPUS *Plantago coronopus* seeds are dispersed by rain in small portions. The range of temperatures required for their germination, the optimal temperature for germination, as well as the percentage of germination at the suitable range of temperatures, are dependent on the length of time since seed maturation as well as dry storage temperatures (Evenari *et al.*, 1982; Gutterman and Shem-Tov, 1996; Gutterman *et al.*, 1998; Shem-Tov *et al.*, 1999).

ASTERISCUS HIEROCHUNTICUS Other serotinous aerial seeds, such as those of *Asteriscus hierochunticus* (Michon) Wikl. (= *A. pygmaeus*) (*Asteraceae*), are dispersed by rain and adhere by their pappus or hairs to the wet soil surface near the dry mother plant. The achenes that remain free are dispersed by wind (anemochorous dispersal) and many of them accumulate in diggings and depressions which act as wind traps for seeds (Gutterman, 1988, 1993). The dispersal of these achenes is delayed and they are released from the protecting lignified, dry plant and inflorescence, whorl after whorl. The nearer the achenes to the centre of the capitulum, the higher their dormancy after being disconnected. Connected achenes never germinate. These aerial seed banks can last for many years until all the seeds are dispersed (Evenari *et al.*, 1982; Gutterman, 1993; Gutterman and Ginott, 1994). *Gymnarhenna micrantha* Desf. (*Asteraceae*) has a similar tactic (Koller and Roth, 1964).

MESEMBRYANTHEMUM NODIFLORUM In other plants with rain-dispersed seeds, such as *Mesembryanthemum nodiflorum*, there are combinations of dispersal strategies and seeds with long-term dormancy. This dormancy may be according to the seed position in the capsules during seed maturation and seed age (afterripening). During the first summer all the seeds are enclosed in capsules on the dry mother plants. During the following winter or winters, the seeds are released during different rain events according to the length of the time of wetting. First to be dispersed are the terminal seeds which are separated from the seed cluster in the capsule after 15 min of wetting. The seeds of the middle part only become separated after 200 min of wetting. The basal seeds are the last and only become separated after 320 min. After one or several winters, almost the whole seed bank of a particular plant is transferred to the soil to become a soil seed bank with a long period of dormancy as afterripening.

The percentage of dormancy is dependent on the position during maturation of the seeds in the capsule on the mother plant, even 28 years after maturation. Even then, the seeds that originated from the terminal part of the capsules germinate in the highest percentages, whereas the seeds originating from the basal part germinate in the lowest (Fig. 10.4) (Gutterman, 1980/81, 1994b, in preparation).

Soil seed banks of dust-like diaspores

Dispersal mechanisms

There is a second group of many plant species in the Negev which have escape strategies of seed dispersal by wind for their numerous dust-like seeds (Evenari and Gutterman, 1976; Loria and Noy-Meir, 1979/80; Gutterman, 1993, 1994a). Some of these seeds are dispersed shortly after maturation, such as in *Schismus arabicus* Nees (*Poaceae*). Others, such as *Spergularia diandra* (Guss.) Heldr. et Sart. (*Caryophyllaceae*), have delayed dehiscence from the capsule or inflorescence (serotiny). The mature seeds remain in their dry capsules until strong gusts of wind disperse the seeds at intervals until they have all been dispersed. All the seeds join the soil seed banks within a few weeks after maturation. The tiny seeds of these two very common plant species enter cracks in the soil surface and become covered with fine soil particles. In this way they may escape from massive seed consumption. The long-living seeds of these plants have developed different tactics and strategies of seed dormancy whereby only small portions of the large soil seed banks are ready for germination after one rainfall.

There are several plant species with dust-like diaspores that occur in other deserts, such as in Namaqualand in southern Africa (Rösch, 1977; Van Rheede van Oudtshoorn and van Rooyen, 1999).

Seed dormancy and germination plasticity in annual plant species occurring in deserts

SPERGULARIA DIANDRA The seeds of *Spergularia diandra* that develop from the first flower that terminates on the main stem are black. These black seeds are the heaviest and germinate in the highest percentages, in comparison with the brown seeds that mature in capsules from flowers on the lateral branches. The flowers that appear when the plant starts to senesce produce yellow seeds which are the lightest with the highest percentage of dormancy (Fig. 10.5). The daylength during seed maturation has a large influence on the germination plasticity of the *S. diandra* seeds (Fig. 10.6). Longer dry storage at higher temperatures reduces seed dormancy (Gutterman, 1994c). In some of the *S. diandra* populations in the Negev there are three genotypes producing seeds with hairs, glabrous seeds or partially hairy seeds. Therefore, in one population there are nine types of seeds with differing colour, size, dispersability and germinability.

In *S. diandra*, the percentage of dormant seeds depends on the position of their capsules on the mother plant, maturation conditions during seed maturation, and post-maturation temperatures (Figs 10.5 and 10.6) (Gutterman, 1993, 1994a,c, 1996a, 1997b). The germination plasticity of the seeds of this plant enable small portions of the seeds in the seed bank to

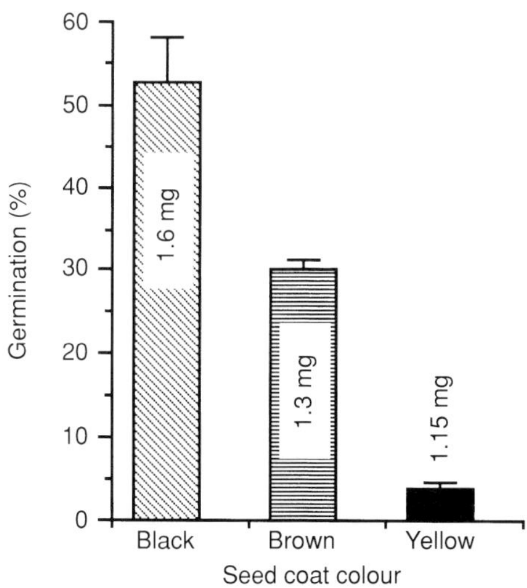

Fig. 10.5. Comparison of germination after 9 days of wetting (average % ± standard error of 4 × 50 seeds) of black, brown and yellow hairy *Spergularia diandra* seeds harvested on 27 June 1989 from a natural population near Sede Boker. Seeds were wetted on 2 May 1994 at 15°C in the dark for 6 days. The average weight (mg) of 5 × 100 seeds is marked (Gutterman, 1994b).

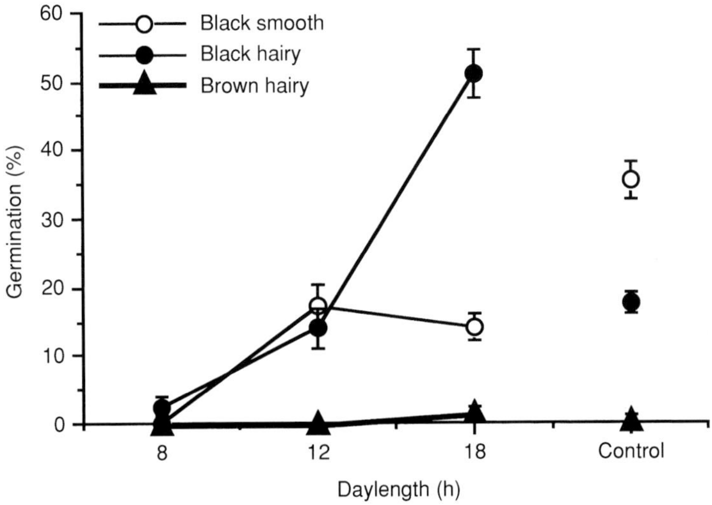

Fig. 10.6. Comparison of germination (% ± standard error of 4 × 50 seeds) of black and brown, hairy, and black, smooth *Spergularia diandra* seeds matured under 8, 12 and 18 h daylengths outdoors and under natural daylength and temperatures in the growing season (control). Seeds were collected between 7 and 9 March 1994 from mother plants that had developed from seeds harvested on 27 June 1989 near Sede Boker. The germination experiment began on 10 March 1994. There were significant differences in germination percentages after 47 days between phenotypes ($P = 0.0001$) and genotypes ($P = 0.0001$) (at 95% in three-way ANOVA) after angular transformation among seeds matured under different daylengths (from Gutterman, 1994c).

germinate in winter even after a light rainfall of 10–15 mm (Loria and Noy-Meir, 1979/80). This opportunistic strategy is very important for the survival of plant species in large desert areas. The plasticity of seed germination is influenced by a combination of factors. These include: the time and position of the appearance of the flowers from which the capsules and seeds later develop, the daylength during seed maturation, storage conditions and duration (Gutterman, 1993, 1994a,b, 1996a, 1997a,b). These strategies influence intermittent germination, reduce the risk to the survival of the species and may even be critical under the unpredictable small amounts and distribution of rain in desert areas.

SCHISMUS ARABICUS In *Schismus arabicus*, a very common annual species, the escape mechanisms of seed dispersal, and opportunistic strategy of seed germination, enables it to be dominant and appear over large areas in deserts. It even emerges, develops and produces seeds in years with less than the annual average rainfall. As little as 10–15 mm of rain in winter may engender germination of a small portion of the large *S. arabicus* soil seed bank. The dormancy of portions of the seed population and the plasticity in their germination are influenced by environmental factors during seed maturation, as well as the dry storage temperatures after seed maturation, in addition to the amount of rain, temperatures and relative humidity (RH) during germination. The daylength during seed maturation has a great influence on the plasticity of seed germination and the percentage of dormant caryopses in the years following maturation. Post-maturation dry storage temperatures also affect primary seed dormancy (afterripening) in this plant (see below) (Table 10.2). The number of seedlings that emerge is influenced by environmental factors such as temperature and RH, and amounts of water. In mild winter temperatures and high RH, some *S. arabicus* seedlings may emerge after 10 mm of rain, or in depressions even after 7.1 mm. In high summer temperatures and low RH, seedlings may only emerge after wetting by an equivalent amount of water to 90 mm of rain. The higher the amounts of water, from 90 to 150 and 200 mm, the higher the number of *S. arabicus* seedlings that emerged (Evenari and

Table 10.2. The influences of dry storage temperatures on the germination percentage (± standard error) of *Schismus arabicus* caryopses at 20°C in light (L) or dark (D) with short illuminations (5 min) during daily observations; 4 × 50 caryopses for each treatment in 50 mm diameter Petri dishes. (A) Freshly harvested mature *S. arabicus* caryopses collected from a natural population near Sede Boker on 18 April 1993; (B) caryopses harvested on 16 May 1993 and stored at constant temperatures of 40°C or −18°C for 70 days, or under naturally fluctuating day/night summer temperatures of 17–40°C in shade; (C) caryopses harvested on 28 April 1991 and stored at room temperatures of 10–27°C. This germination experiment started on 25 July 1993 (adapted from Gutterman, 1996b).

				Germination (%)	
	Treatment no.	Dry storage conditions (°C)	Wetting at 20°C L or D	Time of wetting (days) 1	6
A		—	L	0	0.5
		—	D	0	0
B	I	40	L	0	8.0 ± 1.8
	II	40	D	41 ± 5.3[a]	51.0 ± 2.6[a]
	III	−18	L	0	0.5 ± 0.25
	IV	−18	D	0	0
	V	17–40	L	0	1.5 ± 0.9
	VI	17–40	D	1.0 ± 1.0	7.0 ± 2.0
C	VII	10–27	L	0	28.5 ± 2.3
	VIII	10–27	D	37 ± 1.9[a]	43.5 ± 1.2[a]

[a]Significant at 95% by one factor ANOVA Fisher PLSD and Scheffe *F*-test:II vs. I, III, IV, V, VI,VII; VII vs. VIII; $P = 0.0001$.

Gutterman, 1976; Loria and Noy-Meir, 1979/80; Gutterman, 1982, 1993, 1996b,c; Gutterman and Evenari, 1994).

Seed dormancy and germination plasticity in annual plants that are weeds in cultivated fields

Similar strategies to those developed in desert plants to retain long-term seed banks have also developed in plant species that emerge as weeds in cultivated fields (Sidhu and Cavers, 1977; Kane and Cavers, 1992; Benech-Arnold and Sánchez, 1995; Baskin and Baskin, 1998; Cavers *et al.*, 1998). These weeds also exist in unpredictable environments because of the various attempts by farmers to destroy them.

PORTULACA OLERACEAE *Portulaca oleraceae* L. (*Portulacaceae*) is widely distributed and is one of the eight most common weeds of the tropical and temperate regions of the world (Holm *et al.*, 1977). This annual succulent plant may germinate within 3–6 h of wetting, in a wide range of temperatures from 20 to 44°C in light. The longer the time of dry storage at higher temperatures, the higher and faster their germination. The germinability of the seeds was found to have a quantitative short-day effect. The longer the daylength, from 8 to 18 h per day, during the last 8 days of seed maturation, the higher the percentage of dormant seeds among the mature seeds (Gutterman, 1974, 1985, 1993).

LACTUCA SERRIOLA The date of maturation of *Lactuca serriola* L. (*Asteraceae*) achenes during summer, from July to October, when the natural daylength decreases from 15 h to 12 h per day, has an effect on the percentage of dormant seeds. This was found when groups of *L. serriola* seeds were collected daily and tested in a range of temperatures from 10 to 25°C, in light or dark, every 2 weeks, as well as after different periods of storage. At supra-optimal temperatures during wetting the achenes enter scoto- and thermodormancy which may also regulate the proper time for germination, thus preventing germination in high summer temperatures. This has also been found in *Lactuca sativa* L. (*Asteraceae*) (Gutterman *et al.*, 1972, 1975; Small and Gutterman, 1991, 1992a,b; Gutterman, 1994d).

AMARANTHUS RETROFLEXUS *Amaranthus retroflexus* L. (*Amaranthaceae*) (redroot pigweed) plants produce more than one million seeds m^{-2} which have different degrees of dormancy. This enables them, and seeds of other weeds producing similar seed numbers, to remain in wet soil for several years and only a small percentage of their very large seed bank may germinate from time to time (Karssen, 1980/81; Cavers and Steele, 1984). Kigel *et al.* (1979) found that *A. retroflexus* seed germinability is influenced by temperature and daylength regimes. At the lower temperatures of 22/17°C, the heavier the seeds and the higher the percentage of dormancy among the seed population in comparison with higher temperatures of 27/22°C.

CHENOPODIUM ALBUM Karssen (1970) found in *Chenopodium album* L. (*Chenopodiaceae*) that there is a temperature and daylength effect during seed maturation on seed germination. The higher temperatures during seed maturation (22/22°C) increased the percentage of dormant seeds in comparison with lower temperatures of 22/12°C during seed maturation. The germination percentages achieved in the dark showed that temperature and daylength effects during seed development and maturation are much more pronounced than in seeds that germinate in light. The combination of higher temperatures of 22/22°C and long days during seed maturation resulted in the lowest percentage of seed germination (22%). However, the lower temperatures of 22/12°C and short days during seed maturation resulted in the highest percentage of germination (96%).

CHENOPODIUM POLYSPERMUM *Chenopodium polyspermum* L. (*Chenopodiaceae*) seed germination, weight and seed coat thickness, were found to be affected by the daylength during seed development and maturation. The longer the daylength, from 10 to 24 h light per day, the lower the germination percentage and seed weight. But, under long days of 24 h of light, the seed coat was much thicker (46 μm) than under short days of 8 h light per day (20 μm). The more long days during the 24 days from flowering to seed maturation, the thicker the tegument (seed coat) (Jacques, 1957, 1968; Pourrat and Jacques, 1975). Karssen (1970) found similar daylength effects of tegument thickness in *C. album.*

Post-maturation Dry Storage Temperatures Affecting Primary Dormancy

In the areas of the northern part of the Sahara and Arabian Deserts, rain occurs in winter, followed by long, dry and hot summers. Freshly matured seeds of some of the common plant species have primary dormancy (afterripening) that prevents germination of all the seeds of these plant species shortly after maturation (Evenari *et al.*, 1982). This important strategy prevents germination of these seeds after a late rain that occurs just before the summer.

PLANTAGO CORONOPUS *Plantago coronopus* seeds do not germinate shortly after maturation. Some of these seeds are only capable of germinating after a period of dry storage at high summer temperatures or a long period of mild temperatures. Post-maturation temperatures and duration of dry storage affect the percentage and duration of primary dormancy of

P. coronopus seed. During wetting, the optimal temperatures and the range of temperatures required for germination were found to be affected by the time since seed maturation. Young seeds that have matured during the last growing season germinate in lower temperatures and are inhibited by temperatures higher than 25°C. Older seeds that had matured in previous years germinate at high temperatures of 25 and 30°C but are inhibited by lower temperatures (Evenari *et al.*, 1982; Gutterman *et al.*, 1998).

SCHISMUS ARABICUS Freshly harvested caryopses of *S. arabicus* do not germinate and require a period of about 70 days storage at high temperatures of at least 40°C. Even then, only about 50% of the seeds germinate (Table 10.2). These seeds will not germinate even after 2 years if they are dry stored at −18°C (Gutterman, 1996c). This afterripening or primary dormancy is a very important survival strategy in a desert such as the Negev because, after seed maturation at the beginning of summer, germination is prevented if there is a rainfall soon after seed maturation when temperatures are still mild. However, non-dormant caryopses of *S. arabicus* plants may even germinate in high summer temperatures, but only after receiving much higher amounts of water than in winter (see above) (Gutterman, 1993, 1994a).

HORDEUM SPONTANEUM Post-maturation environmental factors may reduce the percentage of seeds with primary dormancy (afterripening), as found in *Hordeum spontaneum* C. Koch (*Poaceae*). This wild barley of the Mediterranean and Irano-Turanian phytogeographic region is also found in wadis of the Negev Desert highlands. Great differences were found among local and edaphic ecotypes in Israel in the afterripening temperatures and duration affecting caryopses germination after dry storage following maturation (Nevo *et al.*, 1984; Gutterman and Nevo, 1994; Gutterman *et al.*, 1996; Gutterman and Gozlan, 1999; Gozlan and Gutterman, 1999) (Fig. 10.7A and B).

Afterripening has also been found in other cereals such as *Avena fatua* L. (Simpson, 1965), *Bromus rubens* L. (Corbineau *et al.*, 1992), wheat (Corbineau *et al.*, 1981; Côme, 1982; Corbineau and Côme, 1982) and oats (Corbineau *et al.*, 1986), and many other winter annuals, as studied and summarized by Baskin and Baskin (1998).

Secondary Dormancy

Supra-optimal temperatures during wetting of achenes of *Lactuca sativa* L. (*Asteraceae*) and some other species, causes them to enter secondary dormancy, such as thermodormancy (Gutterman *et al.*, 1972). This has also been found in natural populations of *Lactuca scarriola* L. in the Negev Desert. Such high temperatures may prevent germination in a season with

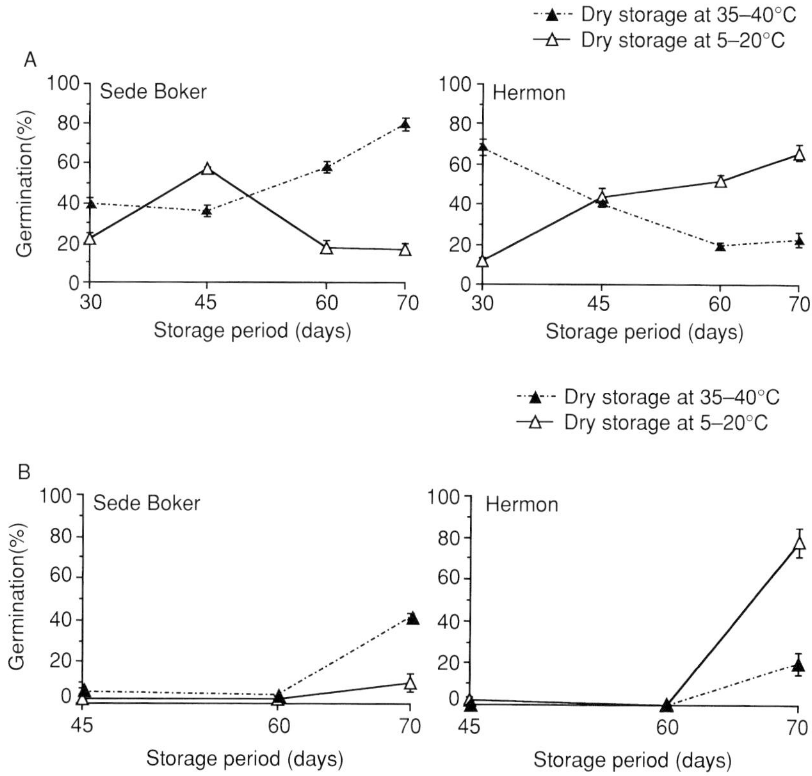

Fig. 10.7. (A) Germination (% ± SE) after 15 days of wetting at 10°C in the dark of *Hordeum spontaneum* caryopses originating from Sede Boker and Mount Hermon ecotypes. Caryopses were dry stored after harvest for 30, 45, 60 and 70 days at 5–20°C or 35–40°C. (B) Germination (± SE) after 15 days of wetting at 20°C in the dark of *H. spontaneum* caryopses originating from Sede Boker and Mount Hermon ecotypes. Caryopses were dry stored after harvest for 45, 60 or 70 days at 5–20°C or 35–40°C (Gozlan and Gutterman, 1999).

supra-optimal temperatures and later increase germination plasticity in the following growing season (Small and Gutterman, 1991, 1992a,b).

The Inhibitory Effect of Desert Soils

In addition to intermittent germination and germination plasticity, which affect seed dormancy, the soil as a substrate for seed germination has also been found to inhibit seed germination in some plant species tested. The inhibitory effect increased when seeds of *Plantago coronopus* germinated under laboratory conditions on soil crusts collected from ten locations in the Negev. These locations were along the rainfall gradient from 50 mm year^{-1} in Machtesh Ramon, in the central part of the Negev, to Lahav with 325 mm year^{-1} in the northern part of the Negev (Shem-Tov *et al.*, 1999). Differing soil structure, physical and biological effects were also found among three plant species tested, *P. coronopus*, *Reboudia pinnata* (Viv.) Taeckh. et Boulos (*Brassicaceae*) and *Carrichtera annua* (Zaady *et al.*, 1997).

Conclusions

The group of plants that includes *Schismus arabicus* and *Spergularia diandra* has opportunistic strategies of seed germination. Because of the germination plasticity and high percentage of dormant seeds according to position, maturation and post-maturation conditions, only small portions of the soil seed bank germinate after one rain event, but even after very small amounts of rain. Germination of seeds on the soil surface may be engendered by as little as 10–15 mm of rain. However, seeds in depressions may germinate when as little as 7 mm of rain falls if the intensity is above 4 mm h^{-1} and causes runoff water that enters the depressions. Most of the other plant species found in deserts do not germinate after such a small amount of rain.

Seedlings of these plants may survive if further rain events occur shortly after germination. If not, the risk to species survival is not great because only a small portion of the seed bank germinates. Even if only an additional small amount of rain does fall, the plant will complete its life cycle and produce many seeds in a year when seeds of other plant species do not germinate because of shortage of water. Therefore, these plants are very common in regions such as the Negev Desert where the main rain events are small and there are only two or three occasions when there may be enough rain in one day for the cautious plant species to germinate.

Cautious plants such as *Blepharis* spp. disperse only a small number of their limited aerial seed bank during an efficient rain, but almost all the dispersed seeds germinate and develop plants which produce mature seeds. The better the growth conditions, the larger the plants and the greater the number of seeds produced. In other plant species, many intermediate tactics were found between the extremes of cautious and opportunistic strategies. Seed dormancy and intermittent germination is essential for the survival of many plant species in the more extreme desert conditions, as has been summarized by Baskin and Baskin (1998).

References

Baskin, C.C. and Baskin, J.M. (1998) *Seeds – Ecology, Biogeography, and Evolution of Dormancy and Germination.* Academic Press, San Diego.

Benech-Arnold, R.L. and Sánchez, R.A. (1995) Modeling weed seed germination. In: Kigel, J. and Galili, G. (eds) *Seed Development and Germination.* Marcel Dekker, New York, pp. 545–566.

Cavers, P.B. and Steele, M.G. (1984) Patterns of change in seed weight over time in individual plants. *American Naturalist* 124, 324–335.

Cavers, P.B., Qaderi, M.M., Downs, P.M., Doucet, C., Manku, R. and Meier, L. (1998) The Thistles: a spectrum of seed banks. *Aspects of Applied Biology* 51, 135–141.

Côme, D. (1982) Germination. In: Mazliak, P. (ed.) *Physiologie Végétale II. Croissance et Développement.* Hermann, Paris, pp. 129–225.

Corbineau, F. and Côme, D. (1982) Evolution des Dormances de deux variétés d'Orge (*Hordeum vulgare* L.) au cours de leur maturation et de leur conservation au sec. *Compte Rendus des Séances de l'Académie des Sciences, Paris* (série III) 294, 967–970.

Corbineau, F., Sanchez, A., Côme, D. and Chaussat, R. (1981) La dormance du caryopse de blé (*Triticum aestivum* L. var *Champlein*) en relation avec la température et l'oxygène. *Comptes Rendus de l'Académie d'Agriculture de France* 9, 826–834.

Corbineau, F., Lecat, S. and Côme, D. (1986) Dormancy of three cultivars of oat seeds (*Avena sativa* L.). *Seed Science and Technology* 14, 725–735.

Corbineau, F., Belaid, D. and Côme, D. (1992) Dormancy of *Bromus rubens* L. seeds in relation to temperature, light and oxygen effects. *Weed Research* 32, 303–310.

Datta, S.C., Evenari, M. and Gutterman, Y. (1970) The heteroblasty of *Aegilops ovata* L. *Israel Journal of Botany* 19, 463–483.

Datta, S.C., Gutterman, Y. and Evenari, M. (1972) The influence of the origin of the mother plant on yield and germination of their caryopses in *Aegilops ovata* L. *Planta* 105, 155–164.

Evenari, M. (1981) Ecology of the Negev desert, a critical review of our knowledge. In: Shuval, H. (ed.) *Development in Arid Zone Ecology and Environmental Quality.* Balaban ISS, Rehovot, pp. 1–33.

Evenari, M. and Gutterman, Y. (1976) Observations on the secondary succession of three plant communities in the Negev desert, Israel. I. *Artemisietum herbae albae.* In: Jacques, R. (ed.) *Hommage au Prof. P. Chouard. Etudes de Biologie Végétale.* CNRS Gif sur Yvette, Paris, pp. 57–86.

Evenari, M., Shanan, L. and Tadmor, N. (1982) *The Negev. The Challenge of a Desert,* 2nd edn. Harvard University Press, Cambridge, Massachusetts.

Friedman, J., Stein, Z. and Rushkin, E. (1981) Drought tolerance of germinating seeds and young seedlings of *Anastatica hieronchuntica* L. *Oecologia* 51, 400–403.

Gozlan, S. and Gutterman, Y. (1999) Dry storage temperatures, duration, and salt concentrations, affect germination of local and edaphic ecotypes of *Hordeum spontaneum* (Poaceae) from Israel. *Biological Journal of the Linnean Society* 67, 163–180.

Gutterman, Y. (1972) Delayed seed dispersal and rapid germination as survival mechanisms of the desert plant *Blepharis persica* (Burm.) Kuntze. *Oecologia* 10, 145–149.

Gutterman, Y. (1974) The influence of the photoperiodic regime and red/far-red light treatments of *Portulaca oleracea* L. plants on the germinability of their seeds. *Oecologia* 17, 27–38.

Gutterman, Y. (1980/81) Annual rhythm and position effect in the germinability of *Mesembryanthemum nodiflorum. Israel Journal of Botany* 29, 93–97.

Gutterman, Y. (1982) Survival mechanisms of desert winter annual plants in the Negev Highlands of Israel. In: Mann, H.S. (ed.) *Scientific Reviews on Arid Zone Research* 1. Scientific Publishers, Jodhpur, India, pp. 249–283.

Gutterman, Y. (1985) Flowering, seed development, and the influences during seed maturation on seed germination of annual weeds, In: Duke, S.O. (ed.) *Weed Physiology* I. CRC Press, Baco Raton, Florida, pp. 1–25.

Gutterman, Y. (1988) An ecological assessment of porcupine activity in a desert biome. In: Ghosh, P.K. and Prakash, I. (eds) *Ecophysiology of Desert Vertebrates.* Scientific Publishers, Jodhpur, India, pp. 289–372.

Gutterman, Y. (1990) Seed dispersal by rain (ombrohydrochory) in some of the flowering desert plants in the deserts of Israel and the Sinai Peninsula. *Mitteilungen aus dem Institut für Allgemeine Botanik Hamburg* 23b, 841–852.

Gutterman, Y. (1993) *Seed Germination in Desert Plants. Adaptations of Desert Organisms.* Springer, Berlin.

Gutterman, Y. (1994a) Strategies of seed dispersal and germination in plants inhabiting deserts. *Botanical Review* 60, 373–425.

Gutterman, Y. (1994b) Long-term seed position influences on seed germinability of the desert annual, *Mesembryanthemum nodiflorum* L. *Israel Journal of Plant Sciences* 42, 197–205.

Gutterman, Y. (1994c) In Memoriam – Michael Evenari and his desert. Seed dispersal and germination strategies of *Spergularia diandra* compared with some other desert annual plants inhabiting the Negev Desert of Israel. *Israel Journal of Plant Sciences* 42, 261–274.

Gutterman, Y. (1994d) Germinability under natural temperatures of *Lactuca serriola* L. achenes matured and collected on different dates from a natural population in the Negev Desert Highlands. *Journal of Arid Environments* 28, 117–127.

Gutterman, Y. (1995) Seed dispersal, germination and flowering strategies of desert plants. In: Nierenberg, W.A. (ed.) *Encyclopedia of Environmental Biology,* vol. 3. Academic Press, San Diego, pp. 293–316.

Gutterman, Y. (1996a) Environmental influences during seed maturation, and storage affecting germinability in *Spergularia diandra* genotypes inhabiting the Negev Desert, Israel. *Journal of Arid Environments* 34, 313–323.

Gutterman, Y. (1996b) Temperatures during storage, light and wetting, affecting caryopses germinability of *Schismus arabicus,* a common desert annual grass. *Journal of Arid Environments* 33, 73–85.

Gutterman, Y. (1996c) Effect of day length during plant development and caryopsis maturation on flowering and germination, in addition to temperature during dry storage and light during wetting, of *Schismus arabicus* (Poaceae) in the Negev Desert, Israel. *Journal of Arid Environments* 33, 439–448.

Gutterman, Y. (1997a) Genotypic, phenotypic and opportunistic germination strategies of some common desert annuals compared with plants with other seed dispersal and germination strategies. In: Ellis, R.H., Black, M., Murdoch, A.J. and Hong, T.O. (eds) *Basic and Applied Aspects of Seed Biology. Proceedings of 5th Workshop on Seeds, Reading, UK, 1995.* Kluwer Academic Publishers, Dordrecht, pp. 611–622.

Gutterman, Y. (1997b) Effect of daylength on flowering and seed morphology of *Spergularia diandra* occurring in the Negev Desert, Israel. *Journal of Arid Environments* 36, 611–622.

Gutterman, Y. (1998a) Ecological strategies of desert annual plants. In: Ambasht, R.S. (ed.) *Modern Trends in Ecology and Environment.* Backhuys Publishers, Leiden, The Netherlands. pp. 203–231.

Gutterman, Y. (1998b) Ecophysiological genotypic and phenotypic strategies affecting seed 'readiness to germinate' in plants occurring in deserts. In: Taylor, A.G. and Huang, X.-L. (eds) *Progress in Seed Research. Proceedings of the 2nd International Conference on Seed Science and Technology.* Communication Services of the New York State Agricultural Experiment Station, Geneva, New York, pp. 10–19.

Gutterman, Y. and Evenari, M. (1994) The influences of amounts and distribution of irrigation during the hot and dry season on emergence and survival of some desert winter annual plants in the Negev Desert. *Israel Journal of Plant Sciences* 42, 1–14.

Gutterman, Y. and Ginott, S. (1994) The long-term protected 'seed bank' in the dry inflorescents, the mechanism of achenes (seeds) dispersal by rain (ombrohydrochory) and the germination of the annual desert plant *Asteriscus pygmaeus. Journal of Arid Environments* 26, 149–163.

Gutterman, Y. and Gozlan, S. (1999) Afterripening, amounts of rain for germination, and seedling drought tolerance, of local and edaphic ecotypes of *Hordeum spontaneum* from Israel. In: Wasser, S. (ed.) *Evolutionary Theory and Processes: Modern Perspectives,* Festschrift in Honour of Professor E. Nevo. Kluwer Academic, Dordrecht, pp. 419–439.

Gutterman, Y. and Nevo, E. (1994) Temperatures and ecological–genetic differentiation affecting the germination of *Hordeum spontaneum* caryopses harvested from three populations: the Negev Desert and opposing slopes on Mediterranean Mount Carmel. *Israel Journal of Plant Sciences* 42, 183–195.

Gutterman, Y. and Shem-Tov, S. (1996) Structure and function of the mucilaginous seed coats of *Plantago coronopus* inhabiting the Negev Desert of Israel. *Israel Journal of Plant Sciences* 44, 125–134.

Gutterman, Y. and Shem-Tov, S. (1997a) Mucilaginous seed coat structure of *Carrichtera annua* and *Anastatica hierochontica* from populations occurring in the Negev Desert highlands of Israel, and its adhesion to the soil crust. *Journal of Arid Environments* 35, 697–705.

Gutterman, Y. and Shem-Tov, S. (1997b) The efficiency of the strategy of mucilaginous seeds of some common annuals of the Negev adhering to the soil crust to delay collection by ants. *Israel Journal of Plant Sciences* 45, 317–327.

Gutterman, Y., Witztum, A. and Evenari, M. (1967) Seed dispersal and germination in *Blepharis persica* (Burm.) Kuntze. *Israel Journal of Botany* 16, 213–234.

Gutterman, Y., Evenari, M. and Heydecker, W. (1972) Phytochrome and temperature relations in *Lactuca sativa* L. Grand Rapids seed germination after thermodormancy. *Nature New Biology* 235, 144–145.

Gutterman, Y., Thomas, T.H. and Heydecker, W. (1975) Effect on the progeny of applying different day length and hormone treatments to parent plants of *Lactuca scariola. Physiologia Plantarum* 34, 30–38.

Gutterman, Y., Corbineau, F. and Côme, D. (1996) Dormancy of *Hordeum spontaneum* caryopses from a population on the Negev Desert highlands. *Journal of Arid Environments* 33, 337–345.

Gutterman, Y., Shem-Tov, S. and Gozlan, S. (1998) The effect of post maturation temperatures and duration on seed germinability of *Plantago coronopus* occurring in natural populations in the Negev Desert highlands. *Journal of Arid Environments* 38, 451–463.

Harper, J.L. (1977) *Population Biology of Plants.* Academic Press, New York.

Holm, L.G., Plucknett, D.L., Pancho, J.V. and Herberger, J.P. (1977) *The World's Worst Weeds: Distribution and Biology.* University Press of Hawaii, Honolulu.

Huang, Z. and Gutterman, Y. (1999) Water absorption by mucilaginous achenes of *Artemisia monosperma*: floating and germination as affected by salt concentrations. *Israel Journal of Plant Sciences* 47, 27–34.

Jacques, R. (1957) Quelques données sur le photoperiodisme de *Chenopodium polyspermum* L., influence sur la germination des graines. *Colloque International sur le Photo-thermo Periodisme,* publ. 34, serie B. IUBS, Parma, pp. 125–130.

Jacques, R. (1968) Action de la lumiere par l'intermediaire du phytochrome sur la germination, la croissance et le developpement de *Chenopodium polyspermum* L. *Physiologie Végétale* 6, 137–164.

Kane, M. and Cavers, P.B. (1992) Patterns of seed weight distribution and germination with time in a weedy biotype of proso millet (*Panicum milliaceum*). *Canadian Journal of Botany* 70, 562–567.

Karssen, C.M. (1970) The light promoted germination of the seeds of *Chenopodium album* L., III. Effect of the photoperiod during growth and development of the plants on the dormancy of the produced seeds. *Acta Botanica Neerlandica* 19, 81–94.

Karssen, C.M. (1980/81) Environmental conditions and endogenous mechanisms involved in secondary dormancy of seeds. *Israel Journal of Botany* 19, 45–64.

Kigel, J., Gibly, A. and Negbi, M. (1979) Seed germination in *Amaranthus retroflexus* L. as affected by the photoperiod and age during flower induction of the parent plants. *Journal of Experimental Botany* 30, 997–1002.

Koller, D. and Roth, N. (1964) Studies on the ecological and physiological significance of amphicary in *Gymnarrhena micrantha* (Compositae). *American Journal of Botany* 51, 26–35.

Lamont, B.B., Le Maitre, D.C., Cowling, R.M. and Enright, N.J. (1991) Canopy seed storage in woody plants. *Botanical Review* 57, 277–317.

Loria, M. and Noy-Meir, I. (1979/80) Dynamics of some annual populations in a desert loess plain. *Israel Journal of Botany* 28, 211–225.

Nevo, E.A., Beiles, Y., Gutterman, Y., Storch, N. and Kaplan, N. (1984) Genetic resources of wild cereals in Israel and vicinity: II. Phenotypic variation within and between populations of wild barley, *Hordeum spontaneum. Euphytica* 33, 737–756.

Pourrat, Y. and Jacques, R. (1975) The influence of photoperiodic conditions received by the mother plant on morphological and physiological characteristics of *Chenopodium polyspermum* L. seeds. *Plant Science Letters* 4, 273–279.

Rösch, M.W. (1977) Enkele plantekologiese aspekte van die Hester Malan-natuurreservaat. MSc. thesis, University of Pretoria, Pretoria, South Africa.

Salisbury, E.J. (1942) *The Reproductive Capacity of Plants.* Bell, London.

Shem-Tov, S., Zaady, E., Groffman, P.M. and Gutterman, Y. (1999) Soil carbon content along a rainfall gradient and inhibition of germination: a potential mechanism for regulating distribution of *Plantago coronopus. Soil Biology and Biochemistry* 31, 1209–1217.

Shmida, A., Evenari, M., and Noy-Meir, I. (1986) Hot desert ecosystems: an integrated view. In: Evenari, M., Noy-Meir, I. and Goodall, D.W. (eds) *Ecosystems of the World*, 12B, *Hot Deserts and Arid Shrublands B*. Elsevier Science, Amsterdam, pp. 379–388.

Sidhu, S.S. and Cavers, P.B. (1977) Maturity–dormancy relationships in attached and detached seeds of *Medicago lupulina* L. (black medick). *Botanical Gazette* 138, 174–182.

Simpson, G.M. (1965) Dormancy studies in seed of *Avena fatua*. 4. The role of gibberellin in embryo dormancy. *Canadian Journal of Botany* 43, 793–816.

Small, J.G.C. and Gutterman, Y. (1991) Evidence for inhibitor involvement in thermodormancy of Grand Rapids lettuce seeds. *Seed Science Research* 1, 263–267.

Small, J.G.C. and Gutterman, Y. (1992a) Effect of sodium chloride on prevention of thermodormancy, ethylene and protein synthesis and respiration in Grand Rapids lettuce seeds. *Physiologia Plantarum* 84, 35–40.

Small, J.G.C. and Gutterman, Y. (1992b) A comparison of thermo- and skotodormancy in seeds of *Lactuca serriola* in terms of induction, alleviation, respiration, ethylene and protein synthesis. *Plant Growth Regulation* 11, 301–310.

Van Rheede van Oudtshoorn, K. and van Rooyen, M.W. (1999) *Dispersal Biology of Desert Plants. Adaptations of Desert Organisms*. Springer, Berlin.

Wurzburger, J. and Koller, D. (1976) Differential effects of the parental photo-thermal environment on development of dormancy in caryopses of *Aegilops kotschyi*. *Journal of Experimental Botany* 27, 43–48.

Zaady, E., Gutterman, Y. and Boeken, B. (1997)The germination of mucilaginous seeds of *Plantago coronopus*, *Reboudia pinnata* and *Carrichtera annua* on cyano-bacterial soil crust from the Negev Desert. *Plant and Soil* 190, 247–252.

Zohary, M. (1937) Die verbreitungsökologischen Verhältnisse der Pflanzen Palästinas. Die antitelechorischen Erscheinungen. *Beihefte zum Botanischen Centralblatt* 56, 1–55.

Zohary, M. (1962) *Plant Life of Palestine*. Ronald, New York.

11

Seasonally Regulated Proteins in Peach (*Prunus persica* L. Batsch): What Are They and What Do They Do?

Michael Wisniewski[1] and Rajeev Arora[2]

[1]USDA-ARS, Appalachian Fruit Research Station, 45 Wiltshire Road, Kearneysville, WV 25430, USA; [2]Department of Plant and Soil Science, University of West Virginia, Morgantown, West Virginia, USA

Introduction

The seasonal process of cold acclimation in woody plants allows them to tolerate and survive freezing (Sakai and Larcher, 1987; Wisniewski and Arora, 1993). Additionally, temperate woody plants undergo a period of dormancy where growth is arrested. Transitions (onset and loss) in dormancy and cold acclimation partially overlap, making it difficult to associate physiological and molecular changes specifically with one or the other phenological events. To overcome this problem, researchers have devised several different approaches. One of these strategies has been the use of sibling genotypes of peach segregating for deciduous (dormant) and evergreen (non-dormant) habits. These genotypes exhibit a significant difference in their ability to cold acclimate while the apical meristems of the evergreen genotype do not form buds or exhibit endodormancy (Arora *et al.*, 1992; Arora and Wisniewski, 1994; Rodriguez *et al.*, 1994). This system has allowed us to identify several seasonally regulated proteins that may be associated with either dormancy or cold acclimation. While these proteins could be classified as bark or vegetative storage proteins (BSP or VSP) based on their abundance and seasonal pattern, protein and/or gene sequence data indicate that they have functions other than simple nitrogen storage. The purpose of this chapter will be: (i) to briefly review various model systems that have been used to study proteins related to dormancy and cold hardiness; and (ii) describe the seasonally regulated proteins we have identified using evergreen and deciduous genotypes of peach and discuss the putative functions of these proteins.

In particular, we draw attention to the problems of identifying proteins as BSPs or VSPs based solely on their abundance and seasonal pattern of accumulation.

Model Systems Used to Study Bud Dormancy and Cold Acclimation in Woody Plants

Besides the use of sibling genotypes of peach, two principal approaches have been used to overcome the problem of the superimposition of dormancy and cold acclimation in woody plants. These approaches have involved either the differential induction of dormancy and cold acclimation or the differential regulation of chill-unit accumulation (chilling requirement) and cold hardiness (Wisniewski and Arora, 2000). In the former approach, Fennell and Hoover (1991) showed that *Vitis labruscana* is able to fully enter endodormancy in response to short photoperiods without cold acclimating. Salzman *et al.* (1996) used this system to study differential expression of proteins during cold hardiness and dormancy.

Arora *et al.* (1997) used the latter approach to differentially modify chill-unit accumulation and cold hardiness transitions in blueberry (*Vaccinium*, section *Cyanococcus*) cultivars that differ in their chilling requirement. By subjecting cold-acclimated plants to temperatures that induced deacclimation to cold without negating chill-unit accumulation, they were able to identify bud proteins that were specifically associated with deacclimation. In particular, they identified a family of dehydrin proteins that were specifically associated with cold hardiness transitions rather than with changes in dormancy status.

In a series of papers, Wisniewski and co-workers used sibling genotypes of deciduous and evergreen peach, the latter lacking endodormancy, to identify proteins that were specifically associated with either dormancy or cold acclimation (Arora *et al.*, 1992; Arora and Wisniewski, 1994, 1996; Arora *et al.*, 1996). In these studies, they identified three major proteins with molecular masses of 16, 19 and 60 kDa. These proteins accumulate to significant levels in the autumn, reach a maximum in midwinter, decline after bud break, and are nearly absent during summer months. This pattern of accumulation was descriptive of a class of proteins referred to as bark or vegetative storage proteins (Kang and Titus, 1987; Wetzel *et al.*, 1989; Coleman *et al.*, 1992; Stepien *et al.*, 1994). By definition, however, this class of proteins serves simply as a source of nitrogen reserves and is not believed to have other functions. Based on our own studies, we believe that defining a storage protein simply by its pattern of expression and abundance may lead to erroneous conclusions regarding its functional role. Instead, detailed functional analyses of seasonally regulated proteins are needed in order to better understand their relationship to physiological processes such as cold acclimation and dormancy.

Seasonally Regulated Proteins in Peach

Seasonal patterns of proteins and cold hardiness were characterized in bark and xylem tissues of genetically related deciduous and evergreen peach (Arora *et al.*, 1992). In contrast to deciduous trees, which entered endodormancy and abscised leaves in the autumn, evergreen trees retained their leaves and exhibited shoot elongation under favourable environmental conditions. A successive increase in the cold hardiness of bark and xylem tissues was observed during the autumn in both genotypes (Fig. 11.1). This

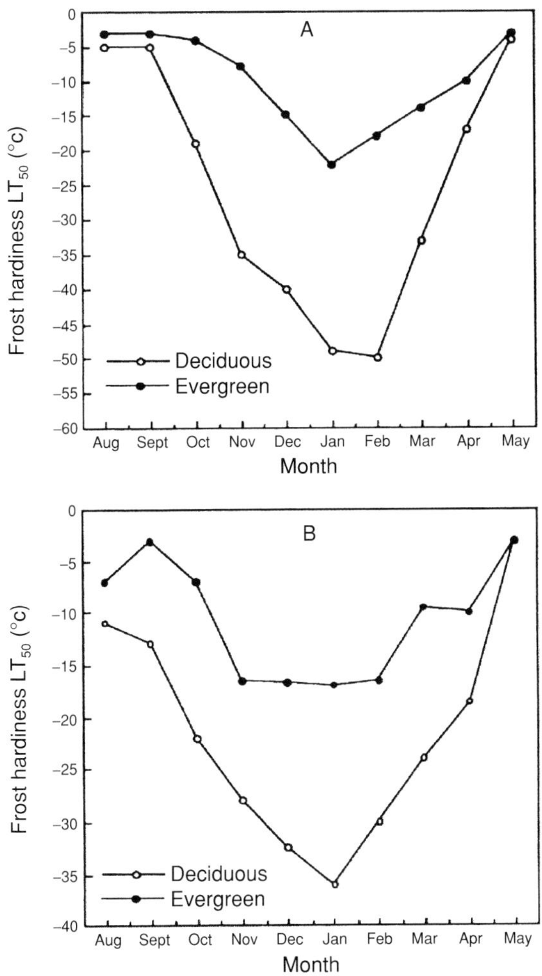

Fig. 11.1. Seasonal changes in the cold hardiness of bark (A) and xylem (B) tissues of sibling deciduous and evergreen peach trees. Reprinted from Arora *et al.* (1992) with permission from the American Society of Plant Physiologists.

was followed by a subsequent decrease in cold hardiness from midwinter to spring. The maximum hardiness level attained in deciduous trees was more than twofold that of evergeens. Seasonal patterns of proteins from bark and xylem of the sibling genotypes was characterized by one-dimensional SDS-PAGE (Figs 11.2 and 11.3). Among other qualitative and quantitative changes, accumulation of a 19 kDa polypeptide in the bark of both genotypes was observed during autumn followed by a decrease in spring. This polypeptide accumulated to higher levels in the deciduous peach compared with the evergreen. Additionally, a 16 kDa protein exhibited a similar pattern of accumulation in the deciduous trees but not the evergreen trees, suggesting that this protein may be more associated with dormancy than with cold acclimation. Based on their seasonal pattern of accumulation, both proteins conformed to the criteria of bark storage proteins.

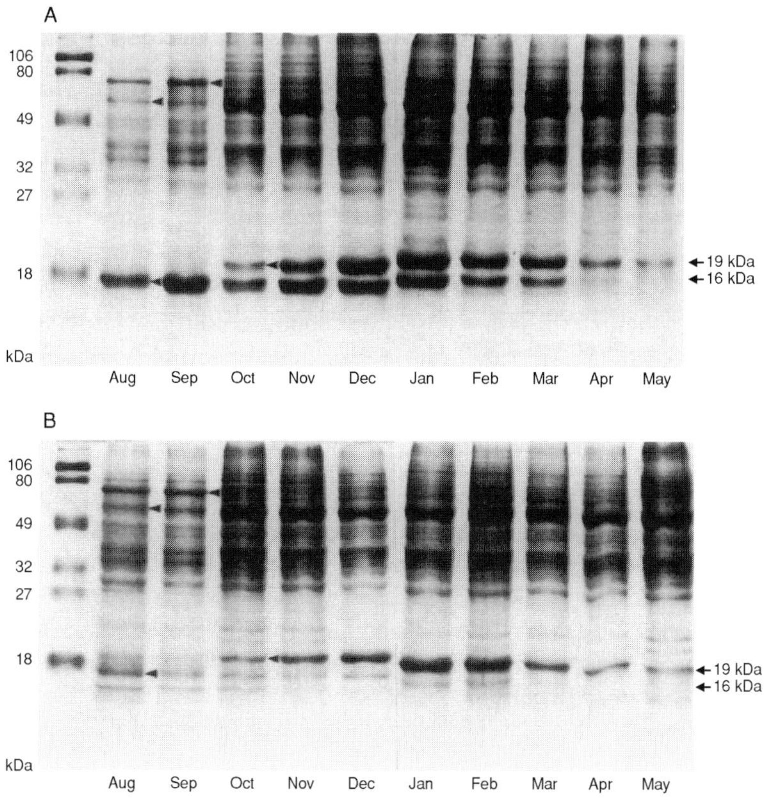

Fig. 11.2. Monthly SDS-PAGE protein profiles of the bark of sibling deciduous (A) and evergreen (B) peach trees. Arrows indicate the position of 78, 60, 19 and 16 kDa polypeptides. Reprinted from Arora *et al.* (1992) with permission from the American Society of Plant Physiologists.

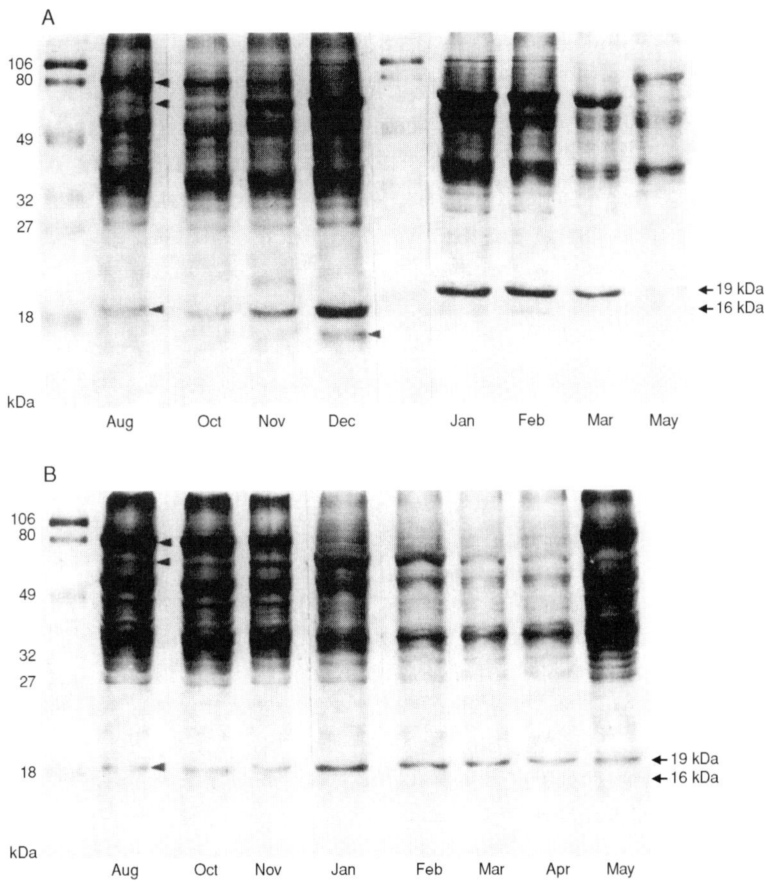

Fig. 11.3. Monthly SDS-PAGE protein profiles of the xylem tissues from sibling deciduous (A) and evergreen (B) peach trees. Arrows indicate the position of 80, 70, 19 and 16 kDa polypeptides. Reprinted from Arora *et al.* (1992) with permission from the American Society of Plant Physiologists.

In subsequent research (Arora and Wisniewski, 1994) we identified a 60 kDa polypeptide (PCA60) in bark tissues that also accumulated during cold acclimation and was also heat stable. Immunological studies indicated that this protein belongs to the dehydrin family of proteins. PCA60 accumulated to higher levels in the bark tissues of the deciduous genotype than in the evergreen genotype (Fig. 11.4). In a later study (Arora and Wisniewski, 1996), this protein was found to have a similar pattern of expression and accumulation in the xylem tissues of the two genotypes (Fig. 11.5).

The dehydrin family of proteins is induced by environmental stresses that result in cellular dehydration (Close, 1997). They are glycine-rich proteins, deficient in tryptophan and cysteine, and remain soluble at

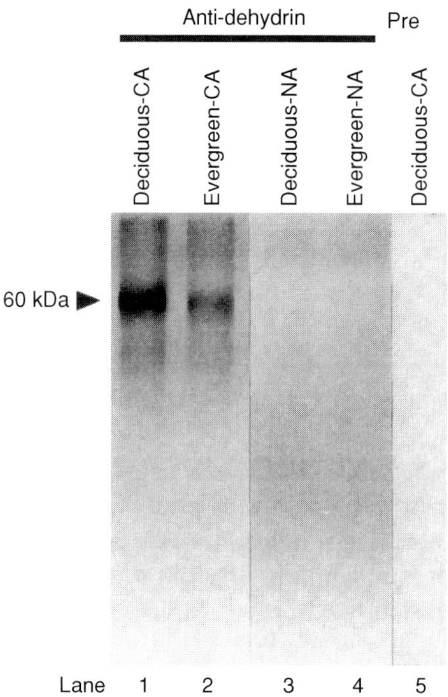

Fig. 11.4. Western blot analysis of heat-stable bark proteins from cold-acclimated (CA) and non-acclimated (NA) tissues of sibling deciduous and evergreen peach genotypes. Lanes 1 through 4 were probed with dehydrin antiserum (anti-dehydrin), whereas lane 5 was probed with preimmune serum (Pre). Reprinted from Arora and Wisniewski (1994).

temperatures approaching 100°C. Additionally, dehydrins are characterized by a lysine rich amino acid sequence (KIKEKLPG) present at the carboxy terminus and often repeated throughout the polypeptide. The polypeptide is believed to form an amphipathic helix and might interact with lipids in a manner similar to the interaction of class A amphipathic α-helices of apolipoproteins with lipids, or with exposed hydrophobic patches of partially denatured proteins (Close, 1997) in a manner similar to chaperones (Mayhew and Hartl, 1996). The presence of particular motifs in dehydrins is often related to the stress that induces expresion of the protein. For example, if the gene is primarily cold-inducible, the corresponding protein is rich in lysine repeats and lacks a serine track (Palva and Heino, 1998). Dehydrins accumulate in a wide array of tissues and have been cellularly localized in the cytosol and nucleus (Neven *et al.*, 1993; Edgerton-Warburton *et al.*, 1997; Danyluk *et al.*, 1998; Wisniewski *et al.*, 1999).

While much has been learned about dehydrins in herbaceous plants, information on dehydrins in woody plants is limited. Seasonal expression of dehydrins has been noted in several species (Wisniewski *et al.*, 1996) including peach (*Prunus persica*) (Arora and Wisniewski, 1994) and blueberry (*Vaccinium*, sec. *Cyanococcus*) (Muthalif and Rowland, 1994). Chilling and

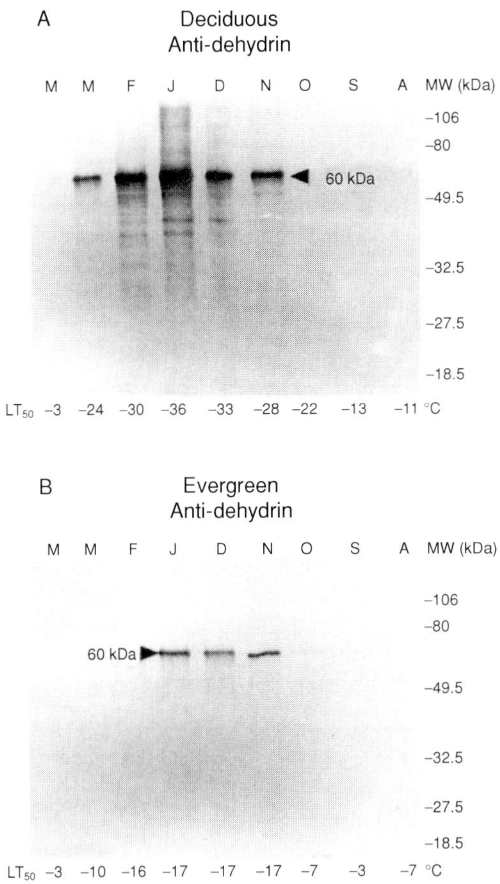

Fig. 11.5. Seasonal variation in the accumulation of 60 kDa dehydrin protein and cold hardiness in the xylem tissues of sibling deciduous (A) and evergreen (B) peach genotypes. Letters above each lane (right to left) denote successive monthly samples from August to May, except for April. Corresponding cold hardiness values (LT_{50}) are listed at the bottom of each lane. Reprinted from Arora and Wisniewski (1996).

desiccation have also been shown to induce dehydrin gene expression in citrus (*Poncirus trifoliata* and *Citrus grandis*) leaves (Cai *et al.*, 1995). The gene (*ppdhn1*) coding for the peach dehydrin PCA60 has been cloned and characterized (Artlip *et al.*, 1997). The *ppdhn1* codes for a protein that contains nine lysine-rich repeats, two DEYGNP motifs at the amino acid terminus, and is immunolocalized primarily in the cytosol and nucleolus of various cell types of bark and xylem tissues of peach (Artlip *et al.*, 1997; Wisniewski *et al.*, 1999). RNA and immunoblot analyses indicated that the gene transcript and protein exhibited a distinct seasonal pattern in both evergreen and deciduous genotypes. Accumulation began in early autumn, reached a maximum in midwinter, declined during early spring, and was absent during the summer months. While overall seasonal patterns were similar, there was a significant difference between the genotypes. Transcript and protein levels increased earlier and to a higher level in the deciduous genotype than in the evergreen and there was also a close correspondence between dehydrin expression and cold hardiness (Arora *et al.*, 1992; Artlip *et al.*, 1997). These results are very similar to those reported for protein (Muthalif and Rowland, 1994) and transcript levels (Levi *et al.*, 1999) of the 65 and 60 kDa dehydrins in the buds of a cold hardy variety of the blueberry 'Bluecrop' and the less cold hardy variety 'Tifblue'. Arora *et al.* (1997), using conditions which caused deacclimation but not the negation of chill-unit accumulation established that the expression of three dehydrins (65, 60 and 14 kDa) was specifically associated with cold hardiness transitions rather than with changes in dormancy status. Induction of *ppdhn1* in response to drought and abscisic acid (ABA) has been documented (Artlip and Wisniewski, 1997) and purified extracts of PCA60 have also been shown to have both cryoprotective and antifreeze activities *in vitro* (Wisniewski *et al.*, 1999). Thus, it appears that PCA60 plays a functional role in cold hardiness rather than dormancy. The role and function of the seasonally expressed 16 and 19 kDa proteins in peach, however, is more ambiguous.

Bark storage (BSPs), or vegetative storage proteins (VSPs) as they are sometimes referred to, are involved in seasonal cycling of nitrogen in a number of woody perennials (Titus and Kang, 1982; Stepien *et al.*, 1994). BSP mRNAs and protein levels of *Populus* have been shown to be induced by short photoperiods and their expression regulated by altered source–sink relationships (Coleman *et al.*, 1992, 1993). Whether storage proteins are causally linked with dormancy and cold acclimation or whether they have functions other than nitrogen storage is an open question.

Seasonal SDS-PAGE profiles of bark proteins from deciduous and evergreen genotypes of peach identified at least three proteins (16, 19 and 60 kDa) that could be classified as storage proteins based on their patterns of accumulation (Arora *et al.*, 1992; Arora and Wisniewski, 1994). Interestingly, two of the proteins (19 and 60 kDa) had sequence homology to proteins with presumed functions other than nitrogen storage (Table 11.1). The third protein (16 kDa) had no sequence similarity to other proteins in the database and thus may truly represent a storage protein. Our observations on the seasonal accumulation of the 16 kDa protein in different-aged tissues of peach support this contention (unpublished data). We noted the greatest accumulation of the 16 kDa protein during the autumn in current year shoots. Lesser amounts were present in the older tissues and roots. BSPs do not accumulate uniformly throughout the tree. Instead, they are present in the greatest amounts in tissues adjacent to sites of subsequent spring growth. This reduces the distance needed to transport the recycled nitrogen and provides for greater efficiency (Coleman, 1997). The 16 kDa protein was either present in low amounts or undetectable in the evergreen genotype. Arora *et al.* (1992) attributed this to an apparent lack of sensitivity of evergreen peach to shorter photoperiods during the

Table 11.1. Sequence comparison of seasonally regulated proteins in peach with other reported protein sequences.

Plant	Polypeptide		Sequence	Identification
Peach	**60 kDa**	1	**RLPGGQKDDQYL**	
		2	**RLPGGQNVDPTTGPYGGGGAAG**	Dehydrins, cold-
		3	**RLPIGQKVD**	acclimation and
Barley	B8		KLPGGAH	drought-induced
	B9		KLPGGAH	proteins
Spinach	CAP 85		KLPG-QH	
Wheat	WCS 120		KLPGGHGDHQQTGGT	
Peach	**19 kDa**	1	**ITFGEGSQYGYVK**	
Apricot	—		ITFGEGSQYGYVK	
Cherry	—		ITFGEGSQYGYVK	Allergens, pathogenesis-
Apple	MalD1		ITFGEGSQYGYVK	related proteins
Pear	—		ITFGEGSQYGYVK	
Hazel	—		ITFGEGSRYKYVK	
Peach	**19 kDa**	2	**HSEILEGDGGPGTIK**	
Apricot	—		HSEILEGDGGPGTIK	
Cherry	—		HSEILEGDGGPGTIK	Allergens, pathogenesis-
Pear	—		HAEILEGNGGPGTIK	related proteins, ABA-
Pea	—		EILEGNGGPGTVK	responsive proteins,
Bean	—		EIVEGNGGPGTIK	fruit-ripening-related
Birch	—		ENIEGNGGPGTIK	proteins
Peach	**19 kDa**	3	**AFVLDADNLVPK**	
Cherry	—		AFVLDADNLVPK	
Apple	MalD1		AFVLDADNLIPK	
Pear	—		AFVLDADNLIPK	Allergens
Apricot	—		AFILDADNLIPK	
Birch	—		SFVLDADNLIPK	
Peach	**16 kDa**	1	**GQAHLIPNVSSGHIK**	
	16 kDa	2	**EKVEYDDENKVATLIGLDGEVFK**	
			No known sequence homology with other proteins	

autumn. These findings, taken together with the effect of short photoperiods on the induction of bud dormancy (Nitsch, 1957; Junttila, 1976) and BSP gene expression (Coleman *et al.*, 1992; Langheinrich and Tischner, 1991), suggest that seasonal accumulation (or lack) of the 16 kDa peach protein may be more closely linked to dormancy rather than cold acclimation.

Establishing a direct role for storage proteins in either cold acclimation or dormancy will require detailed studies on the expression of these genes and proteins under controlled environmental conditions. The availability of transgenics, exhibiting over- or under-expression of BSPs, or BSP mutants would also assist in determining the effect of these genes on cold acclimation and dormancy in woody perennials. Defining BSPs by their pattern of accumulation and by the extent to which they accumulate may not be a good functional definition. It is possible that BSPs might belong to a larger family of proteins, some of which, in addition to nitrogen storage, may have a functional role in cold acclimation, whereas others may be associated more with dormancy.

References

Arora, R. and Wisniewski, M.E. (1994) Cold acclimation in genetically related (sibling) deciduous and evergreen peach (*Prunus persica*). II. A 60-kilodalton bark protein in cold acclimated tissues of peach is heat-stable and related to the dehydrin family of proteins. *Plant Physiology* 105, 95–101.

Arora, R. and Wisniewski, M.E. (1996) Accumulation of 60-kD dehydrin protein in peach xylem tissues and its relationship to cold acclimation. *HortScience* 31, 923–925.

Arora, R., Wisniewski, M.E. and Scorza, R. (1992) Cold acclimation in genetically related (sibling) deciduous and evergreen peach (*Prunus persica*). I. Seasonal changes in cold hardiness and polypeptides of bark and xylem tissues. *Plant Physiology* 99, 1562–1568.

Arora, R., Wisniewski, M.E. and Rowland, L.J. (1996) Cold acclimation and alterations in dehydrin-like and bark storage proteins in the leaves of sibling deciduous and evergreen peach. *Journal of the American Society for Horticultural Science* 121, 915–919.

Arora, R., Rowland, L.J. and Panta, G.R. (1997) Chill-responsive dehydrins in blueberry: are they associated with cold hardiness or dormancy transitions? *Physiologia Plantarum* 101, 8–16.

Artlip, T.S. and Wisniewski, M.E. (1997) Tissue-specific expression of a dehydrin gene in one-year-old 'Rio Oso Gem' peach trees. *Journal of the American Society for Horticultural Science* 122, 784–787.

Artlip, T.S., Callahan, A.M., Bassett, C.L. and Wisniewski, M.E. (1997) Seasonal expression of dehydrin gene in sibling deciduous and evergreen peach (*Prunus persica* L. Batsch). *Plant Molecular Biology* 33, 61–70.

Cai, Q., Moore, G.A. and Guy, C.L. (1995) An unusual group 2 LEA gene family in citrus responsive to low temperature. *Plant Molecular Biology* 29, 11–23.

Close, T.J. (1997) Dehydrins: a commonalty in the response of plants to dehydration and low temperatures. *Physiologia Plantarum* 100, 291–296.

Coleman, G.D. (1997) Seasonal vegetative storage proteins of poplar. In: Klopfenstein, N.B., Chun, Y.W., Kim, M.S. and Ahuja, M.R. (eds) *Micropropagation, Genetic Engineering and Molecular Biology of* Populus. USDA Forest Service General Technical Report, Fort Collins, Colorado, RM-GTR-297, pp. 124–130.

Coleman, G.D., Chen, T.H.H. and Fuchigami, L.H. (1992) Complementary DNA cloning of poplar bark storage protein and control of its expression by photoperiod. *Plant Physiology* 98, 687–693.

Coleman, G.D., Englert, J.M., Chen, T.H.H. and Fuchigami, L.H. (1993) Physiological and environmental requirements for poplar (*Populus deltoides*) bark storage protein degradation. *Plant Physiology* 102, 53–59.

Danyluk, J., Perron, A., Houde, M., Limin, A., Fowler, B., Benhamou, N. and Sarhan, F. (1998) Accumulation of an acidic dehydrin in the vicinity of the plasma membrane during cold acclimation of wheat. *The Plant Cell* 10, 623–638.

Edgerton-Warburton, L.M., Balsamo, R.A. and Close, T.J. (1997) Temporal expression and ultrastructural localization of dehydrins in *Zea mays* L. *Physiologia Plantarum.* 101, 545–555.

Fennell, A. and Hoover, E. (1991) Photoperiod influences growth, bud dormancy and cold acclimation in *Vitis labruscana* and *V. riparia*. *Journal of the American Society for Horticultural Science* 116, 270–273.

Junttila, O. (1976) Apical growth cessation and shoot tip abscission in *Salix*. *Physiologia Plantarum* 38, 278–286.

Kang, S.-M. and Titus, J.S. (1987) Specific proteins may determine maximum cold resistance in apple shoots. *Journal of Horticultural Science* 62, 281–285.

Langheinrich, U. and Tischner, R. (1991) Vegetative storage proteins in poplar induction and characterization of a 32- and a 36-kilodalton polypeptide. *Plant Physiology* 97, 1017–1025.

Levi, A., Panta, G.R., Parmentier, C.M., Muthalif, M.M., Arora, R., Shanker, S. and Rowland, L.J. (1999) Complementary DNA cloning, sequencing and expression of an unusual dehydrin from blueberry buds. *Physiologia Plantarum* 107, 98–109.

Mayhew, M. and Hartl, F.U. (1996) Molecular chaperone proteins. In: Neidhardt, F. (ed.) Escherischia coli *and* Salmonella typhimurium: *Cellular and Molecular Biology*. American Society for Microbiology, Washington, DC, pp. 922–937.

Muthalif, M.M. and Rowland, L.J. (1994) Identification of dehydrin-like proteins responsive to chilling in floral buds of blueberry (*Vaccinium* section Cyanococcus). *Plant Physiology* 104, 1439–1447.

Neven, L.G., Haskell, D.W., Hofig, A., Li, Q.-B. and Guy, C.L. (1993) Characterization of a spinach gene responsive to low temperature and water stress. *Plant Molecular Biology* 21, 291–305.

Nitsch, J.P. (1957) Growth responses of woody plants to photoperiodic stimuli. *Proceedings of the American Society for Horticultural Science* 70, 512–525.

Palva, T. and Heino, P. (1998) Molecular mechanism of plant cold acclimation and freezing tolerance. In: Li, P.H. and Chen, T.H.H. (eds) *Plant Cold Hardiness: Molecular Biology, Biochemistry, and Physiology*. Plenum Press, New York, pp. 3–14.

Rodriguez, A.J., Sherman, W.B., Scorza, R., Wisniewski, M. and Okie, W.R. (1994) 'Evergreen' peach, its inheritance and dormant behavior. *Journal of the American Society for Horticultural Science* 119, 789–792.

Sakai, A. and Larcher, W. (1987) *Frost Survival of Plants: Responses and Adaptation to Freezing Stress.* Springer, Berlin.

Salzman, R.A., Bressan, R.A., Hasegawa, P.M., Ashworth, E.N. and Bordelon, B.P. (1996) Programmed accumulation of LEA-like proteins during desiccation and cold acclimation of overwintering grape buds. *Plant, Cell and Environment* 19, 713–720.

Stepien, V., Sauter, J.J. and Martin, F. (1994) Vegetative storage proteins in woody plants. *Plant Physiology and Biochemistry* 32, 1–8.

Titus, J.S. and Kang, S.-M. (1982) Nitrogen metabolism, translocation and recycling in apple trees. *Horticultural Reviews* 4, 204–246.

Wetzel, S., Demmers, C. and Greenwood, J.S. (1989) Seasonally fluctuating bark proteins are a potential form of nitrogen storage in three temperate hardwoods. *Planta* 178, 275–281.

Wisniewski, M. and Arora, R. (1993) Adaptation and response of fruit trees to freezing temperatures. In: Biggs, A.R. (ed.) *Cytology, Histology, and Histochemistry of Fruit Tree Diseases.* CRC Press, Boca Raton, Florida, pp. 299–320.

Wisniewski, M. and Arora, R. (2000) Structural and biochemical aspects of cold hardiness in woody plants. In: Jain, M. and Minocha, S. (eds) *The Molecular Biology of Woody Plants* (in press).

Wisniewski, M.E., Close, T.J., Artlip, T. and Arora, R. (1996) Seasonal patterns of dehydrins and 70-kDa heat-shock proteins in bark tissues of eight species of woody plants. *Physiologia Plantarum* 96, 496–505.

Wisniewski, M., Webb, R., Balsamo, R., Close, T., Yu, X.-M. and Griffith, M. (1999) Purification, immunolocalization, cryoprotective, and antifreeze activity of PCA60: A dehydrin from peach (*Prunus persica* L. Batsch). *Physiologia Plantarum* 105, 600–608.

12

Aspects of ABA and Fusicoccin Signal Transduction in Barley Grains: I Effect of Fusicoccin on ABA-induced Gene Expression in Embryo and Aleurone; II Spatial and Temporal Expression of 14-3-3 Proteins in Dormant Embryos

Christa Testerink, René M. van der Meulen and Mei Wang

Center for Phytotechnology, Leiden University, The Netherlands Organization for Applied Scientific Research (UL/TNO), Department of Applied Plant Sciences, Wassenaarseweg 64, 2333 AL Leiden, The Netherlands

Introduction

The plant hormone abscisic acid (ABA) plays an important role in induction and maintenance of seed dormancy (Black, 1983; Wang *et al.*, 1994, 1995). Comparative physiological studies with both dormant and non-dormant barley grains demonstrated that a decrease in endogenous ABA content is one of the prerequisites for germination of barley embryos (Wang *et al.*, 1995). The use of a barley mutant that showed a reduced sensitivity to applied ABA further supported that a decrease in endogenous ABA level outside the embryo was necessary for germination (Visser *et al.*, 1996). In our previous studies, we have classified various dormancy-breaking compounds into two classes according to their effects on endogenous ABA content (Wang *et al.*, 1998). Class I compounds are those chemical compounds that have an effect on the level of endogenous ABA, while class II compounds do not have an effect on ABA content.

We are particularly interested in compounds of class II, such as fusicoccin (FC), which show induction of germination, but have no effect on endogenous ABA content. FC is a fungal toxin that has a wide range of effects on plant physiology, such as induction of stomatal opening, stimulation of cell elongation and promotion of seed germination (Lado *et al.*, 1974; Marrè, 1979). Since FC antagonizes ABA-induced physiological responses (Marrè *et al.*, 1979), it is possible that FC can crosstalk with ABA signal transduction pathways. Therefore, we studied the effect of fusicoccin on ABA-induced gene expression.

In addition, we studied the expression of 14-3-3 proteins in barley embryos, since 14-3-3 proteins are involved in the binding of FC. It has recently become clear that the binding site of FC (the FC-receptor) is formed by the association of a 14-3-3 protein with the plasma membrane H^+-ATPase (Jahn *et al.*, 1997; Oecking *et al.*, 1997). 14-3-3 proteins form a protein family that is present in all eukaryotes studied so far, and 14-3-3 proteins appear to have many different functions (Ferl, 1996). Although the biological functions of 14-3-3 proteins are highly diverse, a common theme for their action appears to be interaction with other proteins. These interacting proteins include enzymes, such as nitrate reductase and H^+-ATPase, kinases, phosphatases and other signalling proteins (Palmgren *et al.*, 1998).

Several different 14-3-3 isoforms have been identified in most plant species; at least ten isoforms both in *Arabidopsis* (Wu *et al.*, 1997) and tomato (Roberts and Bowles, 1999). However, there seems to be little functional isoform specificity of these proteins. Instead, tissue-specific and temporal regulation of 14-3-3 proteins seems to be a major way to control 14-3-3 action (reviewed recently by Finnie *et al.*, 1999). In non-dormant barley embryos, differences in spatial and temporal regulation of three barley 14-3-3 isoforms were detected (Testerink *et al.*, 1999). In this study, we investigated expression of these different barley 14-3-3 isoforms in embryos of dormant grains, and show germination-related expression of *14-3-3A* mRNA.

Materials and Methods

Chemicals

[α-^{32}P]-dCTP 110 TBq mmol^{-1} was obtained from Amersham (UK). Gene Screen plus was from Dupont (USA). Dextran sulphate was from Pharmacia Biotech (Uppsala, Sweden). GA_3 (chemical purity 99%) and all other chemicals were from Sigma (USA). FC was obtained from Dr A.H. de Boer (Department of Genetics, Section of Plant Physiology, Vrije Universiteit Amsterdam, The Netherlands).

Barley grains, germination conditions and aleurone isolation

Non-dormant *Hordeum distichum* L. cv. 'Triumph' grains were from Heineken Technical Services (HTS) (Zoeterwoude, The Netherlands).

Dormant grains were obtained by growing genetically identical 'Triumph' plants first for 16 h at 21°C/8 h at 10°C under continuous light for 20 days. Then conditions were changed to 16 h at 14°C/8 h at 10°C, also under continuous light, until the grains were fully ripe. Grains were dried to 6–8% moisture content and stored at –20°C to preserve dormancy.

Twenty intact grains were incubated on two layers of Whatman No. 1 paper (Whatman, Maidstone, Kent, UK) in a Petri dish (9 cm) containing 3 ml water. Plates were incubated at 20°C in the dark. For germination of isolated embryos, embryos were incubated in 300 µl water or FC, as has been described before by Wang *et al.* (1998). Barley aleurone layers were prepared as described before (Wang *et al.*, 1992) and ten aleurone layers from non-dormant barley grains were incubated in 1 ml water, with or without ABA and/or FC in the dark at 20°C for 4 h.

Northern analysis and cDNA probes

Total cellular RNA was isolated, purified and analysed by Northern blotting as described by Wang *et al.* (1992). Hybridization was performed in 1% SDS, 1 M NaCl, 10% dextran sulphate and 0.1 mg ml^{-1} sonicated salmon sperm DNA with random-primed labelled cDNA hybridization probes. The cDNA probes used were rice *Rab-16* cDNA (Mundy and Chua, 1988), barley *gapdh* (glyceraldehyde-3-phosphate dehydrogenase) cDNA (Cojecki, 1986) and barley *14-3-3A*, *14-3-3B* and *14-3-3C* cDNA 3′ UTRs (Testerink *et al.*, 1999).

Barley protein isolation and Western analysis

Protein was isolated from barley embryos at different times of imbibition. 20 µg of soluble protein was analysed on 15% SDS-PAGE and blotted on to nitrocellulose membrane. Blots were incubated overnight with the isoform-specific anti-14-3-3 antibodies (1:20,000, at 4°C) and bands were visualized by alkaline phosphatase (AP)-labelled goat anti-rabbit antibody, followed by incubation with NBT/BCIP substrate (Promega). Production of the isoform-specific antibodies was described by Testerink *et al.* (1999).

In situ immunolocalization studies

Immunolocalization studies were performed exactly as described by Testerink *et al.* (1999). In short, barley grains were imbibed in water for 24 h, and half grains were fixed, embedded in paraffin and cut into 10 µm sections. Rehydrated sections were incubated in 0.4% SDS, 3 mM 2-mercaptoethanol, 12 mM Tris pH 6.8 for 20 min and blocked in 0.9% NaCl, 0.1% BSAc (Aurion) in 20 mM Tris pH 8.2 for 3 min. Incubation with purified anti-14-3-3 antibody overnight at 4°C and secondary AP-labelled goat anti-rabbit antibody at 20°C was done in the same blocking buffer. The signal was visualized using NBT/BCIP substrate (Promega).

Results

Effect of FC on ABA-induced *Rab* gene expression

We previously demonstrated that the expression and maintenance of *Rab* mRNA in barley embryos are ABA-level dependent. In dry embryos of non-dormant barley grains a relatively high level of *Rab* mRNA was detected, which decreased during imbibition and disappeared at the onset of germination (16 h of imbibition, see Wang *et al.*, 1994). In embryos from dormant grains, a higher level of *Rab* mRNA was detected (Wang *et al.*, 1998), which correlates with a relatively higher ABA content in these embryos (Wang *et al.*, 1995, 1998).

During the first 16 h of imbibition, *Rab* mRNA levels stayed rather high in dormant embryos (Fig. 12.1). Addition of FC showed a significant reduction of *Rab* mRNA expression. This inhibitory effect of FC on Rab mRNA expression was not due to the reduction of ABA content in the embryos (Wang *et al.*, 1998), nor did FC treatment affect gene expression of the general metabolic enzyme GAPDH (Fig. 12.1, lower panel). In both the control and the FC-treated embryos, the level of *gapdh* gene expression increased, due to the resumption of the glycolytic pathway during imbibition. The observed inhibition of *Rab* gene expression by FC raises the question of whether FC interacts with ABA signal transduction and thereby represses *Rab* mRNA expression.

In order to know if FC is indeed able to inhibit ABA-induced *Rab* gene expression, we used an isolated aleurone system, in which no endogenous

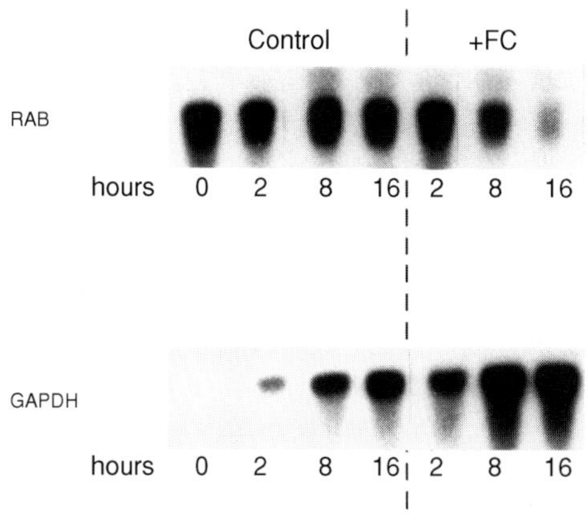

Fig. 12.1. *Rab* mRNA expression in dormant barley embryos treated with fusicoccin. Isolated embryos were incubated with or without FC, and RNA was isolated at different time points. *Rab* gene expression was analysed on Northern blots. One typical example of two independent experiments is presented.

ABA can be detected. In these isolated aleurone layers (from non-dormant barley grains), *Rab* expression is ABA concentration dependent (van Beckum *et al.*, 1993). Two different ABA concentrations (optimal: 5×10^{-7} M ABA and sub-optimal: 5×10^{-8} M ABA) were used for our studies. Figure 12.2 shows that the expression of *Rab* mRNA in isolated aleurone was induced by application of ABA. At the optimal concentration of ABA, FC had no significant effect on *Rab* expression, while at a sub-optimal concentration of ABA, FC was able to inhibit *Rab* expression (Fig. 12.2). The inhibitory effect of FC was concentration dependent; 100 μM FC gave a maximal inhibitory effect. FC alone, without addition of ABA, did not affect *Rab* mRNA expression (Fig. 12.2).

14-3-3 expression in dormant barley embryos

In the second part of our study, we investigated expression of the different 14-3-3 isoforms in dormant barley embryos. Figure 12.3A shows mRNA expression of *14-3-3A*, *14-3-3B* and *14-3-3C* in embryos of dormant intact grains that were incubated in water or FC. In addition, expression of the different 14-3-3 isoforms in 48 h imbibed non-dormant embryos is shown for comparison. In dormant embryos, expression of all three isoforms increased upon imbibition, this increase was delayed compared with the *14-3-3* increase during imbibition of non-dormant embryos (Testerink *et al.*, 1999). The expression level of *14-3-3C* mRNA in dormant embryos was comparable with the level in non-dormant, germinating grains, and *14-3-3B* expression was only slightly lower than in non-dormant embryos. *14-3-3A* expression, however, was considerably lower in the dormant embryos than in non-dormant, germinating embryos (Fig. 12.3A). In addition, when dormancy of dormant grains was broken with FC, expression of *14-3-3A* in the embryos was induced to a level that was comparable with the level in non-dormant, germinated embryos. Expression of *14-3-3B* and *14-3-3C* was not affected by FC treatment of dormant grains.

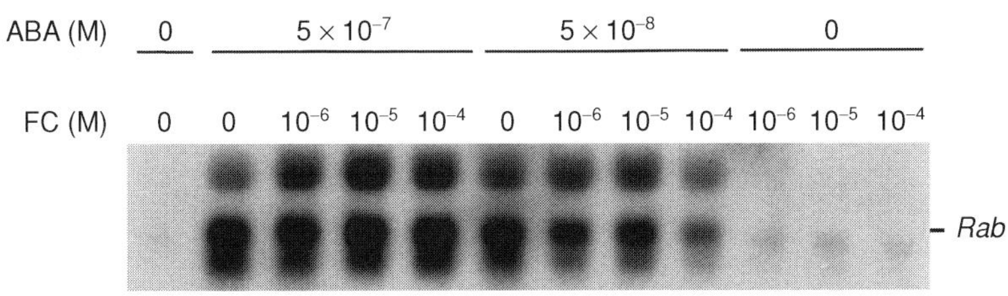

Fig. 12.2. *Rab* mRNA expression in isolated barley aleurone layers. Isolated aleurone layers were incubated in different concentrations of ABA and fusicoccin for 4 h, and *Rab* gene expression was analysed. One typical example of three independent experiments is presented.

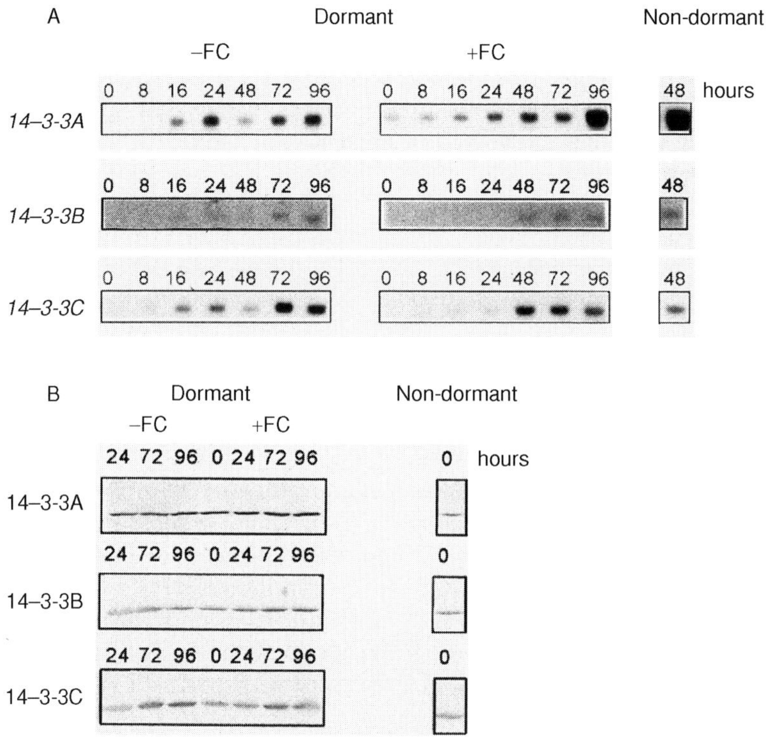

Fig. 12.3. *14-3-3* mRNA and protein expression in embryos of dormant barley grains incubated in water or fusicoccin, compared with expression in non-dormant embryos. (A) mRNA expression of *14-3-3A*, *14-3-3B* and *14-3-3C*. Intact grains were imbibed and at indicated times, the embryos were isolated and used for Northern analysis. (B) 14-3-3A, 14-3-3B and 14-3-3C protein expression. Intact grains were imbibed and at indicated times, the embryos were isolated and used for Western analysis.

Using isoform-specific anti-14-3-3 antibodies, 14-3-3 protein expression was studied in embryos of intact dormant grains (Fig. 12.3B). Like in non-dormant embryos (Testerink *et al.*, 1999), the increase in *14-3-3* mRNA levels in the embryo during imbibition of dormant grains was not reflected in the protein level. During imbibition, the level of 14-3-3A, 14-3-3B and 14-3-3C protein stayed rather constant, only 14-3-3C protein expression slightly increased after 72 h of imbibition. 14-3-3 protein levels in dormant embryos were comparable with 14-3-3 levels in non-dormant grains, and were not affected by FC treatment.

Next, localization of the different 14-3-3 protein isoforms in the barley embryo was studied using *in situ* immunolocalization. In non-dormant embryos, 14-3-3 isoforms were shown to be differentially expressed; whereas 14-3-3A and 14-3-3B were expressed all over the embryo, 14-3-3C protein was predominantly present in the scutellum and shoot apical meristem (Testerink *et al.*, 1999). In 24-h imbibed dormant embryos the same isoform-specific distribution could be observed (Fig. 12.4). 14-3-3A and 14-3-3B protein expression was present in all tissues, while 14-3-3C expression was restricted to the scutellum. Unfortunately, it was not possible to determine whether 14-3-3C was also expressed in the shoot apical meristem, since it proved not to be feasible to obtain high-quality sections exactly through the middle of dormant embryos.

Discussion

We are interested in the mechanism by which the fungal toxin fusicoccin (FC) interferes with the ABA signalling leading to gene expression and inhibition of germination. FC binds to a 14-3-3/H^+-ATPase complex and profoundly alters plasma membrane ion fluxes by stimulating apoplast acidification, hyperpolarization of the membrane potential and the increase of proton-motive force (Marra *et al.*, 1992; De Boer, 1997). When isolated embryos were incubated in FC, a strong decrease in external pH (pHe) was observed (van der Meulen *et al.*, 2000).

However, the antagonistic effect of FC on ABA-induced gene expression (Figs 12.1 and 12.2) is not likely to involve regulation of external pH,

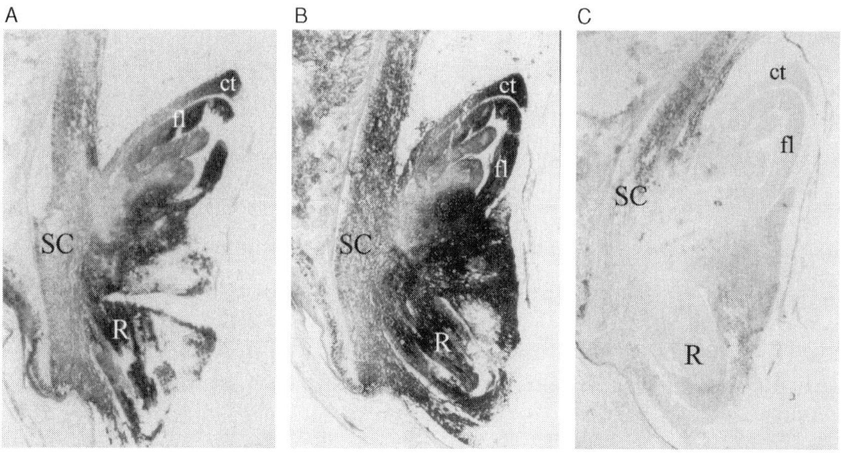

Fig. 12.4. Immunolocalization of 14-3-3 proteins in longitudinal sections of 24-h imbibed dormant barley grains. Protein was detected using anti-14-3-3 antibodies, followed by a secondary AP-conjugated antibody and incubation with NBT/BCIP. (A) Detection of 14-3-3A protein with anti-14-3-3A antibody. (B) Detection of 14-3-3B protein with anti-14-3-3B antibody. (C) Detection of 14-3-3C with anti-14-3-3C antibody. Localization of 14-3-3A, B and C protein was studied each in at least three embryos and one representative example is shown here. SC, scutellum; fl, foliage leaf; ct, coleoptile; R, radicle.

since in isolated aleurone, ABA also induces a decrease in external pH, mainly by stimulation of the secretion of malate (Heimovaara-Dijkstra *et al.*, 1994). However, although the pH of the medium does not seem important for this effect of FC, it is still possible that modulation of H^+-ATPase activity is involved.

Another possible point of crosstalk between FC and ABA signal transduction could be modulation of kinase activity. In barley aleurone protoplasts, ABA is able to induce a transient activation of mitogen-activated protein kinase (MAP kinase). Inhibition of ABA-induced MAP kinase activity is correlated with the inhibition of ABA-induced *Rab* gene expression (Knetsch *et al.*, 1996). In *Saccharomyces cerevisiae* pseudohyphal development and animal mitogenic signalling, 14-3-3 proteins are required for RAS/MAP kinase cascade signalling (Roberts *et al.*, 1997). Although the mode of action is not yet clear, it is possible that the interaction between FC and ABA is via 14-3-3 proteins and MAP kinase activity, thereby regulating ABA-induced responses such as *Rab* gene expression.

In the second part of our study, we investigated expression of 14-3-3 proteins in dormant barley embryos. Three different 14-3-3 isoforms have been identified in barley, 14-3-3A (Brandt *et al.*, 1992), 14-3-3B (GenBank accession no. X93170) and 14-3-3C (Y14200). In non-dormant embryos, these isoforms were shown to be differentially expressed (Testerink *et al.*, 1999). On the mRNA level, *14-3-3A* was induced during germination of non-dormant grains, after which expression decreased again. *14-3-3B* and *C* mRNA expression also increased upon germination, but did not decrease after germination was completed. In the present study, our hypothesis that *14-3-3A* mRNA expression is correlated with germination, is confirmed. *14-3-3A* expression appeared to be lower in dormant grains than in non-dormant grains, and was induced by dormancy-breaking FC treatment. *14-3-3B* and *C*, on the other hand, are expressed in similar amounts in both dormant and non-dormant grains, and were not induced by FC treatment.

Although the differential mRNA expression of *14-3-3A*, *B* and *C* shows germination-related regulation of the barley 14-3-3 isoforms, the role of these differences in temporal expression is not immediately apparent, as protein levels of all three isoforms stay rather constant during imbibition. Both in dormant and in non-dormant embryos, all three 14-3-3 isoforms are already present in the dry grain, and these levels hardly change during imbibition. Thus, the increase in mRNA levels does not result in higher protein levels. Possibly, turnover of the proteins is increased upon imbibition, and *de novo* protein synthesis is required to keep levels of 14-3-3 proteins constant.

In non-dormant barley embryos, differences in spatial expression were observed for the different 14-3-3 isoforms (Testerink *et al.*, 1999). We investigated spatial expression of the 14-3-3 isoforms in dormant embryos and found that the expression pattern of all three 14-3-3 isoforms was similar to the expression in non-dormant embryos. 14-3-3A and 14-3-3B were present in all embryo tissues, while 14-3-3C expression was restricted to the scutellum tissue. Hence, the tissue-specific expression patterns of 14-3-3 proteins in the barley embryo do not seem to be related to the germination process of barley grains. It will be interesting now to study overexpressing and antisense plants of the different isoforms, to elucidate the role of the 14-3-3 isoforms present in the barley embryo.

Acknowledgements

We would like to thank Jan Kijne for stimulating discussion and Bert van Duijn and Sjoukje Heimovaara-Dijkstra for critical reading of the manuscript. This work was partially supported by European Community programme no. PL 962275.

References

Black, M. (1983) Abscisic acid in seed germination and dormancy. In: Addicott, F.T. (ed.) *Abscisic Acid.* Preager Publishers, New York, pp. 337–363.

Brandt, J., Thordal-Christensen, H., Vad, K., Gregersen, P.L. and Collinge, D.B. (1992) A pathogen-induced gene of barley encodes a protein showing high similarity to a protein kinase regulator. *Plant Journal* 2, 815–820.

Cojecki, J. (1986) Identification and characterisation of a cDNA clone for glyceraldehyde-3-phosphate dehydrogenase in barley. *Carlsberg Research Communication* 51, 203–310.

De Boer, B. (1997) Fusicoccin – a key to multiple 14-3-3 locks. *Trends in Plant Science* 2, 60–66.

Ferl, R.J. (1996) 14-3-3 proteins and signal transduction. *Annual Review of Plant Physiology and Plant Molecular Biology* 47, 48–73.

Finnie, C., Borch, J. and Collinge, D.B. (1999) 14-3-3 proteins: eukaryotic regulatory proteins with many functions. *Plant Molecular Biology* 40, 545–554.

Heimovaara-Dijkstra, S., Heistek, J.C. and Wang, M. (1994) Counteractive effects of ABA and GA_3 on extracellular and intracellular pH and malate in barley aleurone. *Plant Physiology* 106, 359–365.

Jahn, T., Fuglsang, A.T., Olsson, A., Brüntrup, I.M., Collinge, D.B., Volkmann, D., Sommarin, M., Palmgren, M.G. and Larsson, C. (1997) The 14-3-3 protein interacts directly with the C-terminal region of the plant plasma membrane H^+-ATPase. *Plant Cell* 9, 1805–1814.

Knetsch, M.L.W., Wang, M., Snaar-Jagalska, B.E. and Heimovaara-Dijkstra, S. (1996) Abscisic acid induces mitogen-activated protein kinase activation in barley aleurone protoplasts. *Plant Cell* 8, 1061–1067.

Lado, P., Rasi-Caldogno, F. and Colombo, R. (1974) Promoting effect of fusicoccin on seed germination. *Physiologia Plantarum* 31, 149–152.

Marra, M., Ballio, A., Fullone, M.R. and Aducci, P. (1992) Some properties of a functional reconstituted plasmalemma H^+-ATPase activated by fusicoccin. *Plant Physiology* 98, 1029–1034.

Marrè, E. (1979) Fusicoccin: a tool in plant physiology. *Annual Review of Plant Physiology* 30, 273–288.

Mundy, J. and Chua, N.-H. (1988) Abscisic acid and water-stress induce the expression of a novel rice gene. *EMBO Journal* 7, 2270–2286.

Oecking, C., Piotrowski, M., Hagemeier, J. and Hagemann, K. (1997) Topology and target interaction of the fusicoccin-binding 14-3-3 homologs of *Commelina communis*. *Plant Journal* 12, 441–453.

Palmgren, M.G., Fuglsang, A.T. and Jahn, T. (1998) Deciphering the role of 14-3-3 proteins. Experimental Biology Online 3: 4 (http://www.link.springer.de/link/service/journals/00898/bibs/8003001/80030004.htm).

Roberts, M.R. and Bowles, D.J. (1999) Fusicoccin, 14-3-3 proteins, and defense responses in tomato plants. *Plant Physiology* 119, 1243–1250.

Roberts, R.L., Mosch, H.U. and Fink, G.R. (1997) 14-3-3 proteins are essential for ras/MAPK cascade signaling during pseudohyphal development in *S. cerevisiae*. *Cell* 89, 1055–1065.

Testerink, C., Van der Meulen, R.M., Oppedijk, B.J., De Boer, A.H., Heimovaara-Dijkstra, S., Kijne, J.W. and Wang, M. (1999) Differences in spatial expression between 14-3-3 isoforms in germinating barley embryos. *Plant Physiology* 121, 81–87.

Van Beckum, J.M.M., Libbenga, K.R. and Wang, M. (1993) Abscisic acid and gibberellic acid regulated responses of embryos and aleurone layers isolated from dormant and non-dormant barley grains. *Physiologia Plantarum* 89, 483–489.

Van der Meulen, R.M., Vredenbregt-Heistek, J.C., Caspers, M.P.M. and Wang, M. (2000) A synergistic effect of fusicoccin and gibberellic acid on germination and α-amylase expression of barley grains. In: Black, M., Bradford K. and Vazquez-Ramos, J. (eds) *Seed Biology*. CAB International, Wallingford, UK.

Visser, K., Vissers, A.P.A., Cagirgan, M.I., Kijne, J.W. and Wang, M. (1996) Rapid germination of a barley mutant is correlated with a rapid turnover of abscisic acid outside the embryo. *Plant Physiology* 111, 1127–1133.

Wang, M., Van Duijn, B., Van der Meulen, R. M. and Heidekamp, F. (1992) Effect of abscisic acid analogues on intracellular calcium level and gene expression in barley aleurone protoplasts. In: Karssen, C.M., Van Loon, L.C. and Vreugdenhill, D. (eds) *Plant Growth Substances*. Kluwer Academic Press, Dordrecht, The Netherlands, pp. 635–642.

Wang, M., Bakhuizen, R., Heimovaara-Dijkstra, S., Van Zeijl, M.J., Vries, M.A., Van Beckum, J.M. and Sinjorgo, K.M.C. (1994) The role of ABA and GA in barley grain dormancy: a comparative study between embryo and aleurone dormancy. *Russian Journal of Plant Physiology* 41 (5), 659–667.

Wang, M., Heimovaara-Dijkstra, S. and Van Duijn, B. (1995) Modulation of germination of embryos isolated from dormant and non-dormant barley grains by manipulation of endogenous abscisic acid. *Planta* 195, 586–592.

Wang, M., Van der Meulen, R.M., Visser, K., Van Schaik, H.P., Van Duijn, B. and De Boer, A.H. (1998) Effects of dormancy-breaking chemicals on ABA levels in barley grain embryos. *Seed Science Research* 8, 129–137.

Wu, K., Rooney, M.F. and Ferl, R.J. (1997) The *Arabidopsis* 14-3-3 multigene family. *Plant Physiology* 114, 1421–1431.

Dormancy of Cereal Seeds as Related to Embryo Sensitivity to ABA and Water Potential

13

Françoise Corbineau and Daniel Côme

Laboratoire de Physiologie Végétale Appliquée, Université Pierre et Marie Curie, Tour 53, 1er étage, 4 Place Jussieu, 75252 Paris cedex 05, France

Introduction

Seed germination is subject to a very precise regulation, the complexity of which originates both in the action of various external factors and in characteristics within the seeds themselves. Seeds are considered to be dormant when they do not germinate, or germinate poorly, in environmental conditions that are apparently favourable for germination. Primary dormancy in cereals and other grasses sets in during seed development on the mother plant (e.g. Gosling *et al.*, 1981). It corresponds to an inability of the seeds to germinate at relatively high temperatures (above about 10–20°C) and is often referred to as relative dormancy. This high temperature dormancy is predominantly due to an inhibitory action of the covering structures (seed coat-imposed dormancy); in other cases it is an embryo dormancy or a combination of the two (Simpson, 1990; Bewley and Black, 1994).

Secondary dormancy usually develops under conditions which do not support germination of seeds (Karssen, 1980/81; Corbineau *et al.*, 1993b); it can occur in seeds which are not dormant or which already possess primary dormancy (Karssen, 1980/81).

During seed development, the inability to germinate of the immature embryo might result from the inhibitory effect of endogenous abscisic acid (ABA) and restricted water uptake because of high external osmolarity (Kermode, 1995). Low water potential of the embryo environment prevents cell elongation. ABA is generally considered to be involved in the onset of dormancy during seed development (Robichaud *et al.*, 1980; Karssen *et al.*, 1983; Le Page-Degivry *et al.*, 1990). ABA-deficient mutants of *Arabidopsis*

(eds J.-D. Viémont and J. Crabbé)

produce non-dormant seeds (Karssen *et al.*, 1983), and application of the ABA-synthesis inhibitor fluridone to developing maize embryos (Fong *et al.*, 1983) or sunflower achenes (Le Page-Degivry *et al.*, 1990) results, respectively, in a precocious germination of immature seeds and in lack of dormancy at harvest. However, there exists no clear relationship between the ABA content of mature seeds of barley (Boivin *et al.*, 1995) or wheat (Walker-Simmons, 1987; Walker-Simmons and Sessing, 1990) and dormancy intensity.

ABA and low water potential of the germination medium inhibit germination of mature seeds (Bewley and Black, 1994). However, their mechanisms of action are different. In tomato, for example, low water potential prevents water uptake and induction of cell cycle (Özbingöl *et al.*, 1999), whereas exogenous ABA does not block the latter (Liu *et al.*, 1994). Moreover, both factors do not have the same effect on gene expression in Douglas fir seeds (Jarvis *et al.*, 1997).

Regulation of dormancy and/or germination by ABA must take into account not only the hormone levels but also the sensitivity of the embryo to ABA. Exogenous ABA inhibits germination of seeds of various species (Bewley and Black, 1994). But responsiveness of the embryo to ABA is highly correlated with dormancy intensity in wheat (Walker-Simmons, 1987; Morris *et al.*, 1989; Corbineau *et al.*, 1993a), oat (Poljakoff-Mayber *et al.*, 1990), barley (Wang *et al.*, 1995; Corbineau and Côme, 1996) and species other than cereal, such as sunflower (Le Page-Degivry *et al.*, 1996). In all cases, embryos from dormant seeds are much more sensitive to ABA than those from non-dormant seeds. ABA might also affect indirectly seed responsiveness to water potential via the state of dormancy. For example, mature tomato seeds are more sensitive to water potential of the germination medium than ABA-deficient *sit* ones (Groot and Karssen, 1992).

The aims of the present work were to study: (i) the sensitivity to water potential of the germination medium and to ABA of cereal embryos as related to primary dormancy of seeds of various genotypes at harvest time, and (ii) its possible modulation during dry storage which breaks dormancy (dry afterripening) or during reinforcement of dormancy at high temperature (thermodormancy).

What is Cereal Seed Dormancy?

Sensitivity of dormant seeds to external factors

Freshly harvested cereal seeds are considered to be dormant because their germination is difficult above 10–20°C depending on the species, the cultivar and the seed batch. This dormancy is a relative phenomenon; it is hardly expressed at low temperatures (5–10°C), but it increases as the temperature rises (Fig. 13.1A). Using barley seeds from numerous cultivars

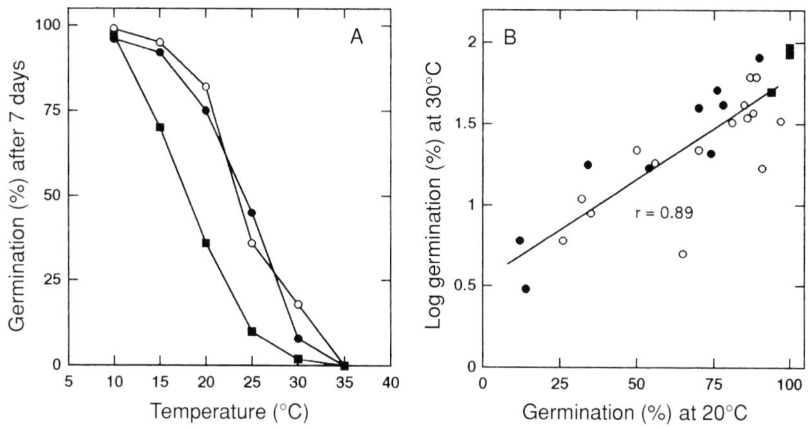

Fig. 13.1. Effects of temperature on germination of freshly harvested cereal seeds. (A) Effects of temperature on the germination percentages obtained after 7 days with seeds from batches of wheat (○), barley (●) and oats (■); (B) relationship between the germination percentages obtained after 3 days at 20°C and the log of the germination percentages obtained after 3 days at 30°C with barley seeds from various cultivars collected in 1992 (●), 1993 (○) and 1995 (■). r, regression coefficient.

collected in 1992, 1993 and 1995, it was shown that there exists an exponential relationship between the germination percentages obtained after 3 days at 20°C and those obtained after the same duration at 30°C (Fig. 13.1B; Corbineau and Côme, 1996). Dormant seeds are also very sensitive to oxygen deprivation (Corbineau and Côme, 1980; Corbineau *et al.*, 1981, 1986) and to a decrease in water potential of the germination medium (Corbineau and Côme, 1996).

Modulation of dormancy by environmental factors

Dormancy is a gene characteristic, however, its intensity is mainly modulated by the agronomical and climatic conditions during seed development on the mother plants (Belderock, 1968; Grahl and Schrodter, 1975; Olsson and Mattson, 1976; Black *et al.*, 1987). Dormancy intensity depends also on the maturity stage of the grain (Gordon, 1980; Corbineau and Côme, 1982; Black *et al.*, 1987). Dormancy is very deep in immature seeds, then it decreases progressively during dehydration of the grains on the mother plant (Gosling *et al.*, 1981; Corbineau and Côme, 1982). This decrease in dormancy during the maturation phase depends on the genotype but is also regulated by environmental factors. In wheat, for example, the maintenance of dormancy on the plant increases with decreasing temperatures higher than 12.5°C to which grains are submitted during the dough stage (Belderock, 1965).

Origin of dormancy

Different studies performed with various cereals have shown that the inability of freshly harvested seeds to germinate at high temperatures is mainly due to an inhibition of germination by the grain covering structures (pericarp and seed coat) and glumellae when they remain around the caryopsis (barley, oat, rice). They inhibit germination because they limit oxygen supply to the embryo by fixing oxygen through polyphenoloxidase-mediated oxidation of phenolic compounds (Lenoir *et al.*, 1983, 1986; Corbineau *et al.*, 1986). This barrier to oxygen diffusion rises with increasing temperature (Lenoir *et al.*, 1986) and decreases during dry afterripening (Corbineau *et al.*, 1986; Lenoir *et al.*, 1986).

Sensitivity of Embryos of Primary Dormant Seeds to ABA and Water Potential

There is no clear correlation between germination index and endogenous ABA level in cereal seeds (Walker-Simmons, 1987; Black, 1991; Boivin *et al.*, 1995), but the responsiveness of embryos to exogenous ABA depends on the dormancy intensity of the seeds. Abscisic acid and a low water potential of the germination medium inhibit more or less the germination of isolated embryos depending on the depth of dormancy. Results obtained with barley seeds from various cultivars exhibiting different degrees of dormancy at harvest show a linear relationship between the germination at 30°C of whole seeds incubated in water and that of isolated embryos incubated in a polyethylene glycol-8000 (PEG) solution at −1.5 MPa (Fig. 13.2A). Such a linear correlation does not exist concerning the sensitivity of embryos to ABA (Fig. 13.2B).

Sensitivity of embryos to ABA depends on the genotype. In barley, for example, embryos isolated from dormant seeds from 'Barberousse' and 'Baraka' cultivars are much more sensitive to ABA than those isolated from seeds from the 'Golf' cultivar which are less dormant at harvest (Corbineau *et al.*, 1993a). In wheat, the responsiveness of isolated embryos to ABA is also highly correlated with the resistance to sprouting; it is much weaker for grains susceptible to sprouting (less dormant ones) than for those which are resistant (more dormant ones) (Walker-Simmons, 1987; Morris *et al.*, 1989; Corbineau *et al.*, 1993a).

Fig. 13.3 shows that for a wheat cultivar susceptible to sprouting ('Thésée') the responsiveness of embryos to 0.5 μM ABA depends on the stage of seed development. It sharply increases during the dough stage and then decreases progressively during the maturation phase which is associated with a decrease in seed dormancy. It remains high until the end of seed maturation for cultivars which are more resistant to sprouting (Walker-Simmons, 1987; Corbineau *et al.*, 1993a). Inhibition of

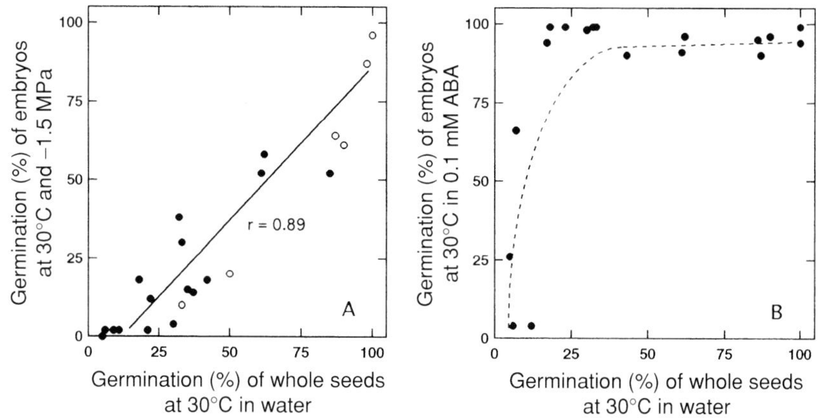

Fig. 13.2. Relationships between the germination percentages obtained after 3 days at 30°C with whole seeds placed in water and isolated embryos placed in a PEG-8000 solution at −1.5 MPa (A) or in a solution of 0.1 mM ABA (B). Barley seeds from various cultivars collected in 1993 (●) and 1995 (○). r, regression coefficient.

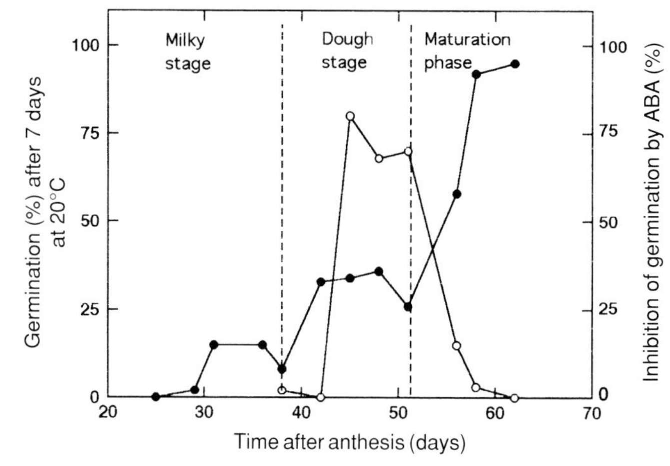

Fig. 13.3. Changes, during development of wheat seeds ('Thésée' cultivar), in the germination percentages obtained after 7 days at 20°C with whole seeds (●), and in the inhibition of germination at 20°C of isolated embryos by 0.5 μM ABA (○). Inhibition of germination is expressed as the difference between the germination percentages obtained after 7 days with embryos placed in water and those placed in the presence of ABA.

germination by ABA and water potential is highly correlated with the degree of dormancy and is also temperature dependent. In oat, the embryos are about 1000 times more sensitive to ABA at 30°C than at 10°C (Corbineau *et al.*, 1991, 1993b).

Changes in Sensitivity of Embryos to ABA and Water Potential during Afterripening

After harvest, cereal seeds become progressively able to germinate at high temperatures during dry storage in the open air at room temperature. This breaking of dormancy is expressed by a widening of the temperature range compatible with good germination (Côme and Corbineau, 1984; Côme *et al.*, 1984). Non-dormant seeds are also less sensitive to oxygen deprivation (Corbineau and Côme, 1980; Lenoir *et al.*, 1986) and to a decrease in water potential of the germination medium (Corbineau and Côme, 1996) than are dormant seeds.

Embryos progressively lose their sensitivity to ABA during afterripening. Those isolated from dormant seeds are 100 to 1000 times more sensitive to ABA than those isolated from non-dormant ones (Corbineau *et al.*, 1993b). In oat, for example (Fig. 13.4), 1 μM ABA is sufficient to inhibit the germination of 50% of the embryos isolated from dormant freshly harvested seeds while embryos isolated from non-dormant seeds (stored dry for 38 days at 30°C) germinate almost completely in 100 μM ABA. Embryos also become less sensitive to water potential of the germination medium during release of primary dormancy (data not shown).

Changes in Sensitivity of Embryos to ABA and Water Potential during Onset and Release of Thermodormancy

Effects of a pre-treatment at high temperature on subsequent germination

Incubation of dormant oat seeds at 30°C for a short period (2–72 h) is associated with a loss of ability to germinate subsequently at 20°C (Fig. 13.5; Corbineau *et al.*, 1993b). This phenomenon can be considered as a reinforcement of primary dormancy, i.e. a thermodormancy. After 24–72 h at 30°C, only 20–25% of seeds remain capable of germinating at 20°C. It is interesting to note that 4-h exposure to 30°C is sufficient to cause a thermodormancy in approximately 20–30% of the seeds. The inductive effect of a 30°C-treatment on thermodormancy falls as treatment time increases, and after 7–8 days at 30°C seeds again show the same behaviour as primary dormant ones, i.e. they are almost all able to germinate at 20°C. Fluridone applied during seed incubation at 30°C prevents the induction of thermodormancy (data not shown), suggesting that ABA synthesis is involved in the onset of the reinforcement of dormancy.

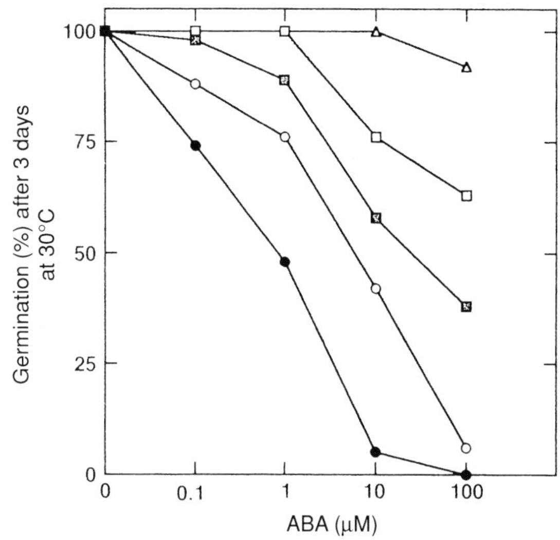

Fig. 13.4. Effects of ABA concentration on the germination percentages obtained after 3 days at 30°C with embryos isolated from oat seeds at harvest time (●) and after 12 (○), 19 (■), 26 (□) and 38 days (Δ) of dry afterripening at 30°C.

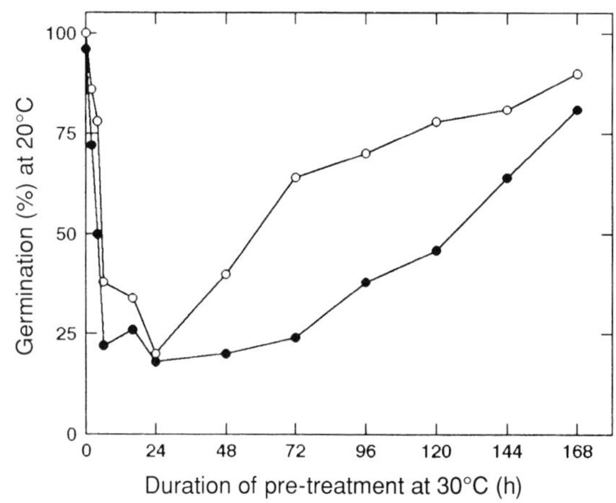

Fig. 13.5. Effects of duration of pre-treatment of oat seeds at 30°C in water on the percentages of germination recorded 3 (●) and 7 days (○) after their transfer to 20°C.

Sensitivity of embryos to ABA and water potential

Figure 13.6 shows the effects of water potential of the germination medium (Fig. 13.6A) and of exogenous ABA (Fig. 13.6B) on the germination at 20°C of embryos isolated from control primary dormant seeds and seeds pre-treated at 30°C in water for 3 days (thermodormant seeds) and 7 days (seeds for which the thermodormancy has almost completely disappeared). Embryos isolated from thermodormant seeds are much more sensitive to water potential and to ABA than those isolated from primary dormant seeds. A long treatment (7 days) at 30°C, which results in release of thermodormancy, is associated with a decrease in sensitivity of embryos to both ABA and water potential of the germination medium.

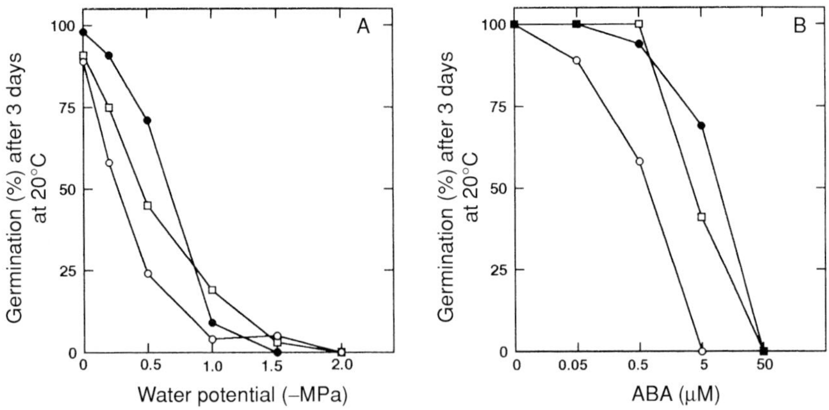

Fig. 13.4. Effects of water potential (A) and ABA concentration (B) of the germination medium on the germination percentages obtained after 3 days at 20°C with embryos isolated from seeds with primary dormancy (●) and seeds pre-treated at 30°C in water for 3 days (thermodormant seeds) (○) or 7 days (seeds for which the thermodormancy has been released) (□).

Discussion and conclusion

Responsiveness of embryos of cereals to ABA and water potential of the germination medium depends on the depth of seed dormancy. The various genotypes of the same species differ widely in their general response to these two factors. However, at least in barley, there is an inverse correlation between the ability of whole seeds to germinate at relatively high temperatures, independently of the genotype, and the sensitivity of the embryos to ABA and water potential. Results obtained with numerous batches of barley seeds show a linear relationship between the germination at 30°C of whole seeds incubated in water and that of isolated embryos incubated in a PEG solution at −1.5 MPa. Such a linear correlation does not exist for the sensitivity of embryos to ABA, suggesting that both factors act differently and probably independently. Nevertheless, Schopfer and Plachy (1984) have reported that ABA inhibits embryonic water uptake in *Brassica napus.*

The great difference observed in the responsiveness to ABA of embryos isolated from dormant and non-dormant cereal seeds (Walker-Simmons, 1987; Corbineau *et al.*, 1993a; Van Beckum *et al.*, 1993; Wang *et al.*, 1994; Benech-Arnold *et al.*, 1998) may explain in part the lack of a clear relationship between ABA levels in mature embryos and seed dormancy intensity (Black, 1991). During cereal seed development, the sensitivity of the embryo to ABA is at its highest at the dough stage (Walker-Simmons, 1987; Corbineau *et al.*, 1993a). It decreases then during the maturation phase on the mother plant and during dry storage after harvest, i.e. during dormancy release. This decrease in responsiveness to ABA is probably associated with changes in ABA perception and/or in the ABA signal transduction pathway in which protein phosphorylation through ABA-regulated protein kinases might play a key role (Walker-Simmons, 1998).

If germination of dormant seeds is associated with a reduction in ABA embryonic content through a reduced *de novo* ABA synthesis during imbibition (Wang *et al.*, 1995; Le Page-Degivry *et al.*, 1996) and with a higher diffusion of the hormone in the germination medium (Wang *et al.*, 1995), it also requires a decrease in the ABA sensitivity of the embryo. The role of changes in embryo responsiveness to ABA in regulating dormancy has been suggested for numerous cereals, like maize (Robichaud *et al.*, 1980), wheat (Walker-Simmons, 1987; Corbineau *et al.*, 1993a), barley (Wang *et al.*, 1995; Benech-Arnold *et al.*, 1998), oat (Corbineau *et al.*, 1991) and sorghum (Steinbach *et al.*, 1995), and for other species (Groot and Karssen, 1992; Le Page-Degivry *et al.*, 1996; Jarvis *et al.*, 1997).

Moreover, temperature affects embryo sensitivity to ABA and to water potential. In cereals, the inhibitory effects of these two factors are more effective at 30°C than at 10°C. In oat, induction of thermodormancy by incubating seeds in water at 30°C for about 1–3 days seems to require ABA synthesis since it is prevented by fluridone. It also results in an increase in sensitivity of embryos to ABA and to water potential of the germination medium. However, this increase is revealed only at 20°C; at 30°C, the dose–response relationships to ABA are similar for both embryos isolated from primary dormant seeds and from thermodormant ones (Corbineau *et al.*, 1993b). This downward shift in the temperature at which dormancy and the sensitivity to ABA are expressed emphasizes the great importance of temperature in the regulation of ABA signal transduction events. As for primary dormancy, release of thermodormancy after 7–8 days at 30°C is associated with a decrease in ABA sensitivity at 20°C and 30°C.

References

Belderock, B. (1965) Influence of preharvest weather on wheat dormancy and its tendency to sprout in the ear. *Zeitschrift für Acker- und Pflanzenbau* 122, 297–313.

Belderock, B. (1968) Seed dormancy problems in cereals. *Field Crop Abstract* 26, 697–760.

Benech-Arnold, R.L., Giallorenzi, M.C., Frank, J. and Rodriguez, V. (1998) Termination of hull-imposed dormancy in developing barley grains is correlated with changes in embryonic ABA levels and sensitivity. *Seed Science Research* 9, 39–47.

Bewley, J.D. and Black, M. (1994) *Seeds. Physiology of Development and Germination*, 2nd edn. Plenum Press, New York.

Black, M. (1991) Involvement of ABA in the physiology of developing and mature seeds. In: Davies, W.J. (ed.) *Abscisic Acid Physiology and Biochemistry*. BIOS Scientific Publishers, Oxford, pp. 99–124.

Black, M., Butler, J. and Hugues, M. (1987) Control and development of dormancy in cereals. In: Mares, O.P. (ed.) *Fourth International Symposium on Preharvest Sprouting in Cereals*. Westview Press, Boulder, Colorado, pp. 379–392.

Boivin, P., Kohl, S. and Clamagirand, V. (1995) Barley endogenous phytohormones and malting performance. *Bios, Boissons Conditionnement* 255, 119–124.

Côme, D. and Corbineau, F. (1984) La dormance des semences des céréales et son élimination. I. Principales caractéristiques. *Comptes Rendus de l'Académie d'Agriculture de France* 70, 709–715.

Côme, D., Lenoir, C. and Corbineau, F. (1984) La dormance des céréales et son élimination. *Seed Science and Technology* 12, 629–640.

Corbineau, F. and Côme, D. (1980) Quelques caractéristiques de la dormance du caryopse d'orge (*Hordeum vulgare* L., variété Sonja). *Comptes Rendus de l'Académie des Sciences, Paris* (série D) 280, 547–550.

Corbineau, F. and Côme, D. (1982) Evolution de la dormance des semences de deux variétés d'orge (*Hordeum vulgare* L.) au cours de leur maturation et de leur conservation au sec. *Comptes Rendus de l'Académie des Sciences, Paris* (série III) 294, 967–970.

Corbineau, F. and Côme, D. (1996) Barley seed dormancy. *Bios, Boissons Conditionnement* 261, 113–119.

Corbineau, F., Sanchez, A., Côme D. and Chaussat R. (1981) La dormance du caryopse de blé (*Triticum aestivum* L., var. Champlein) en relation avec la température et l'oxygène. *Comptes Rendus de l'Académie d'Agriculture de France* 9, 826–834.

Corbineau, F., Lecat, S. and Côme, D. (1986) Dormancy of three cultivars of oat seeds (*Avena sativa* L.). *Seed Science and Technology* 14, 725–735.

Corbineau, F., Poljakoff-Mayber, A. and Côme, D. (1991) Responsiveness to abscisic acid of embryos of dormant oat (*Avena sativa*) seeds. Involvement of ABA-inducible proteins. *Physiologia Plantarum* 83, 1–6.

Corbineau, F., Benamar, A., Couvreur, F., Gate, P. and Côme, D. (1993a) La germination sur pied du blé tendre: recherche de critères prévisionnels. *Comptes Rendus de l'Académie d'Agriculture de France* 79, 47–54.

Corbineau, F., Black, M. and Côme, D. (1993b) Induction of thermodormancy in *Avena sativa* seeds. *Seed Science Research* 3, 111–117.

Fong, F., Smith, J.D. and Koehler, D.E. (1983) Early events in maize seed development. *Plant Physiology* 73, 899–901.

Gordon, I.L. (1980) Germinability, dormancy and grain development. *Cereal Research Communications* 8, 115–129.

Gosling, P.G., Butler, R.A., Black, M. and Chapman, J. (1981) The onset of germination ability in developing wheat. *Journal of Experimental Botany* 32, 621–627.

Grahl, A. and Schrodter, H. (1975) Weather conditions before maturity and dormancy in wheat in relation to prediction of sprouting. *Seed Science and Technology* 3, 815–826.

Groot, S.P.C. and Karssen, C.M. (1992) Dormancy and germination of abscisic acid-deficient tomato seeds: studies with the *sitiens* mutant. *Plant Physiology* 99, 952–958.

Jarvis, S.B., Taylor, M.A., Bianco, J., Corbineau, F. and Davies, H.V. (1997) Dormancy-breakage in seeds of Douglas fir (*Pseudotsuga menziesii* (Mirb.) Franco). Support for the hypothesis that LEA gene expression is essential for this process. *Journal of Plant Physiology* 152, 457–464.

Karssen, C.M. (1980/81) Environmental conditions and endogenous mechanisms involved in secondary dormancy of seeds. *Israel Journal of Botany* 29, 59–64.

Karssen, C.M., Brinkhorst-Van der Swan, D.L.C., Breekland, A.E. and Koorneef, M. (1983) Induction of dormancy during seed development by endogenous abscisic acid: studies on abscisic acid deficient genotypes of *Arabidopsis thaliana* (L.). *Planta* 157, 158–165.

Kermode, A.R. (1995) Regulatory mechanisms in the transition from seed development to germination: interactions between the embryo and the seed environment. In: Kigel, J. and Galili, G. (eds) *Seed Development and Germination.* Marcel Dekker, New York, pp. 273–332.

Lenoir, C., Corbineau, F. and Côme, D. (1983) Rôle des glumelles dans la dormance des semences d'orge. *Physiologie Végétale* 21, 633–643.

Lenoir, C., Corbineau, F. and Côme, D. (1986) Barley (*Hordeum vulgare*) seed dormancy as related to glumella characteristics. *Physiologia Plantarum* 68, 301–307.

Le Page-Degivry, M.T., Barthe, P. and Garello, G. (1990) Involvement of endogenous abscisic acid in onset and release of *Helianthus annuus* embryo dormancy. *Plant Physiology* 92, 1164–1168.

Le Page-Degivry, M.T., Bianco, J., Barthe, P. and Garello, G. (1996) Changes in hormone sensitivity in relation to onset and breaking of sunflower embryo dormancy. In: Lang, G.A. (ed.) *Plant Dormancy. Physiology, Biochemistry and Molecular Biology.* CAB International, Wallingford, UK, pp. 221–231.

Liu, Y., Bergervoet, J.H.W., de Vos, C.H., Hilhorst, H.W.M., Kraak, H.L., Karssen, C.M. and Bino, R.J. (1994) Nuclear replication activities during imbibition of abscisic- and gibberellin-deficient tomato (*Lycopersicon esculentum* Mill.) seeds. *Planta* 194, 368–373.

Morris, F., Moffati, J.M., Sears, R.G. and Paulsen, G.M. (1989) Seed dormancy and responses of caryopses, embryos and calli to abscisic acid. *Plant Physiology* 90, 643–647.

Olsson, G. and Mattson, B. (1976) Seed dormancy in wheat under different weather conditions. *Cereal Research Communications* 4, 181–185.

Özbingöl, N., Corbineau, F., Groot, S.P.C., Bino, R.J. and Côme, D. (1999) Activation of the cell cycle in tomato (*Lycopersicon esculentum* Mill.) seeds during osmoconditioning as related to temperature and oxygen. *Annals of Botany* 84, 245–251.

Poljakoff-Mayber, A., Corbineau, F. and Côme, D. (1990) A possible mechanism of high-temperature dormancy regulation in seeds of *Avena sativa* L. (cv. Moyencourt). *Plant Growth Regulation* 9, 147–156.

Robichaud, C.S., Wong, J. and Sussex, I.M. (1980) Control of *in vitro* growth of viviparous embryo mutants of maize by abscisic acid. *Developmental Genetics* 1, 325–330.

Schopfer, P. and Plachy, C. (1984) Control of seed germination by abscisic acid. II. Effect on embryo water uptake in *Brassica napus* L. *Plant Physiology* 76, 155–160.

Simpson, G.M. (1990) *Seed Dormancy in Grasses.* Cambridge University Press, Cambridge.

Steinbach, H.S., Benech-Arnold, R.L., Kristof, G., Sanchez, R.A. and Marcucci Poltri, S. (1995) Physiological basis of pre-harvest sprouting resistance in *Sorghum bicolor* (L.) Moench. ABA levels and sensitivity in developing embryos of sprouting resistant and susceptible varieties. *Journal of Experimental Botany* 45, 701–709.

Van Beckum, J., Libbenga, K. and Wang, M. (1993) Abscisic acid and gibberellic acid-regulated responses of embryos and aleurone layers isolated from dormant and non dormant barley grains. *Physiologia Plantarum* 89, 483–489.

Walker-Simmons, M. (1987) ABA levels and sensitivity in developing wheat embryos of sprouting resistant and susceptible cultivars. *Plant Physiology* 84, 61–66.

Walker-Simmons, M. (1998) Protein kinases in seeds. *Seed Science Research* 8, 193–200.

Walker-Simmons, M. and Sessing, J. (1990) Temperature effects on embryonic abscisic acid levels during development of wheat grain dormancy. *Journal of Plant Growth Regulation* 9, 51–56.

Wang, M., Bakhuizen, R., Heimovaara-Dijkstra, S., Zeijl, M.J., De Vries, M.A., Van Beckum, J.M. and Sinjorgo, K.M.C. (1994) The role of ABA and GA in barley grain dormancy: a comparative study between embryo dormancy and aleurone dormancy. *Russian Journal of Plant Physiology* 41, 577–584.

Wang, M., Heimovaara-Dijkstra, S. and Van Duijn, B. (1995) Modulation of germination of embryos isolated from dormant and non-dormant barley grains by manipulation of endogenous abscisic acid. *Planta* 195, 586–592.

14

Abscisic Acid Control of Seed Dormancy Expression in *Nicotiana plumbaginifolia* and *Arabidopsis thaliana*

Marc Jullien,[1,2] Denise Bouinot,[2] Sonia Ali-Rachedi,[2] Bruno Sotta[3] and Philippe Grappin[1,2]

[1]UMR INRA-INA PG Biologie des Semences, Route de Saint Cyr, F-78026 Versailles cedex, France; [2]UMR INRA-INA PG Biologie des Semences, Chaire de Physiologie Végétale, INA P-G, 16 rue Claude Bernard, F-75231 Paris cedex 05, France; [3]Laboratoire de Physiologie du Développement des Plantes, UMR de Physiologie Cellulaire et Moléculaire des Plantes, Université Pierre et Marie Curie (PVI), Tour 53 (E5, casier 156), 4 Place Jussieu, F-75252 Paris cedex 05, France

Introduction

In a large range of species, while completing their maturation on the mother plant, developing seeds become potentially competent to germinate, but develop an intrinsic germination block, entering a physiological state called primary dormancy or dormancy. Dormant (D) seeds temporarily fail to germinate in conditions which are adequate for germination, whereas non-dormant (ND) seeds will germinate easily. Generally, primary dormancy may be broken by afterripening and/or stratification which consist respectively of a warm temperature treatment to dry seeds and a low temperature treatment to imbibed seeds (Bewley and Black, 1994). Hypotheses have been proposed to describe the mechanism of seed dormancy, including the hormone balance theory (Wareing and Saunders, 1971; revised by Karssen and Laçka, 1986), which explains dormancy by the opposing action of hormones inhibiting (abscisic acid) and stimulating (mainly gibberellins) germination, and by the metabolic theory (Bewley and Black, 1994) which postulates a specific perturbation of respiration in dormant seeds.

Over the past decades, results obtained from application of abscisic acid (ABA), gibberellic acid (GA), inhibitors of their synthesis, and the isolation of mutant plants impaired in GAs and/or ABA synthesis or sensitivity (reviewed in Hilhorst and Karssen, 1992; Bewley and Black, 1994; Hilhorst, 1995), have supported the hormone theory, demonstrating definitively the essential roles of ABA and GAs in seed dormancy and germination. Although ABA synthesis and sensitivity to ABA have clearly been shown to be involved in the onset of dormancy during seed maturation (Koornneef *et al.*, 1982, 1984; Karssen *et al.*, 1983; Walker-Simmons, 1987; Le Page-Degivry *et al.*, 1990; Groot and Karssen, 1992), controversy still exists concerning the antagonistic role of ABA and GAs in dormancy breaking. For instance, decreasing levels of ABA and increasing levels of GAs have rarely been correlated with the efficiency of dormancy breakage treatments (Wareing and Saunders, 1971; Bewley and Black, 1994).

Results obtained with *Arabidopsis* have shown that afterripening and stratification alleviate the dormancy of GA deficient mutants by increasing their sensitivity to GAs (Karssen and Laçka, 1986; Karssen *et al.*, 1989). This has led to the suggestion that GAs are probably not directly active in the breaking of primary dormancy. In species like sunflower and barley where afterripening does not reduce the ABA content of seeds it has been shown that it does suppress the capacity for ABA synthesis in imbibed embryos (Bianco *et al.*, 1994; Wang *et al.*, 1995). These latter results indicate an essential role for ABA synthesis in dormancy control in imbibed seeds. Furthermore, Toyomasu *et al.* (1994) reported that in dark imbibed lettuce seeds red light or GA treatments both promoted germination and substantially lowered endogenous ABA levels. It appears, therefore, that the expression of dormancy is strongly dependent on ABA synthesis in imbibed seeds, which can be counteracted either by afterripening or in some species by GA production.

This study concerns the involvement of ABA in maintaining dormancy in imbibed seeds of *Nicotiana plumbaginifolia* and *Arabidopsis thaliana.* Firstly, the germination behaviours of freshly harvested and afterripened seeds were characterized. Using the delay in seed germination as the criterion of dormancy expression, we then compared the effects of ABA, GA and fluridone on D and ND imbibed seeds. Finally, ABA content in D and ND seeds dry or imbibed with the dormancy-breaking compounds was measured. The dormancy-breaking effect of afterripening, GA and fluridone was found to correlate strongly with a decrease in endogenous ABA contents during early imbibition and we discuss how GA and ABA might interact to control the expression of seed dormancy.

Materials and Methods

Plant material

The seeds and plants of *Nicotiana plumbaginifolia* cv. 'Viviani' used were derived from the haplodiploidized line PbH1D (Bourgin *et al.*, 1979). The seeds and plants of *Arabidopsis thaliana* ecotype CVi came from the 8th generation of selfing of a single seed descent and were given by Dr M. Koornneef (Wageningen).

Seed production

N. plumbaginifolia plants were grown in a greenhouse at 21°C with 16 h photoperiod of natural light supplemented with artificial light (700 W mercury vapour lamps) and watered with standard nutrient solution (Coïc and Lesaint, 1971). Seeds were harvested at physiological maturity, that is just at the natural dehiscence of the capsules, over a 15-day period. *A. thaliana* plants were grown in a culture cabinet maintained at 20°C in a 16 h photoperiod of artificial light (Osram L 58/31830 luminux plus Wanton Wan White 45 μM m^{-2} s^{-1}) with 65% relative humidity (RH) and watered with pond water. After the full fructification, mature plants were progressively dehydrated for 3 weeks by deprivation of water before harvesting the seeds.

The collected seeds were generally stored for 14 days at room temperature, reaching a moisture content of around 5–7%, and in one experiment divided into samples which were stored for 2–16 months at 24°C, 50% RH, before sowing. Other experiments were done with a seedlot harvested at the same time and treated in two ways: storage at 7°C immediately after harvest to slow down dormancy release and to keep seeds in dormant (D) state or afterripening for 10–12 months at 24°C to yield non-dormant (ND) seeds.

Imbibition and germination assays

Seeds used in each experiment originated from the same seedlot. For all germination kinetics presented, similar results were observed in five independent experiments. Germination is expressed as the cumulative percentage of germinated seeds. Generally each value was the mean ± SE of four measurements of at least 100 seeds.

All germination experiments were performed with seeds surface sterilized in an aqueous solution of the commercial disinfectant Domestos (Lever) 20% v/v for 1 h, then washed with distilled water and dried (*N. plumbaginifolia*). *Arabidopsis* seeds were sterilized by 10-min immersion in 95% ethanol containing 10% of a commercial Bayrochlore solution, rinsed with 95% ethanol then dried. These treatments had no evident effect on seed germination behaviour. Seeds were plated in 10 cm Petri dishes either directly on a medium containing distilled water buffered with Mes (3 mM, pH 5.7) and solidified with agar (7 g l^{-1} Noble agar, Difco) or on nylon filters (Nycom 5 μm, Buisine) covering this same medium. Imbibition took place with or without addition of the growth regulators (±)-ABA (Sigma), GA (Sigma), or the herbicide fluridone (Dowelanco, Le Chesnay, France). Fluridone, 1-methyl-3-phenyl-5-(3-trifluoromethyl-(phenyl))-4-(1*H*)-pyridinone, is an inhibitor of carotenogenesis (Bartels and Watson, 1978) and has been repeatedly demonstrated to suppress the production of the carotenoid-derived ABA in plants and seeds (e.g. Gamble and Mullet, 1986; Xu and Bewley, 1995; Popova and Riddle, 1996; Yoshioka *et al.*, 1998). These chemicals were prepared by dissolving in the minimum volume of DMSO and then diluting in water, and were added to warm medium just prior to agar solidification. The final DMSO concentration used had no effect on seed germination. Dishes were sealed with Parafilm (American National Can, Neenah, Wisconsin) and germination was conducted in a controlled culture room under the following conditions: 8 h of white light (Philips TRM HOW/33 RS tubes, 85 μM m^{-2} s^{-1}) at 25°C, and 16 h of dark at 17°C and a constant 90% RH (*N. plumbaginifolia*). The conditions were the same for *A. thaliana* but the photoperiod was 16 h of light. Germination measurements were scored under lens. Seeds were considered to have germinated when the radicle had pierced through the testa.

ABA measurements

Abscisic acid was extracted from dry or imbibed seeds. At each period of imbibition a seed sample (400 mg) was transferred to microfuge tubes, frozen in liquid nitrogen (N_2) and stored at −80°C. The samples were ground to a powder in liquid N_2. Hormone extraction, purification, quantification by ELISA, and identification of immunoreactive molecules has been previously described (Juillard *et al.*, 1994; Kraepiel *et al.*, 1994). We used a monoclonal anti-ABA antibody (LPDP 229, Jussieu, France) labelled with peroxidase-conjugated goat antibody to mouse immunoglobulins (Sigma). Hormone content was determined five times for each sample.

Results

Characteristics of seed dormancy in *N. plumbaginifolia* and *A. thaliana*

Germination kinetics of seeds freshly harvested (14 days old) and of afterripened seeds (stored dry at 24°C for increasing time) were compared on agar medium in order to characterize seed dormancy (Figs 14.1 and

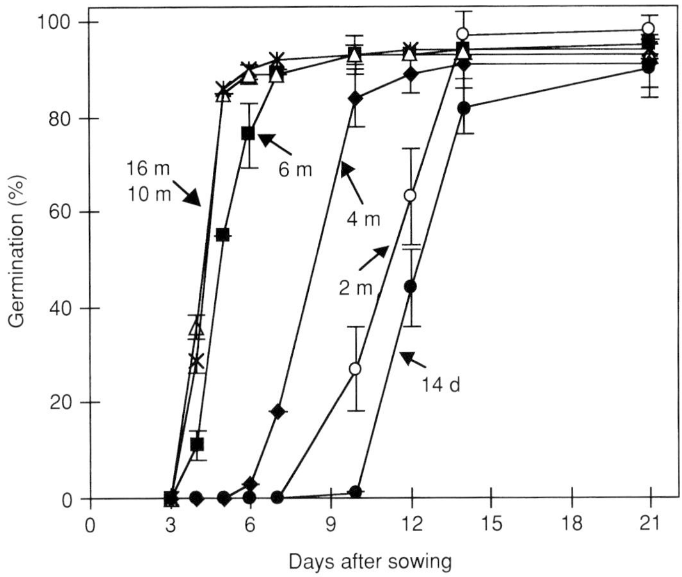

Fig. 14.1. Effect of afterripening on the time course of germination in *N. plumbaginifolia* line PbH1D. Mature seeds were afterripened for 14 days (●), 2 months (○), 4 months (◆), 6 months (■), 10 months (**X**), 16 months (△) and were sown on basal medium in the conditions defined in Materials and Methods; germination percentage was scored as radicle emergence. Data are means of four repeats and the SE of the difference is indicated. Statistical significance was assessed by Student's *t*-test ($P < 0.001$).

14.2). In *N. plumbaginifolia* (Fig. 14.1), 14-day-old seeds needed more than 12 days to reach 50% germination (gt 50%). In comparison, the time required to reach gt 50% was about 4–5 days for the 6–16 month afterripened seeds. Shorter afterripening time (2 and 4 months) yields intermediate values. In *A. thaliana*, 14-day-old seeds needed more than 30 days to reach gt 50% (Fig. 14.2). This value decreased rapidly with increasing afterripening to a gt 50% around 2 days after a stockage of 5–7 months. Roughly, afterripening appeared to alleviate dormancy by the same rate in the two species, but ND-seeds germinate faster in *Arabidopsis* than in *Nicotiana*. Therefore, freshly harvested *Arabidopsis* and *Nicotiana* seeds display a typical dormancy, which is manifested by a pronounced delay in germination. Six months of afterripening treatment at 24°C largely removes primary dormancy.

Hormone and fluridone effects on germination of dormant seeds

The effects of exogenous GA on germination of D seeds were investigated. Germination of *N. plumbaginifolia* D seeds showed a marked stimulation by

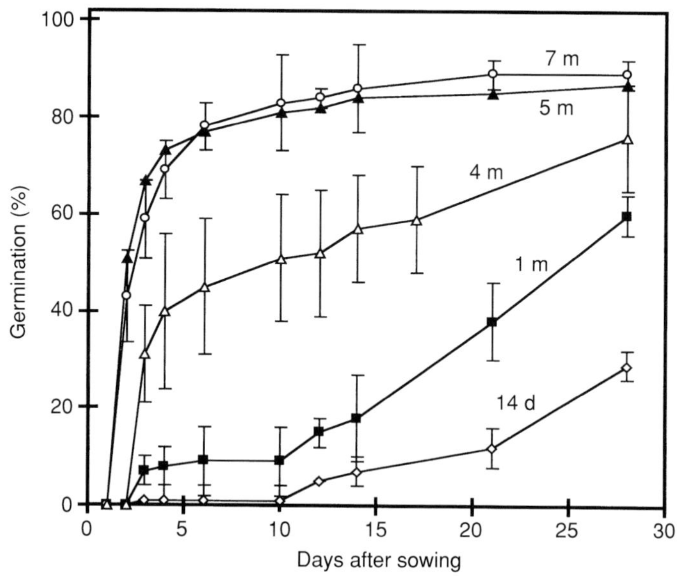

Fig. 14.2. Effect of afterripening on the time course of germination in *A. thaliana* line CVi. Mature seeds were afterripened for 14 days (◇), 1 month (■), 4 months (△), 5 months (▲) or 7 months (○). Data are the mean of four repeats and statistical significance was assessed as in Fig. 14.1.

GA and a strong inhibition by ABA (Fig. 14.3). When imbibed on the optimal GA concentration (100 μM), the D seed germination kinetic was similar to ND seeds (compare Fig. 14.3 and Fig. 14.1). The inhibitory effect of ABA (10 μM) could only be partially reversed by GA added at 100 μM, a concentration which strongly stimulated the germination of D seeds (not shown). Therefore, when exogenous ABA and GA were applied together to D seeds during imbibition, the ABA response (germination inhibited) dominated the GA response (germination advanced).

The germination of D seeds appears to be stimulated very effectively by fluridone, even at low concentrations (1–10 μM), whereas no effect on ND seeds was observed (not shown). At the optimal concentration of 100 μM, the effects of fluridone treatment on the germination of D seeds were quite similar to those observed on GA treatment (Fig. 14.3), but fluridone (gt 50% 6.5 days) was slightly less efficient than GA_3 (gt 50% 5 days) in breaking dormancy. Moreover, taken together, the two compounds have an additive effect on dormancy breakage as gt 50% with the mixture decreased to less than 4 days. By comparison, the gt 50% for the control D seeds in this experiment was 11 days. Finally, the stimulating effect of fluridone (100 μM) on germination was totally suppressed in the presence of 30 μM ABA (Fig. 14.3). This indicates that the stimulating effect of fluridone on the germination of D seeds is probably related to the inhibition of ABA

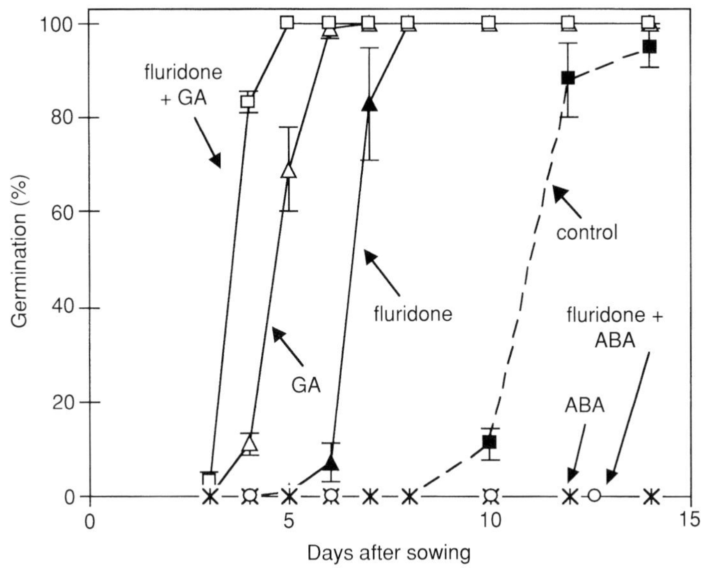

Fig. 14.3. Combined effect of fluridone with GA and ABA on the time course of germination of *N. plumbaginifolia* dormant seeds. Dormant seeds were imbibed on control basal medium (■), with addition of fluridone 10 μM (▲), ABA 10 μM (**X**), GA 100 μM (△), fluridone 10 μM + ABA 10 μM (○), fluridone 10 μM + GA 100 μM (□). Data are means of four repeats and statistical significance was assessed as in Fig. 14.1.

synthesis, and that this synthesis may, therefore, be essential for the expression of seed dormancy.

In the same experiment (but the effects of ABA were not tested) the responses of *A. thaliana* D seeds were found to be very similar with the notable exception of the very low efficiency of GA in breaking the dormancy of D seeds (Fig. 14.4). Fluridone stimulated strongly the germination of *Arabidopsis* D seeds and the additive effects of fluridone with GA on dormancy breaking were spectacular (Fig. 14.4). This last result indicates clearly that GA and ABA interact in the control of seed dormancy in *Arabidopsis* imbibed seeds.

Content of ABA in dormant and afterripened seeds during imbibition

The amounts of endogenous ABA in both D and ND seeds were analysed in dry seeds and every 4 h during the first day of imbibition in *N. plumbaginifolia* (Fig. 14.5) and during some days after sowing in *A. thaliana* (Fig. 14.6). In D dry seeds, ABA content was two- or threefold higher than in ND afterripened dry seeds (Figs 14.5 and 14.6). During the imbibition of

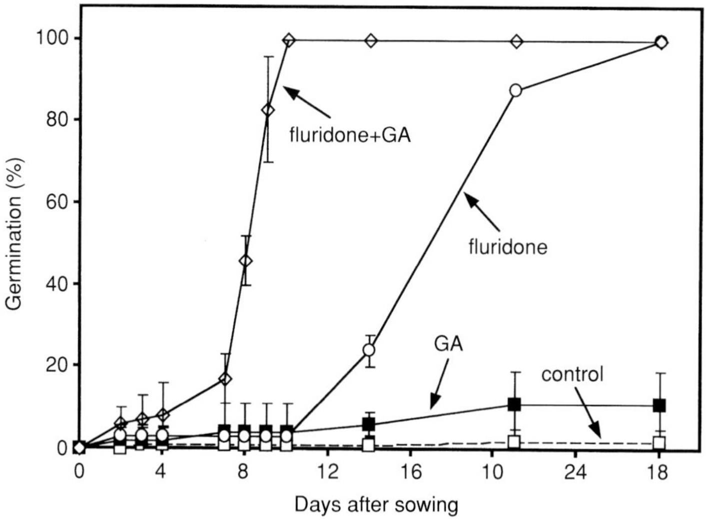

Fig. 14.4. Combined effect of fluridone with GA on the time course of germination of *A. thaliana* dormant seeds. Dormant seeds were imbibed on control basal medium (□), with addition of fluridone 10 μM (○), GA 100 μM (■), fluridone 10 μM + GA 100 μM (◇). Data are means of four repeats and statistical significance was assessed as in Fig. 14.1.

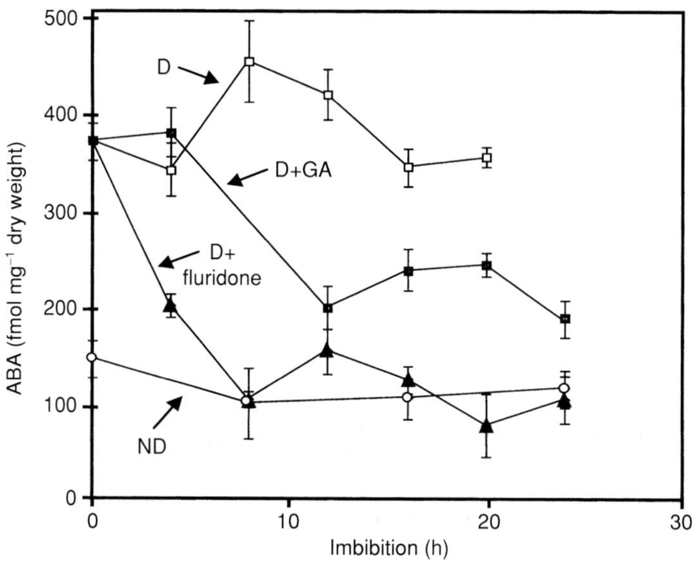

Fig. 14.5. Change in endogenous ABA level in *N. plumbaginifolia* dormant and non-dormant seeds during early imbibition. Dormant seeds were imbibed on water (□), fluridone 10 μM (▲), or GA 100 μM (■). Non-dormant seeds were imbibed on water (○). Data are the means ± SE of five independent measurements. On water the gt 50% was reached in 13 days for D seeds and in 5 days for ND seeds.

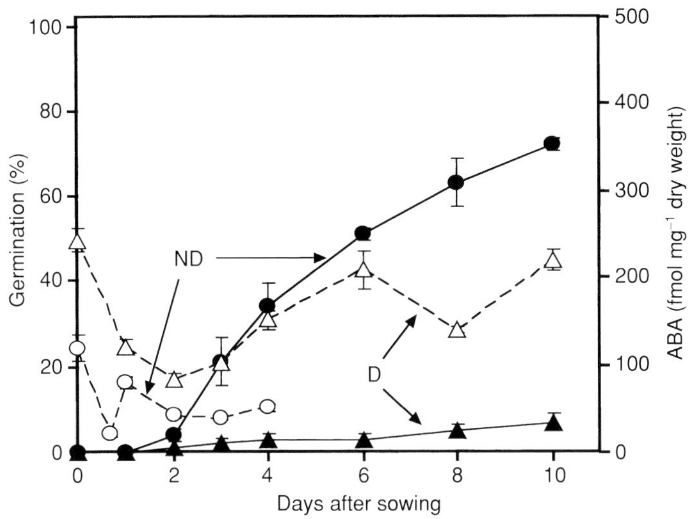

Fig. 14.6. Change in endogenous ABA level in *A. thaliana* dormant and non-dormant seeds during the time course of imbibition and germination. D and ND seeds were imbibed on water. Black symbols and full lines: time course of germination of D and ND seeds. Open symbols and dotted lines: ABA content in corresponding seeds. Data are the means ± SE of five independent measurements.

Nicotiana ND seeds the amount of endogenous ABA decreased slightly, while the ABA content from D seeds decreased transiently during the first 4 h of imbibition but thereafter increased markedly, reaching a peak value (at about 10 h) and then decreasing again slightly (Fig. 14.5). Imbibition of D seeds was, therefore, characterized by a transient, but significant accumulation of ABA, which was not found in ND seeds. This suggests that D seeds may synthesize ABA during early imbibition.

During the imbibition of *Arabidopsis* ND seeds the amount of endogenous ABA decreased the first day after sowing and was stabilized thereafter at a low level when germination began at around day 2 (Fig. 14.6). In D seeds the ABA content also decreased on the first day but was stabilized at a higher level than in ND seeds before increasing again at around day 3–4 to a steady state fourfold higher than in ND seeds.

To determine whether ABA is indeed synthesized *de novo* in the D seeds and to relate this endogenous ABA accumulation to the processes of dormancy maintenance during imbibition, we measured ABA contents from *Nicotiana* D seeds imbibed on fluridone 10 μM and on GA_3 100 μM (Fig. 14.5). The increased level of ABA in imbibed D seeds was not observed in the presence of fluridone, and the ABA content decreased after 8 h to the low level observed in ND seeds. This indicates that some of the endogenous ABA measured in imbibed D seeds is indeed a result of *de novo* synthesis during early imbibition. The imbibition of D seeds on GA had a similar effect to fluridone treatment on the ABA level in seeds. However, the decrease in ABA concentration was less rapid and appeared to reach a steady state level after 20 h that was more elevated than that observed in ND or D fluridone-treated seeds, although clearly lower than that of D seeds. The same experiments are under way with *Arabidopsis*. In *N. plumbaginifolia*, and very probably in *A. thaliana*, D seeds appear to be characterized by a specific ABA synthesis during the first hours of imbibition which can be inhibited (at least in *Nicotiana*) by two of the compounds most effective in breaking seed dormancy: fluridone, an inhibitor of carotenoid and ABA biosynthesis, and GA, which is not generally accepted to act in this way.

Discussion and Conclusion

In *N. plumbaginifolia* and *A. thaliana* freshly harvested seeds of respectively the PbH1D and CVi lines are dormant, in that they exhibit a pronounced germination delay. The complete release of this primary dormancy needs a prolonged period of afterripening (6 months to 1 year) which renders the seeds able to germinate quickly in a well-synchronized manner (Figs 14.1 and 14.2). In these two lines, D and ND seeds can therefore be distinguished easily on the basis of their germination time, making them a useful model for investigations on the control of dormancy during imbibition.

The hormones ABA and GAs are both implicated in the control of seed dormancy

1. In *N. plumbaginifolia,* the ABA deficient mutant *aba1* (Jullien and Bouinot, 1997) and *aba2* (Marin *et al.*, 1996) produced seeds which, even freshly harvested, presented no dormancy. In other respects, it is well known that the ABA-deficient and -responsive mutants in *A. thaliana* produced non-dormant seeds (Hilhorst, 1995). So it appears that in these two species ABA deficiency suppresses any dormancy expression and mimics the effect of natural afterripening.
2. In *N. plumbaginifolia* application of GA to D seeds suppresses dormancy, reducing the germination time to that observed for ND seeds (Figs 14.1 and 14.3). In *A. thaliana,* D seeds are almost insensitive to GA application (Fig. 14.4), which was known for a long time (Karssen and Laçka, 1986).
3. Fluridone, which inhibits carotenoid biosynthesis and therefore ABA biosynthesis (Gamble and Mullet, 1986), is very efficient in breaking dormancy in both species (Figs 14.3 and 14.4). Moreover, fluridone and GA show an additive effect on the suppression of dormancy, which has previously only been demonstrated with sunflower seeds (Le Page-Degivry *et al.* 1990). Although the inhibiting effect of fluridone on the dormancy induction of immature seeds is well documented (Hole *et al.*, 1989; Le

Page-Degivry *et al.*, 1990; Oishi and Bewley, 1990; Steinbach *et al.*, 1997), its dormancy breaking property on mature seeds has rarely been described: for sunflower (Le Page-Degivry and Garello, 1992), and more recently for some other species: *Fagus sylvatica* (Le Page-Degivry *et al.*, 1997), Douglas fir (Bianco *et al.*, 1997) and lettuce (Yoshioka *et al.*, 1998).

Overall, these data support the hypothesis that ABA and GAs play a fundamental and antagonistic role in the expression of seed dormancy in *N. plumbaginifolia* and *A. thaliana*.

Dormant and non-dormant seeds can be discriminated by their ABA biosynthetic capacity

The measurements of ABA content in dry D and ND seeds showed that during a dry storage period, a decrease in ABA content occurred (Figs 14.5 and 14.6) suggesting ABA degradation during afterripening. The early germination of ND seeds could be related to this decrease in ABA level, but as fluridone treatment during imbibition released dormancy (Figs 14.3 and 14.4), it appears that *de novo* ABA synthesis during early imbibition is necessary to maintain dormancy. This was confirmed by the direct measurement of ABA content during early imbibition (Figs 14.5 and 14.6) which showed an accumulation of ABA in D seeds, that was absent in afterripened seeds and was inhibited in fluridone imbibed D seeds in *N. plumbaginifolia* (Fig. 14.5). Very similar results have been reported for lettuce (Yoshioka *et al.*, 1998), where the inhibitory effect of elevated temperature on seed germination can be reversed by fluridone, which in these conditions rapidly decreased the ABA content to the low level observed in seeds imbibed at permissive temperature. Interestingly, these authors found a difference of about fourfold in the ABA content between the non-germinating (D) and germinating (ND) seeds. These data are very similar to those reported here (Figs 14.5 and 14.6). In *N. plumbaginifolia* and *A. thaliana*, as in lettuce (above), sunflower (Bianco *et al.*, 1994) and barley (Wang *et al.*, 1995), dormancy expression results from an active mechanism involving *de novo* ABA synthesis.

GAs may play an active role in the control of dormancy expression by affecting ABA accumulation

In *N. plumbaginifolia,* surprisingly, GA inhibited ABA accumulation in dormant imbibed seeds (Fig. 14.5), suggesting an active role of GAs in the control of this process. In some other species it has been considered that GAs are probably not involved in the control of dormancy *per se*, but act to promote germination after the ABA-mediated inhibition of germination has been overcome (Karssen and Laçka, 1986; Bewley, 1997). More precisely, Le Page-Degivry *et al.* (1996) proposed that ABA is the primary hormone involved at any step during dormancy maintenance and release and that GAs are present at sufficient levels to promote germination as soon as ABA synthesis is inhibited. However, in agreement with our results, an active role for GAs in dormancy release by antagonizing ABA accumulation could be hypothesized: (i) from the selection of ABA-deficient mutants of *Arabidopsis* by screening for the restoration of germination capacity of non-germinating GA-deficient mutant seeds (Koornneef *et al.*, 1982); and (ii) from the germination of the *Arabidopsis* ABA deficient/insensitive mutants on the GA biosynthesis inhibitors uniconazol (Nambara *et al.*, 1991) or paclobutrazol (Léon-Kloosterziel *et al.*, 1996).

To our knowledge, our results are only the second report of an inhibitory effect of GA on ABA accumulation in dormant seeds. Previously Toyomasu *et al.* (1994) reported that endogenous ABA content in lettuce seeds was lowered by red-light irradiation and exogenous GA_3 treatment, which both stimulate germination in the dark. As red-light induces an increase in the level of GA_1, the active gibberellin in lettuce, Toyomasu *et al.* (1994) hypothesized that the photostimulation of germination is mediated by a prerequisite GA-controlled effect which decreases the endogenous ABA content. In *N. plumbaginifolia,* phytochrome-deficient mutant seeds exhibit dormancy in the light and high ABA contents (Kraepiel *et al.*, 1994). Kraepiel *et al.* proposed that in wild-type plants, the phytochrome-mediated light signal enhances ABA degradation. This degradation might be a prerequisite necessary for germination and it should be interesting to test whether this is mediated by GAs. These results lead us to propose that the hypothesized effect of GAs on ABA content may be mediated by stimulating a degradation process rather than by inhibiting ABA synthesis.

In conclusion, we propose a model (Fig. 14.7) where the key step for predisposing seeds to germinate is the inhibition of the capacity for ABA synthesis, or the stimulation of ABA degradation, or both. This predisposition takes place during afterripening. Our observation of the inhibitory effect of exogenous GA on ABA accumulation during early imbibition of *N. plumbaginifolia* seeds allows us to hypothesize that the earliest role of GAs in the dormancy breakage process is by a negative effect on ABA accumulation in seeds. In this model, the content of active GAs or the capacity for GA synthesis or seed perception to GAs increase during afterripening and break seed dormancy by its consequently inhibitory effect on ABA accumulation in imbibed afterripened seeds. The additive effect of fluridone and GA on dormancy breakage (Figs 14.3 and 14.4) may be due to a greater efficiency of the mixture in reducing the ABA level in imbibed seeds. However, this latter aspect of the model remains to be improved, because: (i) GA, which is more efficient in breaking dormancy than fluridone in *N. plumbaginifolia,* induced a slower decrease of ABA than did fluridone (Figs 14.3 and 14.5). This could be easily explained if GAs also acted on

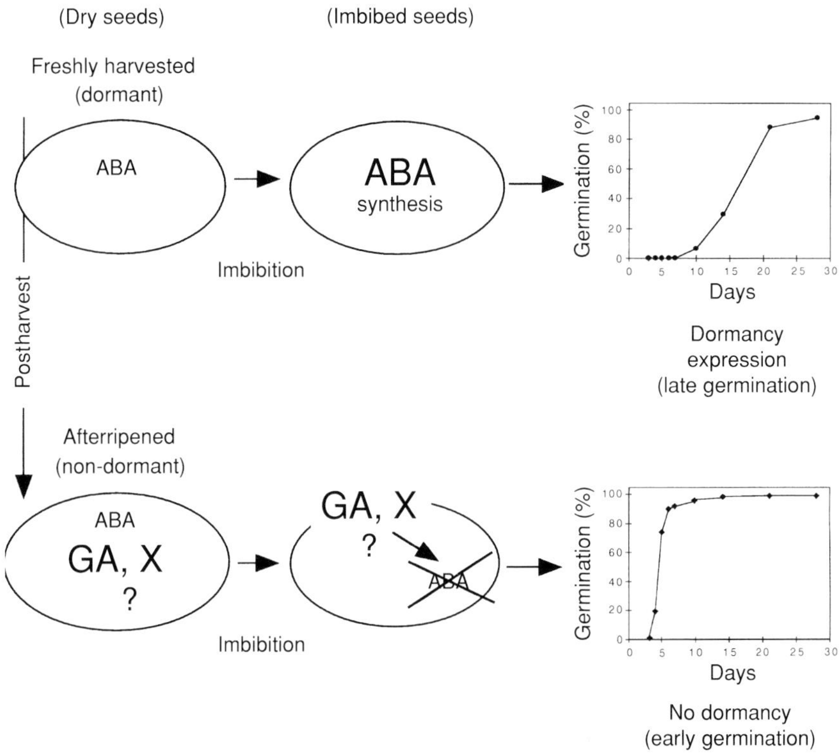

Fig. 14.7. A model for the implication of ABA and GAs in the control of dormancy in imbibed seeds. Upper row: dry D seeds contained a high level of ABA, which increased by neosynthesis with imbibition, leading to late germination. Lower row: dry ND seeds contained a low level of ABA which decreased on imbibition in response to the action of GAs (*N. plumbaginifolia*) or other factors (X), (*A. thaliana*), leading to early germination. Late and early germination curves are *N. plumbaginifolia* examples.

dormancy breaking during imbibition through additional mechanisms rather than just the control of ABA level; (ii) GA was almost ineffective in breaking the dormancy of *A. thaliana* seeds. This indicates that the reduction of ABA level in imbibed *Arabidopsis* ND seeds was mainly controlled by other factor(s) (named X) than GAs which would be activated in *Arabidopsis* by afterripening or stratification.

Acknowledgements

This work was supported by grants from the Ministère de l'Agriculture et de la Pêche, AIP 4767 Biologie du Développement. We are grateful to Christina Gofron for growing the plants, to Malgorzata Bonnet for helping with ELISA tests and to Dr Helen North for correcting the English.

References

Bartels, P.G. and Watson, C.W. (1978) Inhibition of carotenoid synthesis by fluridone and norflurazon. *Weed Science* 26, 198–203.

Bewley, J.D. (1997) Seed germination and dormancy. *Plant Cell* 9, 1055–1066.

Bewley, J.D. and Black, M. (1994) *Seeds: Physiology of Development and Germination.* Plenum Press, New York.

Bianco, J., Garello, G. and Le Page-Degivry, M.T. (1994) Release of dormancy in sunflower embryos by dry storage: involvement of gibberellins and abscisic acid. *Seed Science Research* 4, 57–62.

Bianco, J., Garello, G. and Le Page-Degivry, M.T. (1997) *De novo* ABA synthesis and expression of seed dormancy in gymnosperm: *Pseudotsuga menziesii. Plant Growth Regulation* 21, 115–119.

Bourgin, J.P., Chupeau, Y. and Missonier, C. (1979) Plant regeneration from mesophyll protoplasts of several *Nicotiana* species. *Physiologia Plantarum* 45, 288–292.

Coïc, Y. and Lesaint, C. (1971) Comment assurer une bonne nutrition en eau et ions minéraux en horticulture. *Horticulteur Français* 8, 11–14.

Gamble, P.E. and Mullet, J.E. (1986) Inhibition of carotenoid accumulation and abscisic acid biosynthesis in fluridone-treated dark-grown barley. *European Journal of Biochemistry* 160, 117–121.

Groot, S.P.C. and Karssen, C.M. (1992) Dormancy and germination of abscisic acid-deficient tomato seeds. Studies with the sitiens mutant. *Plant Physiology* 99, 952–958.

Hilhorst, H.W.M. (1995) A critical update on seed dormancy. I. Primary dormancy. *Seed Science Research* 5, 61–73.

Hilhorst, H.W.M. and Karssen, C.M. (1992) Seed dormancy and germination: the role of abscisic acid and gibberellins and the importance of hormone mutants. *Plant Growth Regulation* 11, 225–238.

Hole, D.J., Smith, J.D. and Cobb, B.G. (1989) Regulation of embryo dormancy by manipulation of abscisic acid in kernels and associated cob tissue of *Zea mays* L. cultured *in vitro. Plant Physiology* 91, 101–105.

Juillard, J., Maldiney, R., Sotta, B., Miginiac, E., Kerhoas, L. and Einhorn, J. (1994) HPLC-ELISA and MS as complementary techniques to study plant hormone metabolites. *Analusis* 22, 483–489.

Jullien, M. and Bouinot, D. (1997) Seed dormancy and responses of seeds to phytohormones in *Nicotiana plumbaginifolia.* In: Ellis, R.H., Black, M., Murdoch, A.J. and Hong, T.D. (eds) *Basic and Applied Aspects of Seed Biology,* Kluwer Academic Publishers, Dordrecht, pp. 215–224.

Karssen, C.M. and Laçka, E. (1986) A revision of the hormone balance theory of seed dormancy: studies on gibberellin and/or abscisic acid-deficient mutants of *Arabidopsis thaliana.* In: Bopp, M. (ed.) *Plant Growth Substances.* Springer-Verlag, Berlin, pp. 315–323.

Karssen, C.M., Brinkhorst-Van der Swan, D.L.C., Breekland, A.D. and Koornneef, M. (1983) Induction of dormancy during seed development by endogenous abscisic acid: studies on abscisic acid deficient genotypes of *Arabidopsis thaliana* L. Heynh. *Planta* 157, 158–165.

Karssen, C.M., Zagorski, S., Kepczynski, J. and Groot, S.P.C. (1989) Keyrole for endogenous gibberellins in the control of seed germination. *Annals of Botany* 63, 71–80.

Koornneef, M., Jorna, M.L., Brinkhorst-Van der Swan, D.L.C. and Karssen, C.M. (1982) The isolation of abscisic acid (ABA) deficient mutants by selection of induced revertants in non-germinating gibberellin sensitive lines of *Arabidopsis thaliana* (L.) Heynh. *Theoretical and Applied Genetics* 61, 385–393.

Koornneef, M., Reuling, G. and Karssen, C.M. (1984) The isolation and characterisation of abscisic acid-insensitive mutants of *Arabidopsis thaliana*. *Physiologia Plantarum* 61, 377–383.

Kraepiel, Y., Rousselin, P., Sotta, B., Kerhoas, L., Einhorn, J., Caboche, M. and Miginiac, E. (1994) Analysis of phytochrome- and ABA-deficient mutants suggests that ABA degradation is controlled by light in *Nicotiana plumbaginifolia*. *Plant Journal* 6, 665–672.

Le Page-Degivry, M.T. and Garello, G. (1992) In situ abscisic acid synthesis: a requirement for induction of embryo dormancy in *Helianthus annuus*. *Plant Physiology* 98, 1386–1390.

Le Page-Degivry, M.T., Barthe, P. and Garello, G. (1990) Involvement of endogenous abscisic acid in onset and release of *Helianthus annuus* embryo dormancy. *Plant Physiology* 92, 1164–1168.

Le Page-Degivry, M.T., Bianco, J., Barthe, P. and Garello, G. (1996) Change in hormone sensitivity in relation to onset and breaking of sunflower embryo dormancy. In: Lang, G.A. (ed.) *Plant Dormancy: Physiology, Biochemistry and Molecular Biology*. CAB International, Wallingford, UK, pp. 221–231.

Le Page-Degivry, M.T., Garello, G. and Barthe, P. (1997) Changes in abscisic acid biosynthesis and catabolism during dormancy breaking in *Fagus sylvatica* embryo. *Journal of Plant Growth Regulation* 16, 57–61.

Léon-Kloosterziel, K.M., Alvarez, G.M., Ruijs, G.J., Jacobsen, S.E., Olszewski, N.E., Schwartz, S.H., Zeevaart, J.A.D. and Koornneef, M. (1996) Isolation and characterisation of abscisic acid-deficient *Arabidopsis* mutants at two new loci. *Plant Journal* 10, 655–661.

Marin, E., Nussaume, L., Quesada, A., Gonneau, M., Sotta, B., Hugueney, P., Frey, A. and Marion-Poll, A. (1996) Molecular identification of zeaxanthin epoxidase of *Nicotiana plumbaginifolia*, a gene involved in abscisic acid biosynthesis and corresponding to the *ABA* locus of *Arabidopsis thaliana*. *EMBO Journal* 15, 2331–2342.

Nambara, E., Akazawa, T. and McCourt, P. (1991) Effects of the gibberellin biosynthetic inhibitor Uniconazol on mutants of *Arabidopsis*. *Plant Physiology* 97, 736–738.

Oishi, M.Y. and Bewley, J.P. (1990) Distinction between the responses of developing maize kernels to fluridone and desiccation in relation to germinability, α-amylase, and abscisic acid content. *Plant Physiology* 94, 592–598.

Popova, L.P. and Riddle, K.A. (1996) Development and accumulation of ABA in fluridone-treated and drought-stressed *Vicia faba* plants under different light conditions. *Physiologia Plantarum* 98, 791–797.

Steinbach, H.S., Benech-Arnold, R.L. and Sanchez, R.A. (1997) Hormonal regulation of dormancy in developing sorghum seeds. *Plant Physiology* 113, 149–154.

Toyomasu, T., Yamane, H., Murofushi, N. and Inoue, Y. (1994) Effects of exogenously applied gibberellin and red light on the endogenous levels of abscisic acid in photoblastic lettuce seeds. *Plant Cell Physiology* 35, 127–129.

Walker-Simmons, M. (1987) ABA levels and sensitivity in developing wheat embryos of sprouting resistant and susceptible cultivars. *Plant Physiology* 84, 61–66.

Wang, M., Heimovaara-Dijkstra, S. and Van Duijn, B. (1995) Modulation of germination of embryos isolated from dormant and non dormant barley grains by manipulation of endogenous abscisic acid. *Planta* 195, 586-592.

Wareing, P.F. and Saunders, P.F. (1971) Hormones and dormancy. *Annual Reviews of Plant Physiology* 22, 261–288.

Xu, N. and Bewley, J.D. (1995) The role of abscisic acid in germination, storage protein synthesis and desiccation tolerance in alfalfa (*Medicago sativa* L.) seeds, as shown by inhibition of its synthesis by fluridone during development. *Journal of Experimental Botany* 46, 687–694.

Yoshioka, T., Endo, T. and Satoh, S. (1998) Restoration of seed germination at supraoptimal temperature by fluridone, an inhibitor of abscisic acid biosynthesis. *Plant Cell Physiology* 39, 307–312.

15

The Role of Endogenous Hormones in Potato Tuber Dormancy

Jeffrey C. Suttle

US Department of Agriculture, Agricultural Research Service, Northern Crop Science Laboratory, PO Box 5677, Fargo, ND 58105-5677, USA

Introduction

Indigenous to Central and South America, the potato (*Solanum tuberosum* L.) was initially domesticated by pre-Columbian Indians and has been extensively cultivated since that time. In terms of total production, the potato ranks among the top four food crops and in many countries constitutes a major source of human nutrition. Global production of potatoes exceeds 290 Mt (FAO estimate). Annual production in the United States exceeds 22 Mt (1996 USDA estimate). Over 70% of the total US potato crop is placed in short- to long-term storage for a stable year-round supply. Unlike many crops, potatoes are stored in a fully hydrated and highly perishable form. As such, postharvest losses from both physiological and pathological processes can be severe. Of the physiological processes that affect the quality of stored potatoes, one of the most important is postharvest sprouting.

General Dormancy Characteristics

From a developmental standpoint, the potato is a highly compressed stem. The apical bud represents the terminal bud of a stem and the eyes are analogous to axillary buds. Dormancy is thought to begin at or about the time of tuber initiation (Burton, 1963). Potato tubers exhibit all types of dormancy as defined by Lang *et al.* (1987). At harvest and for an indeterminate period thereafter, all buds (eyes) of the potato are endodormant. After a period of time, endodormancy is lost and, with the exception of (typically) the most apical eye, paradormancy (apical dominance) of the lateral meristems commences. In nature or when stored under cool (*c.* 3°C) temperatures, bud growth is then prevented by ecodormancy. The length of the endodormant (hereafter dormant) period is influenced by both genetics and environment. Environmental conditions during tuber formation affect the length of dormancy; cool and wet conditions extend the dormant period and hot and dry conditions shorten dormancy (Krijthe, 1962; Burton, 1989). Between 3 and 25°C, tuber dormancy duration is inversely related to postharvest storage temperature (Schippers, 1956). Temperatures below 3°C are stressful and result in premature sprouting (Wurr and Allen, 1976). Exposure to a wide variety of conditions or chemical agents results in the rapid termination of dormancy (Coleman, 1987). Aside from acting as stressors, these treatments have essentially nothing in common and their mechanisms of action are unknown. As would be expected from its importance as a survival mechanism, the inheritance pattern of tuber dormancy is quite complex. Quantitative trait loci (QTL) analyses of tuber dormancy have indicated that at least nine distinct loci are associated with tuber dormancy (Van den Berg *et al.*, 1996). Interestingly, despite their distinct developmental origins, a correlation between the lengths of seed and tuber dormancy in potatoes has been noted (Simmonds, 1964).

Cell Biology of Dormancy

Detailed morphological analysis has shown that immediately after harvest, when tubers are most dormant, no bud growth can be detected (Van Ittersum *et al.*, 1992). At the cellular level, very few mitotic figures occur in dormant tuber bud meristems and the resumption of bud growth following the termination of dormancy is accompanied by a large increase in the frequency of metaphase cells (Lesham and Clowes, 1972). Cytological analyses using microdensitometry demonstrated an accumulation of 2C nuclei in dormant meristems (Macdonald and Osborne, 1988). These same authors observed low but detectable rates of ^{3}H-thymidine incorporation into dormant nuclei which they attributed to DNA repair processes. Subsequent studies using flow cytometry of nuclei isolated from tuber buds confirmed these observations and demonstrated that dormant meristematic cells are blocked in the G1 phase of the cell cycle prior to DNA replication (Campbell *et al.*, 1996). Cell cycle progression in all eukaryotes studied is dependent on the activities of a highly conserved group of protein kinases typified by the cdc-2 kinase (Nigg, 1995). The catalytic activity of these heterodimeric kinases is regulated by both their association with another class of proteins known collectively as cyclins and by phosphorylation (Mironov *et al.*, 1999). As judged by Northern and Western analyses, both cdc-2 kinase mRNA and protein were present in dormant meristems (Campbell *et al.*, 1996). Thus, cell cycle arrest in dormant potato buds results from unidentified biochemical factors, possibly mitotic cyclin deficiency and/or inappropriate phosphorylation status of the kinase subunit. This question remains unresolved.

Although mitotically quiescent, dormant meristems are nevertheless metabolically active. Many studies have shown that dormant meristems readily take up and incorporate various precursors into RNA and protein (Macdonald and Osborne, 1988). The resumption of sprout growth is consistently associated with an increase in the rates of incorporation of these precursors (Korableva *et al.*, 1976; Macdonald and Osborne, 1988). The interpretation of these studies is complicated by the fact that dormant tubers readily respond to a variety of stimuli such as wounding and pathogen challenge with a host of sustained biochemical alterations most of which require *de novo* RNA and protein synthesis (Choi and Bostock, 1994; Dixon *et al.*, 1994). It would therefore appear that the attenuated levels of macromolecular synthesis in dormant meristems do not reflect a scarcity of precursors nor a gross metabolic insufficiency, but rather, simply reflect the quiescent state of the meristem. In addition, several studies have attempted to relate specific RNA or protein species with dormancy status; thus far without success (Désiré *et al.*, 1995). As such, the molecular process(es) that regulate entry into and exit from dormancy remain unknown.

Hormonal Regulation of Tuber Dormancy

Plant hormones have been implicated in the regulation of every aspect of plant development from germination to senescence. Potato tuber dormancy is no exception and hypotheses concerning the hormonal regulation of this process have continued to evolve as new information is obtained. As with other aspects of plant hormone research, studies with potatoes have been hindered by both technological and biological factors.

The technological factors that have impeded tuber dormancy research are not unique to this system and are related to the methods used to quantify hormone levels in extracts. Early studies (and in several cases the only studies published to date) relied exclusively on bioassays for hormone quantitation. Aside from more general concerns that apply to all bioassay-derived data, studies with potatoes were further complicated by the presence (see below) of a diverse array of substances that interfere with many bioassay responses. More recent studies have employed immunological and/or mass spectral methodologies to measure hormone levels. Both of these techniques have their own set of benefits and drawbacks but in general, yield far more reliable data.

The biological factors hindering plant hormone studies have proven to be somewhat more intractable. The potato is an underground organ that exhibits considerable variation in induction, growth and final size. In addition, the meristematic portion of the tuber constitutes only a small fraction of the total tuber mass. For example, from a 500 g 'Russet Burbank' tuber, approximately 10–15 mg of eye tissue can be harvested for experimental use. Furthermore, as mentioned above, tuber dormancy begins at (or shortly after) the start of tuberization. Many published studies were begun after harvest. While this may be acceptable for enquiry into the factors affecting dormancy release, it clearly is too late for studies designed to explore early events in dormancy regulation.

The development and use of selected model systems would seem to be the best option to address these issues. In our research, extensive use has been made of *in vitro* tuberization systems. These systems offer several advantages including sterility, the ability to synchronize tuber formation, and easy feeding of chemical agents that influence hormone levels or action. However, the amount of tissue available for experimentation is limited and this requires the use of extremely sensitive analytical techniques for hormone quantitation.

In this review, dormany is divided into three phases: initiation, maintenance and termination. While useful for discussion, this division is arbitrary. At present, it is unclear if or to what extent the physiological mechanisms regulating each of these stages overlap. Nevertheless, this distinction facilitates discussion and will be used for the remainder of this review.

Auxins

The first class of endogenous hormones discovered was also the first to be studied as potential regulators of potato tuber dormancy. The early data were somewhat contradictory. Using *Avena* bioassays for detection, early workers found no clear correlation between endogenous 'acidic' auxin levels and tuber dormancy status (see Hemberg, 1985 for review). An increase in endogenous auxin was generally observed in tubers that had sprouted, a fact which suggested that while auxin is not involved in dormancy regulation, it is required for subsequent sprout growth. Given the high levels of inhibitory substances in potato extracts (see below), interpretation of these bioassay data is equivocal at best.

Adding to the confusion were reports describing the effects of exogenous auxin-like substances such as IAA and 1-naphthalene acetic acid. Early studies by researchers at the Boyce Thompson Institute demonstrated that exogenous auxin was an effective sprout inhibitor (Guthrie, 1939). It was subsequently shown that lower and (presumably) more physiological levels of IAA stimulated the growth of sprouts on eye pieces isolated from non-dormant tubers but had no effect on eyes excised from dormant potatoes (Hemberg, 1949). Interestingly, the weak auxin phenylacetic acid was reported to stimulate sprout growth (i.e. break dormancy) when applied to dormant tubers (Zimmerman and Hitchcock, 1939).

Recent studies using more advanced analytical techniques have not clarified this issue. Using HPLC coupled with fluorescence detection, no change in free IAA levels were observed during 8 months of cold storage (Sukhova *et al.*, 1993). Free IAA levels declined markedly in sprouting tubers. In another study, both free IAA and base-labile (conjugated) IAA contents increased substantially in tuber eyes prior to the onset of sprout growth and under storage conditions (3°C) where sprouting would not occur due to temperature-imposed ecodormancy (C. Sorce, Italy, 1999, personal communication). In this latter study, care was taken to analyse specific tissues. This may explain the apparent inconsistencies in the literature.

At present, it is not possible to assign a specific role for IAA and its derivatives in tuber dormancy regulation. The lack of effect of exogenous auxins on dormancy duration coupled with inconsistent reports on endogenous levels during tuber storage (hence dormancy progression) suggest that auxins do not directly participate in dormancy regulation *per se* but may instead be more involved with subsequent sprout growth.

Abscisic Acid

Initial attempts to quantify endogenous auxin levels in plant tissues by bioassay were routinely hampered by the presence of inhibitory substances in plant extracts. Even following paper chromatography a broad zone of inhibition (inhibitor-β) interfered with the bioassay response. Hemberg (1949) was the first to recognize the potential significance of these inhibitors in dormancy regulation in potatoes and other plants. In a series of thoughtful, carefully designed and interpreted studies, Hemberg demonstrated that extracts prepared from dormant potatoes contained both neutral and acidic inhibitory activity and that the content of the acidic (but not the neutral) inhibitor(s) correlated with tuber dormancy status (Hemberg, 1952). Following the identification of ABA as a component of the β-inhibitor complex, attention focused on examining the involvement of ABA in dormancy regulation (see Hemberg, 1985).

Exogenous ABA temporarily inhibits sprout growth in non-dormant tubers (El-Antably *et al.*, 1967; Van Es and Hartmans, 1969). In order to be effective, ABA must be continuously supplied or repeatedly applied. In no case does exogenous ABA induce dormancy when applied to non-dormant potatoes. Using a variety of detection methods and different potato cultivars, several studies have found a rough correlation between endogenous ABA levels and tuber dormancy (see Coleman, 1987). However, the correlation is not perfect and discrepancies have been reported (i.e. Sorce *et al.*, 1996). In general, in a given cultivar, ABA content declines with increased storage time. In a study examining three cultivars with varying periods of dormancy, the onset of sprouting coincided with a decline in ABA levels in two cultivars (Coleman and King, 1984). ABA levels in periderm/apical bud tissues isolated from 'Russet Burbank' potatoes stored under two temperatures declined with storage time but the absolute levels at which 50% sprouting occurred varied widely (Suttle, 1995). Exposure to dormancy terminating conditions or chemical agents also results in a reduction in endogenous ABA content (Ji and Wang, 1988; Van den Berg *et al.*, 1991). In all of these studies, no threshold ABA level (above which sprouting would not occur) could be identified, suggesting that a loss of ABA sensitivity rather than a decline in total ABA levels heralds the loss of dormancy.

Taken together, these results are consistent with, but not proof of, a role for ABA in potato tuber dormancy. Unequivocal evidence for such a role was obtained only recently (Suttle and Hultstrand, 1994). Critical to the success of these studies was the use of an *in vitro* micro-tuberization system which permitted the precise feeding of fluridone, an inhibitor of *de novo* ABA biosynthesis. Untreated microtubers contain substantial amounts of endogenous ABA and are dormant for several months after formation. Exposure to fluridone resulted in > 95% inhibition of ABA accumulation and precocious sprouting. Exogenous ABA restored internal ABA levels and suppressed premature sprouting. Application of fluridone to pre-formed and fully dormant microtubers also resulted in a reduction in ABA content and stimulated precocious sprout growth. These results demonstrated that the sustained presence of ABA is essential for both the induction and maintenance of tuber dormancy.

Ethylene

Rosa (1925) was the first to report that ethylene treatment results in shortening of the natural rest period of potatoes. However, subsequent studies failed to corroborate these findings (see Denny, 1926). More recent studies have shown that under certain conditions, short-term treatment with ethylene can indeed shorten the period of tuber dormancy (Rylski *et al.*, 1974; Alam *et al.*, 1994). This effect apparently depends not only on concentration and treatment duration but also on the cultivar. Despite numerous attempts using a wide range of concentrations and treatment times, significant stimulation of sprouting of dormant 'Russet Burbank' tubers has never been observed in this laboratory (Suttle, unpublished data). Long-term or continuous exposure to ethylene or treatment with ethylene-releasing agents inhibits sprout growth (Rylski *et al.*, 1974; Cvikrova *et al.*, 1994). Generally, sprout growth resumes following the termination of ethylene treatment. Thus, ethylene does not influence dormancy *per se* but rather acts as a general growth inhibitor. Nevertheless, ethylene has shown promise as a sprout suppressant when continuously supplied in commercial potato storages (Prange *et al.*, 1998).

Intact and undamaged tubers produce low but detectable amounts of ethylene (Rylski *et al.*, 1974). Ethylene production increases dramatically following wounding, chemical treatment or pathogen attack (Creech *et al.*, 1973; Okazawa, 1974). An increase in ethylene production also accompanies the onset of sprout growth (Poapst *et al.*, 1968; Okazawa, 1974). As a whole these observations were not consistent with a role for endogenous ethylene in tuber dormancy regulation. Nevertheless, such a role was recently identified.

Isolated, single-node explants undergoing tuberization produce ethylene throughout the period of *in vitro* culture (Suttle, 1998b). The maximum rate of ethylene production occurs during the initial period of culture and falls to a low but stable rate thereafter. Treatment of tuberizing explants with inhibitors of ethylene action such as silver nitrate or 2,5-norbornadiene results in premature sprouting of the developing microtubers. Exogenous ethylene reverses the effect of norbornadiene. To be effective, these inhibitors must be present from the start of *in vitro* culture. Delaying their application for even 1 week abrogates their efficacy. These results indicate that endogenous ethylene plays a critical role in the induction (but not maintenance) of microtuber dormancy.

With regard to dormancy termination, the role of ethylene is unresolved. A number of the dormancy terminating agents/treatments stimulate endogenous ethylene production by tubers (Poapst *et al.*, 1968; Suttle, unpublished data). Whether endogenous ethylene mediates or participates in the action of these agents has not been critically evaluated.

Gibberellins

Exogenous GA (typically GA_3) has been reported to promote the sprouting of dormant potatoes (Rappaport and Wolf, 1969; Van Es and Hartmans, 1969). In fact, GA_3 is often used in seed potato certification trials where tubers are replanted within several months of harvest to evaluate disease prevalence. As with other aspects of potato tuber dormancy, the efficacy of exogenous GA varies (see Turnbull and Hanke, 1985a). This variation may depend on cultivar and depth of dormancy at the time of treatment.

Many GA-responsive bioassays are very sensitive to endogenous inhibitors. Because of this, early attempts to quantify endogenous GA levels in potatoes were hampered by the abundance of these inhibitors in tuber extracts (see Hemberg, 1985). However, some successes were reported. A transient increase in total GA content was detected prior to the onset of sprout growth (Hemberg, 1985). Similar results were reported by Bialek and Bielinska-Czarnecka (1975). Others have found that endogenous GA levels increase after the onset of sprouting (Smith and Rappaport, 1960).

All of these studies describe total GA levels as determined by several different bioassays. Consequently, the levels reported vary greatly and it is not clear which GAs were being measured.

More recently, GC-MS techniques have been applied to this problem and have resolved some of these uncertainties. In an early study, both GA_{20} and GA_1 were identified (but not quantified) in both tuber and sprout extracts (Jones *et al.*, 1988). These results suggest that, like other solanaceous genera examined to date, the early 13-hydroxylation pathway with GA_1 as the active hormone occurs in potato tissues. In another study, GA_4 and GA_9 were also identified in extracts prepared from tuberizing stolons (Xu *et al.*, 1998). No studies describing changes in endogenous GA levels during dormancy have been reported. This deficiency is currently being addressed in this laboratory.

Thus, while it is tempting to conclude that endogenous GAs are involved in the termination of tuber dormancy, evidence gathered to date is equivocal and does not fully support this hypothesis. Additional studies are needed before the role of GAs in tuber dormancy can be fully understood.

Cytokinins

Hemberg (1970) demonstrated that both synthetic and naturally occurring cytokinins can break potato tuber dormancy and stimulate bud growth. He observed that the effectiveness of exogenous cytokinins was greater in tubers that had been stored prior to treatment. Subsequent studies confirmed this observation and demonstrated that immediately after harvest exogenous cytokinins have little effect on subsequent sprouting but their efficacy increases with storage time and the gradual weakening of dormancy (Turnbull and Hanke, 1985a; Suttle, 1998a). The nature of this refractory period was examined (Suttle, 1998c) and it was found that the change in cytokinin efficacy was not a result of metabolic inactivation but appeared to represent a true change in sensitivity (i.e. receptor availability and/or activity).

Early studies using bioassays demonstrated that an increase in cytokinin-like activity occurred in sprouting tubers (Van Staden and Brown, 1979; Banas *et al.*, 1984; Koda, 1982). Physical wounding of dormant tubers, which often induces sprouting, also results in an increase in cytokinin-like activity (Koda, 1982). Subsequent studies using immunological detection methods have confirmed these results and have also suggested that the increase in cytokinin content may precede the onset of sprouting (Turnbull and Hanke, 1985b; Sukhova *et al.*, 1993). However, interpretation of these studies is not straightforward because of inherent methodological limitations including the lack of sufficient fractionation of extracts prior to analysis and the number of cytokinins quantified.

These issues were subsequently addressed in a recent study in which the endogenous levels of eight cytokinins were determined by side-chain specific mAbs in HPLC-fractionated extracts after specified periods of postharvest storage and known tuber dormancy status (Suttle, 1998a). In this study, an increase in total bioactive cytokinins was found to precede the loss of tuber dormancy and the onset of sprouting. This increase also occurred in tubers prevented from sprouting by storage temperature. This demonstrated that the increase in cytokinin content was not a result of sprout growth but more likely was a cause of sprout growth. In a follow-up study (Suttle *et al.*, 1997), a similar increase in endogenous *cis*-zeatin was also found prior to the onset of sprout growth. Further, exogenous *cis*-zeatin was as effective as *trans*-zeatin in terminating tuber dormancy. Only 10% of exogenous *cis*-[^{14}C]-zeatin was converted to the *trans*-isomer, which indicated that the *cis*-isomer was most likely the biologically active species. Most investigators have considered *cis*-zeatin and its derivatives to be biologically inactive. However, these observations suggest that both cytokinin isomers participate in potato tuber dormancy regulation. Further studies will be required to determine the relative roles of each of these cytokinin isomers in tuber dormancy.

Other Endogenous Factors

If knowledge concerning the roles of the major classes of plant hormones is limited, information concerning the roles of other endogenous regulators is almost completely lacking. It is now clear that the spectrum of plant growth regulating substances is no longer confined to the five traditional groups of hormones but also includes numerous other classes of low molecular weight metabolites, oligosaccharides and peptides. In this review, the potential involvement of only three classes of 'non-traditional' regulators will be discussed.

As originally (functionally) defined, inhibitor-β isolated from potato peel extracts was a heterogeneous mixture of many compounds, some of which remain uncharacterized (Housley and Taylor, 1958). With the identification of ABA as a component of this complex (Milborrow, 1967), attention focused on its involvement in tuber dormancy and little subsequent effort was made to determine the roles of the remaining components. Studies that established the correlation between tuber dormancy and endogenous inhibitor(s) levels relied on two *Avena* bioassays, neither of which is very sensitive to authentic ABA (Holst, 1971; Hemberg, 1985). Using thin-layer chromatography, Holst (1971) also demonstrated that inhibitor-β from potatoes was composed of at least three substances only one of which could be ABA. In a further study using similar chromatographic methods (Franklin and Hemberg, 1980), it was shown that the levels of all inhibitory substances declined as tubers exited dormancy.

Phenolic acids could certainly be a component of this inhibitory complex. Many are growth inhibitors and potatoes are known to contain a number of them (Lewis *et al.*, 1998). It has been reported that the content of several of these phenolics declines as tubers exit dormancy (Cvikrova *et al.*, 1994). However, the exact nature and role of this inhibitory complex in dormancy regulation remain to be determined.

Intact and undamaged potato tubers produce a wide array of volatile compounds. Burton and Meigh (1971) first reported on the accumulation of sprout-inhibiting volatiles in potato storage atmospheres but they were unable to identify the chemical nature of the growth-inhibiting substances. Subsequent studies resulted in the identification of several bioactive volatiles, including the 1,4- and 1,6-isomers of dimethylnaphthalene (Meigh *et al.*, 1973). The extent and persistence of growth inhibition elicited by these isomers is of sufficient magnitude that they are being evaluated as potential sprout suppressants in commercial storage facilities (Lewis *et al.*, 1997). To date, there have been no reports describing rates of production of these endogenous inhibitors or their biosynthetic origins in potatoes during dormancy. Thus, while certainly candidates for endogenous dormancy regulators, the nature and extent of involvement of these volatiles in tuber dormancy regulation is currently uncertain.

The fatty acid-derived hormone jasmonic acid (JA) is produced by all potato tissues studied and elicits a broad range of biological activities (Van den Berg and Ewing, 1991). A derivative of JA, 13-hydroxyjasmonic acid (tuberonic acid) has been posited to be the natural tuberization stimulus in potatoes and other tuber-forming species (Yoshihara *et al.*, 1989). Conflicting reports describing the effects of JA on various types of dormancy have appeared. Application of JA to recently harvested potatoes has been reported to prolong the period of sprout inactivity (E.C. Lulai, North Dakota, 1999, personal communication). No reports describing changes in JA levels during tuber dormancy have appeared. Hence, JA is a possible but unproven regulator of tuber dormancy.

A Look Ahead

From the foregoing, it is clear that progress in determining the roles of endogenous plant hormones in the regulation of potato tuber dormancy is being made. Some of the technical and biological difficulties that have hindered progress in this area have been successfully overcome. Others remain and it is clear that future progress in understanding this highly integrated developmental process will depend on the combined use of appropriate models that circumvent the inherent complexities of this system and advanced analytical instrumentation.

The powerful tools of genetic analysis have not been utilized in tuber dormancy research to the extent that they have in other areas of plant science. To date, only a handful of biochemically characterized mutants of potato have been identified. Two of these mutants ('Droopy' and a dwarf phenotype) have been shown to result from hormone imbalances in ABA biosynthesis and GA metabolism, respectively (Quarrie, 1982; Bamberg and Hanneman, 1991; Van den Berg *et al.*, 1995). In spite of their availability, the two mutants have not been successfully used in dormancy studies. The 'Droopy' mutant occurs in a genetic background that possesses limited dormancy and in the dwarf mutant tuberization has proven difficult to control. With time, other more useful mutants will be identified and there is no doubt future studies will make greater use of this type of analysis.

Plant transformation studies have been conducted using various *Agrobacterium* sp. genes whose products either directly or indirectly affect plant hormone levels. Potatoes have been successfully transformed with the *ipt* (*tmr*) gene of *Agrobacterium tumefaciens* as well as the *rol* A, *rol* B and *rol* C genes of *Agrobacterium rhizogenes* alone and in various combinations using both constitutive and inducible promoters (Schmülling *et al.*, 1993; Walden *et al.*, 1997). The *ipt* (*tmr*) gene codes for a bacterial enzyme that catalyses the formation of IPA-5′-monophosphate from adenosine 5′ monophosphate (AMP) and isopentenyl pyrophosphate (IPP) (Klee *et al.*, 1987; Walden *et al.*, 1997). Although plants transformed by the *rol* genes exhibit numerous morphological abnormalities suggestive of certain hormonal imbalances, the biochemical bases for these phenotypic alterations are currently unknown (Schmülling *et al.*, 1993).

With respect to tuber development, the expression of several of these genes (notably *ipt* and *rol* C) resulted in abnormal tuber morphology (Ooms and Lenton, 1985; Fladung, 1990; Aksenova *et al.*, 1999). In most cases, the phenotypic alterations reported in these studies have not included tuber dormancy characteristics. However, in potatoes transformed with the *ipt* gene, precocious sprouting of tubers was noted (Ooms and Lenton, 1985). This observation is consistent with the effects of exogenous cytokinins on tuber dormancy and lends further support to the hypothesis that endogenous cytokinins are cognate dormancy-terminating agents. To date, studies examining the effects of over- and under-expression of other genes involved in hormone metabolism or action on tuber dormancy have not been reported.

Once the principal hormones involved in tuber dormancy have been identified, the task will turn to understanding the biochemical factors regulating hormone content and turnover. A more difficult task will then be to determine the cell and molecular biology of hormone action in this system. The rapid progress in this area using model systems such as *Arabidopsis* will certainly facilitate this effort.

Although the challenges are great, the potential rewards are even greater. As a developmental process, dormancy affects nearly every aspect of agriculture. A greater understanding of the internal factors that regulate this process will certainly assist in the development of novel strategies to manipulate dormancy for the benefit of producers and consumers alike.

References

Aksenova, N.P., Konstantinova, S.A., Golyanovskaya, S.A., Schmülling, T., Kossmann, J., Willmitzer, L. and Romanov, G.A. (1999) *In vitro* growth and tuber formation by transgenic potato plants harboring *rolC* or *rolB* genes under control of the patatin promoter. *Russian Journal of Plant Physiology* 46, 513–519.

Alam, S.M.M., Murr, D.P. and Kristof, L. (1994) The effect of ethylene and of inhibitors of protein and nucleic acid syntheses on dormancy break and subsequent sprout growth. *Potato Research* 37, 25–33.

Bamberg, J.B. and Hanneman, R.E. (1991) Characterization of a new gibberellin related dwarfing locus in potato (*Solanum tuberosum* L.). *American Potato Journal* 68, 45–52.

Banas, A., Bielinska-Czarnecka, M. and Klocek, J. (1984) Activity of endogenous cytokinins in potato tubers during dormancy and sprouting. *Bulletin Polish Academy of Sciences* (*Biological Science*) 32, 37–40.

Bialek, K. and Bielinska-Czarnecka, M. (1975) Gibberellin-like substances in potato tubers during their growth and dormancy. *Bulletin Academy of Polish Sciences* (*Biological Science*) 23, 213–218.

Burton, W.G. (1963) Concepts and mechanism of dormancy. In: Ivins, J.D. and Milthorpe, F.L. (eds) *The Growth of the Potato.* Butterworth, London, pp. 17–41.

Burton, W.G. (1989) *The Potato,* 3rd edn. John Wiley & Sons, New York.

Burton, W.G. and Meigh, D.F. (1971) The production of growth-suppressing volatile substances by stored potato tubers. *Potato Research* 14, 96–101.

Campbell, M.A., Suttle, J.C. and Sell, T.W. (1996) Changes in cell cycle status and expression of $p34^{cdc2}$ kinase during potato tuber meristem dormancy. *Physiologia Plantarum* 98, 743–752.

Choi, D. and Bostock, R.M. (1994) Involvement of de novo protein synthesis, protein kinase, extracellular Ca^{2+}, and lipoxygenase in arachadonic acid induction of 3-hydroxy-3-methylglutaryl coenzyme A reductase genes and isoprenoid accumulation in potato (*Solanum tuberosum* L.). *Plant Physiology* 104, 1237–1244.

Coleman, W.K. (1987) Dormancy release in potato tubers: a review. *Potato Research* 14, 96–101.

Coleman, W.K. and King, R.R. (1984) Changes in endogenous abscisic acid, soluble sugars and proline levels during tuber dormancy in *Solanum tuberosum* L. *American Potato Journal* 61, 437–449.

Crech, D.L., Workman, M. and Harrison, M.D. (1973) Influence of storage factors on endogenous ethylene production by potato tubers. *American Potato Journal* 64, 145–150.

Cvikrova, M., Sukhova, L.S., Eder, J. and Korableva, N.P. (1994) Possible involvement of abscisic acid, ethylene and phenolic acids in potato tuber dormancy. *Plant Physiology and Biochemistry* 32, 685–691.

Denny, F.E. (1926) Hastening the sprouting of dormant potato tubers. *American Journal of Botany* 13, 118–125.

Desire, S., Couillerot, J.-P., Hilbert, J.-L. and Vasseur, J. (1995) Protein changes in *Solanum tuberosum* during storage and dormancy breaking of *in vitro* microtubers. *Plant Physiology and Biochemistry* 33, 479–487.

Dixon, R.A., Harrison, M.J. and Lamb, C.J. (1994) Early events in the activation of plant defense responses. *Annual Reviews of Plant Pathology* 32, 479–501.

El-Antably, H.M.M., Wareing, P.F. and Hillman, J. (1967) Some physiological responses to D,L- abscisin (dormin). *Planta* 73, 74–90.

Fladung, M. (1990) Transformation of diploid and tetraploid potato clones with the *rol* C gene of *Agrobacterium rhizogenes* and characterization of transgenic plants. *Plant Breeding* 104, 295–304.

Franklin, J. and Hemberg, T. (1980) Seasonal changes in the different components of the inhibitor β complex in potato tubers. *Physiologia Plantarum* 50, 227–232.

Guthrie, J.D. (1939) Inhibition of the growth of buds of potato tubers with the vapor of the methyl ester of naphthaleneacetic acid. *Contributions of the Boyce Thompson Institute* 11, 261–270.

Hemberg, T. (1949) Significance of growth-inhibiting substances and auxins for the rest-period of the potato tuber. *Physiologia Plantarum* 2, 24–36.

Hemberg, T. (1952) The significance of the acid growth-inhibiting substances for the rest-period of the potato tuber. *Physiologia Plantarum* 5, 115–129.

Hemberg, T. (1970) The action of some cytokinins on the rest-period and the content of acid growth-inhibiting substances in potato. *Physiologia Plantarum* 23, 850–858.

Hemberg, T. (1985) Potato rest. In: Li, P.H. (ed.) *Potato Physiology*. Academic Press, New York, pp. 353–388.

Holst, U.-B. (1971) Some properties of inhibitor β from *Solanum tuberosum* compared to abscisic acid. *Physiologia Plantarum* 24, 392–396.

Housley, S. and Taylor, W.C. (1958) Studies on plant-growth hormones. VI. The nature of inhibitor β in potato. *Journal of Experimental Botany* 9, 458–471.

Ji, Z.L. and Wang, S.Y. (1988) Reduction of abscisic acid content and induction of sprouting in potato, *Solanum tuberosum* L. by thidiazuron. *Journal of Plant Growth Regulation* 7, 37–44.

Jones, M.G., Horgan, R. and Hall, M.A. (1988) Endogenous gibberellins in the potato, *Solanum tuberosum. Phytochemistry* 27, 7–10.

Klee, J., Horsch, R. and Rogers, S. (1987) *Agrobacterium*-mediated plant transformation and its further applications to plant biology. *Annual Review of Plant Physiology* 38, 467–486.

Koda, Y. (1982) Effects of storage temperature and wounding on cytokinin levels in potato tubers. *Plant Cell and Physiology* 23, 851–857.

Korableva, N.P., Ladyzhenskaya, E.P. and Metlitskii, L.V. (1976) Synthesis of protein and RNA during the transition of meristematic tissues of potato tubers from dormancy to growth. *Biokhimiya* 44, 1181–1187.

Krijthe, N. (1962) Observations on the sprouting of seed potatoes. *European Journal of Potato Research* 5, 316–333.

Lang, G.A., Early, J.D., Martin, G.C. and Darnell, R.L. (1987) Endo-, para-, and ecodormancy: physiological terminology and classification for dormancy research. *HortScience* 22, 371–377.

Lesham, B. and Clowes, F.A.L. (1972) Rates of mitosis in shoot apices of potatoes at the beginning and end of dormancy. *Annals of Botany* 36, 687–691.

Lewis, C.E., Walker, J.R.L., Lancaster, J.E. and Sutton, K.H. (1998) Determination of anthocyanins, flavonoids and phenolic acids in potatoes. I. Coloured cultivars of *Solanum tuberosum* L. *Journal Science of Food and Agriculture* 77, 45–57.

Lewis, M.D., Kleinkopf, G.E. and Shetty, K.K. (1997) Dimethylnaphthalene and diisopropylnaphthalene for potato sprout control in storage: 1. Application methodolgy and efficacy. *American Potato Journal* 74, 183–197.

Macdonald, M.M. and Osborne, D.J. (1988) Synthesis of nucleic acids and protein in tuber buds of *Solanum tuberosum* during dormancy and early sprouting. *Physiologia Plantarum* 73, 392–400.

Meigh, D.F., Arthur, A., Filmer, E. and Self, R. (1973) Growth-inhibitory volatile aromatic compounds produced by *Solanum tuberosum* tubers. *Phytochemistry* 12, 987–993.

Milborrow, B.V. (1967) The identification of (+) abscisin II ((+) dormin) in plants and measurement of its concentrations. *Planta* 76, 93–113.

Mironov, V., De Veylder, L., Van Montagu, M. and Inze, D. (1999) Cyclin-dependent kinases and cell division in plants – the nexus. *Plant Cell* 11, 509–521.

Nigg, E.A. (1995) Cyclin-dependent protein kinases: key regulators of the eukaryotic cell cycle. *Bioessays* 17, 471–480.

Okazawa, Y. (1974) A relation between ethylene evolution and sprouting of potato tuber. *Journal Faculty of Agriculture Hokkaido University* 57, 443–454.

Ooms, G. and Lenton, J.R. (1985) T-DNA genes to study plant development: precocious tuberisation and enhanced cytokinins in *A. tumefaciens* transformed potato. *Plant Molecular Biology* 5, 205–212.

Poapst, P.A., Durkee, A.B., McGugan, W.A. and Johnston, F.B. (1968) Identification of ethylene in gibberellic-acid-treated potatoes. *Journal Science of Food and Agriculture* 19, 325–327.

Prange, R.K., Kalt, W., Daniels-Lake, B.J., Liew, C.L., Page, R.T., Walsh, J.R., Dean, P. and Coffin, R. (1998) Using ethylene as a sprout control agent in stored Russet Burbank potatoes. *Journal of the American Society of Horticultural Science* 123, 463–469.

Quarrie, S.A. (1982) Droopy: a wilty mutant of potato deficient in abscisic acid. *Plant Cell and Environment* 5, 23–26.

Rappaport, L. and Wolf, N. (1969) The problem of dormancy in potato tubers and related structures. *Symposium Society of Experimental Biology* 23, 219–240.

Rosa, J.T. (1925) Shortening the rest period of potatoes with ethylene gas. *Potato Association of America, Potato News Bulletin* 2, 363–365.

Rylski, I., Rappaport, L. and Pratt, H.K. (1974) Dual effects of ethylene on potato dormancy and sprout growth. *Plant Physiology* 53, 658–662.

Schippers, P.A. (1956) *De Duur and de Rustperiode van een Veertigtall Aardappelrassen, Bewaard bij Verschillende Constante Temperaturen.* Stichting Aardappelbewaring, Serie A, no. 112, Wageningen.

Schmülling, T., Fladung, M., Grossmann, K. and Schell, J. (1993) Hormonal content and sensitivity of transgenic tobacco and potato plants expressing single *rol* genes of *Agrobacterium rhizogenes* T-DNA. *The Plant Journal* 3, 371–382.

Simmonds, N.W. (1964) The genetics of seed and tuber dormancy in the cultivated potato. *Heredity, London* 19, 489–504.

Smith, O.E. and Rappaport, L. (1960) Endogenous gibberellins in resting and sprouting potato tubers. *Advances Chemistry Series* 28, 42–48.

Sorce, C., Piaggesi, A., Ceccarelli, N. and Lorenzi, R. (1996) Role and metabolism of abscisic acid in potato tuber dormancy and sprouting. *Journal of Plant Physiology* 149, 548–552.

Sukhova, L.S., Machackova, I., Eder, J., Bibik, N.D. and Korableva, N.P. (1993) Changes in the levels of free IAA and cytokinins in potato tubers during dormancy and sprouting. *Biologia Plantarum* 35, 387–391.

Suttle, J.C. (1995) Postharvest changes in endogenous ABA levels and ABA metabolism in relation to dormancy in potato tubers. *Physiologia Plantarum* 95, 233–240.

Suttle, J.C. (1998a) Postharvest changes in endogenous cytokinins and cytokinin efficacy in potato tubers in relation to bud endodormancy. *Physiologia Plantarum* 103, 59–69.

Suttle, J.C. (1998b) Involvement of ethylene in potato microtuber dormancy. *Plant Physiology* 118, 843–848.

Suttle, J.C. (1998c) Effects of dormancy status on cytokinin efficacy and metabolism. *Plant Physiology* (Suppl.), Abstract 599, 128.

Suttle, J.C. and Hultstrand, J.F. (1994) Role of endogenous abscisic acid in potato microtuber dormancy. *Plant Physiology* 105, 891–896.

Suttle, J.C., Banowetz, G.M. and Huckle, L.L. (1997) Changes in *cis*-zeatin/*cis*-zeatin riboside levels and biological activities during postharvest storage of potato tubers. *Plant Physiology* 114, S-164.

Turnbull, C.G.N. and Hanke, D.E. (1985a) The control of bud dormancy in potato tubers. Evidence for the primary role of cytokinins and a seasonal pattern of changing sensitivity to cytokinin. *Planta* 165, 359–365.

Turnbull, C.G.N. and Hanke, D.E. (1985b) The control of bud dormancy in potato tubers. Measurement of the seasonal pattern of changing concentrations of zeatin-cytokinins. *Planta* 165, 366–376.

Van den Berg, J.H. and Ewing, E.E. (1991) Jasmonates and their role in plant growth and development, with special reference to the control of potato tuberization: a review. *American Potato Journal* 68, 781–794.

Van den Berg, J.H., Vreugdenhil, D., Ludford, P.M., Hillman, L.L. and Ewing, E.E. (1991) Changes in starch, sugar and abscisic acid contents associated with second growth in tubers of potato (*Solanum tuberosum* L.) one-leaf cuttings. *Journal of Plant Physiology* 139, 86–89.

Van den Berg, J.H., Davies, P.J., Ewing E.E. and Halinske, A. (1995) Metabolism of gibberellin A_{12} and A_{12}-aldehyde and the identification of endogenous gibberellins in potato (*Solanum tuberosum* ssp. *Andigena*) shoots. *Journal of Plant Physiology* 146, 459–466.

Van den Berg, J.H., Ewing, E.E., Plaisted, R.L., McMurry, S. and Bonierbale, M.W. (1996) QTL analysis of potato tuber dormancy. *Theoretical and Applied Genetics* 93, 317–324.

Van Es, A. and Hartmans, K.J. (1969) The influence of abscisin II and gibberellic acid on the sprouting of excised potato buds. *European Potato Journal* 12, 59–63.

Van Ittersum, M.K., Aben, F.B. and Keijzer, C.J. (1992) Morphological changes in tuber buds during dormancy and initial sprout growth of seed potatoes. *Potato Research* 35, 249–260.

Van Staden, J. and Brown, N.A.C. (1979) Investigation into the possibility that potato buds synthesize cytokinins. *Journal of Experimental Botany* 30, 391–397.

Walden, R., Reiss, B., Koncz, C. and Schell, J. (1997) The impact of Ti-plasmid-derived gene vectors on the study of the mechanism of action of phytohormones. *Annual Review of Phytopathology* 35, 45–66.

Wurr, D.C.E. and Allen, E.J. (1976) Effects of cold treatment on the sprout growth of three potato varieties. *Journal of Agricultural Science Cambridge* 86, 221–224.

Xu, X., van Lammeren, A.A.M., Vermeer, E. and Vreugdenhil, D. (1998) The role of gibberellin, abscisic acid, and sucrose in the regulation of potato tuber formation *in vitro*. *Plant Physiology* 117, 575–584.

Yoshihara, T., Omer, E.A., Koshino, H., Sakamura, S., Kekuta, Y. and Koda, Y. (1989) Structure of a tuber-inducing stimulus from potato leaves (*Solanum tuberosum* L.). *Agricultural and Biological Chemistry* 53, 2835–2837.

Zimmerman, P.W. and Hitchcock, A.E. (1939) Experiments with vapors and solutions of growth substances. *Contributions of the Boyce Thompson Institute* 10, 481–508.

Role of ABA, Gibberellins and Auxin in Dormancy and Dormancy Release of Tulip Bulbs

16

Marian Saniewski,[1] Ludwika Kawa-Miszczak,[1] Elzbieta Wegrzynowicz-Lesiak[1] and Hiroshi Okubo[2]

[1]Research Institute of Pomology and Floriculture, Pomologiczna 18, 96-100 Skierniewice, Poland; [2]Laboratory of Horticultural Science, Faculty of Agriculture, Kyushu University 46-01, Fukuoka 812-8581, Japan

Introduction

During the development of tulip flowers three phases can be distinguished (Boonekamp *et al.*, 1990): (i) the initiation and formation of a new sprout with flower (at high temperature); (ii) the internal preparation for stem elongation (at low temperature); (iii) the rapid elongation of the sprout (at high temperature). Tulip bulbs, with a terminal bud containing a complete flower, require 12–16 weeks of low temperature treatment for floral stalk elongation. This suggests a kind of dormancy that can be released by exposure to low temperature (Kamerbeek *et al.*, 1972). The duration of the cold treatment is a major factor determining stalk growth and flowering. Increasing the duration of low temperature treatment decreases the number of days from planting to flowering. Stem and leaf extension of cooled tulip bulbs is due almost entirely to the elongation of cells produced early in the development of the flower bud (Gilford and Rees, 1973).

In this chapter, most attention is paid to hormonal control of dormancy, dormancy release and shoot growth after bulb cooling. The status of endogenous hormones and the effect of exogenously applied plant growth regulators, gibberellins, auxins and abscisic acid (ABA), under *in vivo* and *in vitro* conditions, during dormancy and dormancy release are summarized. The possible role of auxins, gibberellins and ABA in dormancy and dormancy release of tulip bulbs is proposed.

Endogenous Plant Hormones during Dormancy and Dormancy Release of Tulip Bulbs

Hormonal status during dormancy and dormancy release in tulip bulbs was widely investigated by many authors and is generally presented in Fig. 16.1. It is well known that during the cooling of the tulip bulbs the amount of free gibberellins increases (Aung and De Hertogh, 1967, 1968; Aung *et al.*, 1971; Van Bragt, 1971; Syrtanova and Rakhimbayev, 1973; Aung and Rees, 1974; Rakhimbayev *et al.*, 1978; Hanks and Rees, 1980). Hanks and Rees (1980) found two peaks of activity of gibberellin-like substances in tulips sampled at intervals from October (planting time) until the following April (flowering time). The first one occurred in December or in early-January samples, before the cold requirement was completed, and the second peak occurred around the time of rapid shoot extension and flowering found in roots, scales, leaves, stems, daughter bulbs and flowers. Gibberellins, GA_1, GA_4, GA_9, GA_{12}, GA_{24} and GA_{34} were identified in both non-cold and cooled tulip bulb sprouts (Rebers, 1992; Rebers *et al.*, 1994b, 1995). GA_4 was the major gibberellin, while GA_1, GA_9 and GA_{34} were present in lower amounts. Rebers *et al.* (1995) suggested that there was no direct correlation between cold-stimulated growth and a change in endogenous gibberellin status in sprouts or basal plates of tulip bulbs during cold storage. Rebers *et al.* (1995) showed the higher level of GA_4, its inactivation product GA_{34} was present only in growing floral stalks of cooled bulbs, and the absence of a significant quantitative accumulation of GA_1. They suggested that GA_4 is probably involved in floral stalk elongation in tulip.

The presence of ABA in tulips is well documented (Syrtanova *et al.*, 1975; Aung and De Hertogh, 1979; Singh *et al.*, 1979; Terry *et al.*, 1982; Franssen and Voskens, 1992). Low temperature treatment decreases the

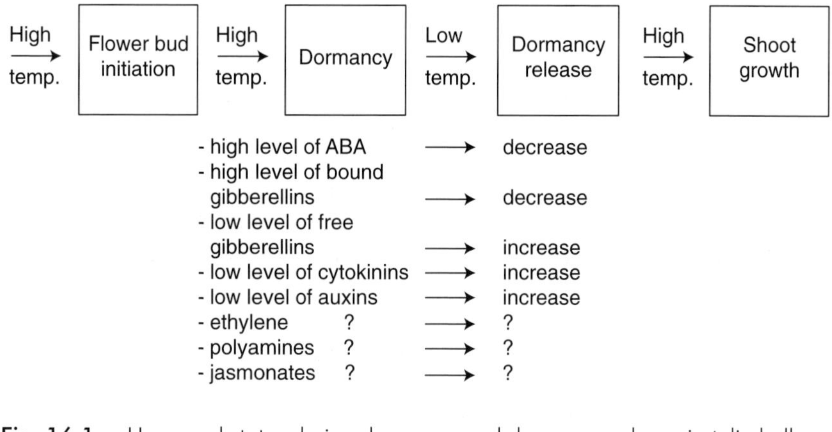

Fig. 16.1. Hormonal status during dormancy and dormancy release in tulip bulbs.

amount of ABA (Syrtanova *et al.*, 1973; Rakhimbayev *et al.*, 1978; Aung and De Hertogh, 1979). Aung and De Hertogh (1979) suggested that probable sites of ABA biosynthesis in the tulip bulb are the developing bulblets, basal plate and roots, since the ABA level is high in these organs and low in the scales and shoot.

Studies on the effect of low-temperature treatment on the level of auxins in the tulip bulb are very limited and preliminary (Franssen and Voskens, 1992).

Dormancy Release of Tulip Bulbs by Gibberellins and Prevention of the Process by ABA

Exogenously applied gibberellins partially substituted for the cold treatment of tulip bulbs and stimulated shoot growth and flowering (Van Bragt and Zijlstra, 1971; Van Bragt and Van Ast, 1976; Rudnicki *et al.*, 1976; Bylov and Smirnova, 1979; Cocozza Talia and Stellacci, 1979; Jones and Hanks, 1984; Hanks, 1984, 1985; Rebers *et al.*, 1994a). Thus, in tulips gibberellins seem to act as a dormancy-breaking agent.

The stimulatory effect of gibberellin on tulip shoot growth and flowering was well documented in non-cooled derooted tulip bulbs (Saniewski *et al.*, 1999a). In control non-cooled derooted bulbs of cv. 'Apeldoorn' and cv. 'Gudoshnik', flower bud blasting occurred and no growth of stem was observed. GA_3 at a concentration of 200 mg l^{-1} greatly induced shoot growth and flowering. The stimulatory effect of gibberellin on shoot growth and flowering in non-cooled derooted bulbs was much stronger than in the case of non-cooled rooted bulbs. It is probable that excision of all roots in non-cooled bulbs decreases the level of ABA, and exogenous gibberellic acid is more effective in comparison with non-cooled rooted tulip bulbs. Or it is also possible that better penetration and uptake of gibberellin take place after removal of roots.

Simultaneous application of ABA at a concentration of 20 mg l^{-1} with GA_3 showed that ABA greatly inhibited shoot growth (stem and leaves) and flowering induced by the gibberellin in non-cooled derooted tulip bulbs (Fig. 16.2) (Saniewski *et al.*, 1999b). Gibberellic acid treatment of cooled derooted tulip bulbs also evidently stimulated elongation of all internodes and flowering, when the leaves were both intact and excised during the duration of the experiments (Kawa-Miszczak *et al.*, 1992; Saniewski *et al.*, 1999c). Thus, it seems that roots in cooled tulip bulbs are the biosynthetic sites of gibberellins or they provide the necessary precursors for their biosynthesis. ABA applied together with GA_3 substantially inhibited shoot growth and delayed flowering of tulip in comparison with GA_3 treatment (Fig. 16.3) (Saniewski *et al.*, 1999c).

These results suggest that hormonal balance between gibberellins and ABA is the important factor for tulip shoot growth.

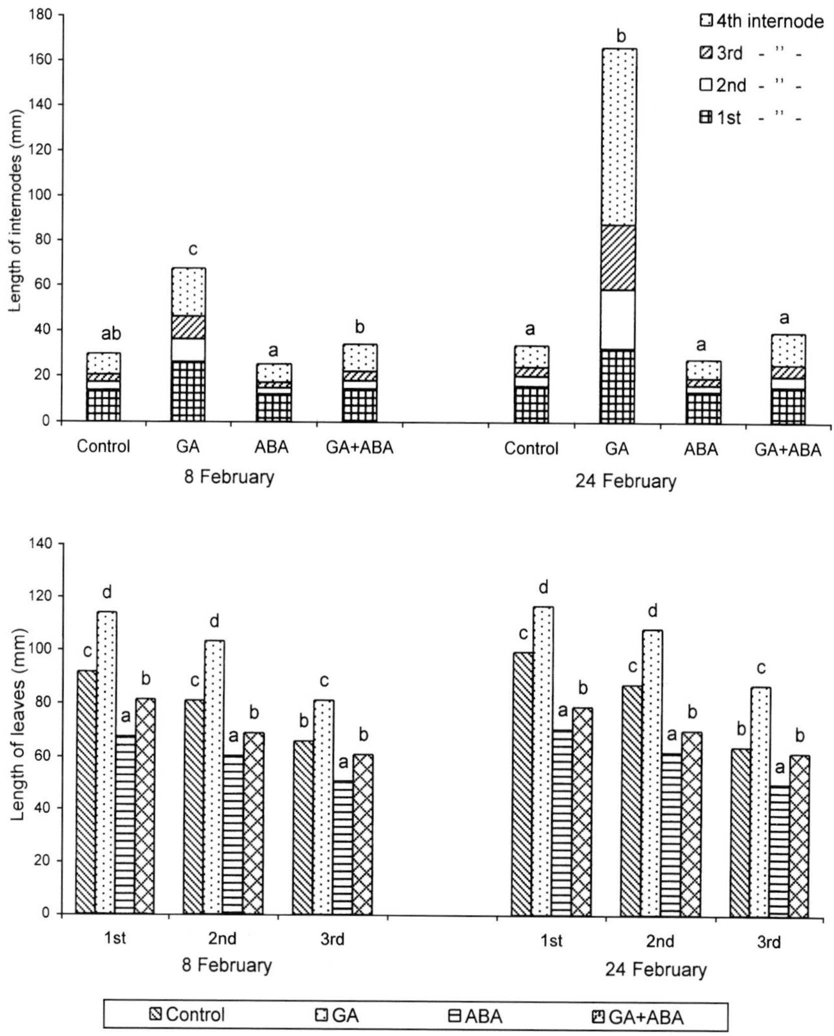

Fig. 16.2. Effect of GA_3 200 mg l^{-1}, ABA 20 mg l^{-1} and their mixture on stem (upper) and leaves (lower) growth of non-cooled derooted bulbs cv. 'Apeldoorn'. Treatments were made on 22 January and the measurements were made on 8 and 24 February. Different letters indicate significant differences according to Duncan's *t*-test ($P = 0.05$), values are calculated separately for each day of measurements; for total stem length (upper), and separately for each leaf (lower) (based on data in Saniewski *et al.*, 1999b).

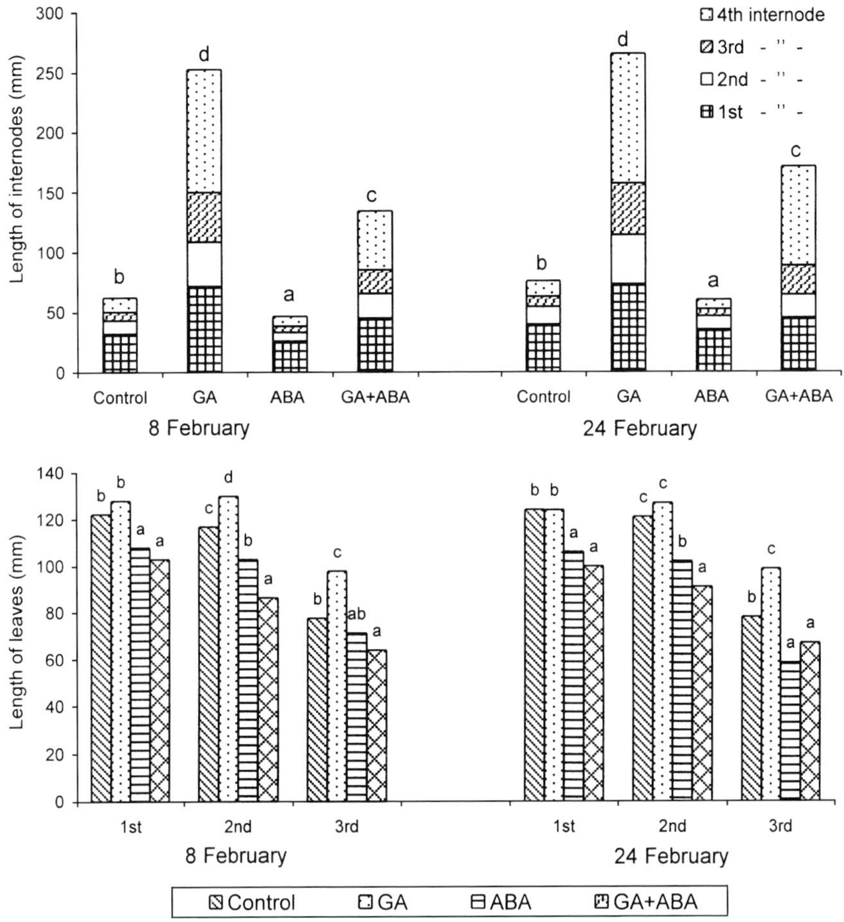

Fig. 16.3. Effect of GA_3 200 mg l^{-1}, ABA 20 mg l^{-1} and their mixture on stem (upper) and leaves (lower) growth of cooled derooted bulbs cv. 'Apeldoorn'. Treatments were made on 22 January and the measurements were made on 8 and 24 February. Different letters indicate significant differences according to Duncan's *t*-test ($P = 0.05$), values are calculated separately for each day of measurements; total stem length (upper) and separately for each leaf (lower) (based on data in Saniewski *et al.*, 1999c).

Auxin Induces Stem Growth in Cooled and Non-cooled Tulip Bulbs: the Role of Gibberellins and ABA in the Process

The leaves and gynoecium provide auxins which control the elongation of the stem in tulip (Op den Kelder *et al.*, 1971; Hanks and Rees, 1977; Saniewski and De Munk, 1981; Banasik and Saniewski, 1985). Excision of all leaves and flower buds in cooled bulbs almost totally inhibited stem growth in tulips. The elongation of all internodes in cooled tulip bulbs is induced by application of auxins in the place of the removed flower bud and after excision of all leaves (Fig. 16.4) (Saniewski and De Munk, 1981; Banasik and Saniewski, 1985). A similar auxin treatment also induced stem growth in cooled derooted tulip bulbs (Saniewski and Okubo, 1997). Saniewski and Wegrzynowicz-Lesiak (1993) documented that the continuous supply of exogenous auxin is necessary for tulip stem elongation after removal of the flower bud and leaves. It has been suggested that the elongation of all internodes in tulips is controlled by the interaction of endogenous auxins and gibberellins (Okubo and Uemoto, 1985, 1986; Saniewski, 1989; Saniewski and Kawa-Miszczak, 1992). Saniewski (1989) has also suggested that gibberellins produced during the cooling of bulbs play an important role in the flower bud development; other gibberellins are synthesized

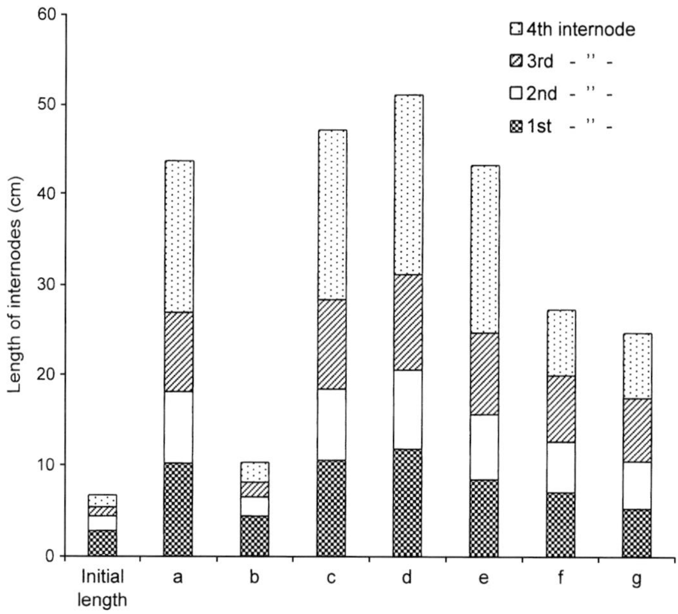

Fig. 16.4. The effect of gibberellins and IAA on tulip stem elongation. Treatments, after excision of all leaves and flower bud, were made on 12 February, when the initial length of stem was about 8.0 cm. Measurements were made on 25 February. a, intact plants; b–g, all leaves and flower bud were removed; b, control, lanolin only in the place of flower bud; c, IAA 0.1% in the place of flower bud; d, IAA 0.1% in the place of flower bud and soaked in GA_3 cotton wick wrapped around entire stem; e, IAA 0.1% in the place of flower bud and soaked in GA_{4+7} cotton wick wrapped around entire stem; f, soaked in GA_3 cotton wick wrapped around entire stem; g, soaked in GA_{4+7} cotton wick wrapped around entire stem; LSD (P=0.05) values for internodes (treatments a–g) are as follows: 1st 0.97 cm, 2nd 0.84 cm; 3rd 0.85 cm, 4th 1.38 cm (based on data in Saniewski *et al.*, 1990b).

during shoot growth and, together with auxin, control stem elongation in tulips. Tulip stem growth, induced by auxin (IAA) applied in the place of the removed flower bud after excision of all leaves, was substantially inhibited by paclobutrazol, an inhibitor of gibberellin biosynthesis (Saniewski, 1989).

It was found that GA_3 and GA_{4+7} applied in soaked cotton wick wrapped around all the internodes of the tulip stem, after excision of all leaves and the flower bud, evidently induced stem elongation (Fig. 16.4) (Saniewski *et al.*, 1990b). The growth of the tulip stem induced by the gibberellins was less than after IAA application in the place of the removed flower bud. It is suggested that exogenous gibberellins taken up by stem tissues, together with endogenous auxin present in the stem or synthesized under the influence of gibberellins, induce elongation of all internodes.

The stem growth of tulips induced by IAA applied in the place of the removed flower bud after excision of all leaves was evidently inhibited by ABA applied by soaking cotton wick wrapped around the entire stem (Fig. 16.5) (Saniewski *et al.*, 1990a). IAA applied in the place of the removed flower bud and after excision of leaves also induced stem growth in non-cooled rooted and derooted tulip bulbs, but to a much smaller degree than in the case of cooled bulbs (Saniewski and De Munk, 1981; Saniewski and Okubo, 1997). Auxin transport inhibitors, 2,3,5-triiodobenzoic acid (TIBA) and naphthylphthalamic acid (NPA) substantially inhibited stem growth induced by IAA both in cooled and non-cooled tulip bulbs (Saniewski and Okubo, 1997, 1998a,b).

The Effect of Auxin, Gibberellins and ABA on the Growth of Leaf, Stem and Pistil Isolated from Non-cooled and Cooled Tulip Bulbs

For better understanding of the hormonal control of growth processes in tulip, studies *in vitro* seem to be a very useful tool. In our experiments the effect of different growth regulators on the stem, leaf and pistil growth were investigated. These organs were excised from non-cooled and cooled tulip bulbs of cv. 'Gudoshnik'. In this system, apical dominance and correlations between organs are eliminated and we can observe the direct effect of growth substances on the growth of isolated organs.

Stalk explants were cultured in normal (basal end down) or inverted (basal end up) positions. Distinct stalk growth of explants isolated from non-cooled and cooled bulbs was observed when they were cultured in the inverted position, with the presence of IAA in the medium (Fig. 16.6) (Gabryszewska and Saniewski, 1983; Kawa and Saniewski, 1990c). Thus, only basipetal transport of IAA or its metabolites seems to induce stalk elongation in tulips. Also, distinct stalk growth of explants isolated from cooled bulbs was observed when they were cultured in normal and inverted

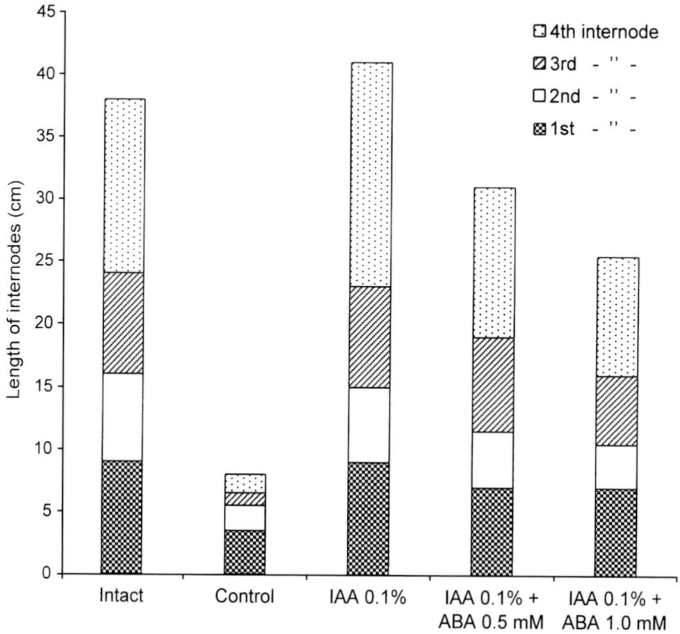

Fig. 16.5. The effect of ABA on stem elongation induced by IAA 0.1%. Treatments, after excision of all leaves and flower bud, were made on 22 January, when the initial length of stem was about 8.0 cm. Measurements were made on 2 February, when extension growth was completed. Control, lanolin only in the place of removed flower bud; IAA 0.1% in the place of removed flower bud; IAA 0.1% in the place of removed flower bud and ABA at concentrations 0.5 and 1.0 mM (twice) was applied through soaking cotton wick wrapped around entire stem; LSD ($P = 0.05$) values for internodes are as follows: 1st 0.96 cm, 2nd 0.76 cm, 3rd 0.88 cm, 4th 1.56 cm (based on data in Saniewski *et al.*, 1990a).

positions with the presence of GA in the medium (Kawa and Saniewski, 1990c). The fact that only stalk explants isolated from cooled bulbs are sensitive to GA treatment may indicate that gibberellins stimulate the auxin-release or response system after low temperature treatment. The addition of ABA on the medium alone and together with IAA or GA inhibited the growth of stalk explants (Fig. 16.6) (Saniewski and Gabryszewska, 1983; Kawa and Saniewski, 1990c). The growth of stalks isolated from cooled bulbs was much greater than the growth of stalks isolated from non-cooled tulip bulbs.

Leaf explants isolated from non-cooled bulbs on the earlier dates (September) and cultured on the control medium (without plant growth regulators) showed slight growth. However, the addition of gibberellic acid to the medium strongly stimulated an increase of length and fresh and dry weight of the explants, cultured in both normal and inverted positions. The

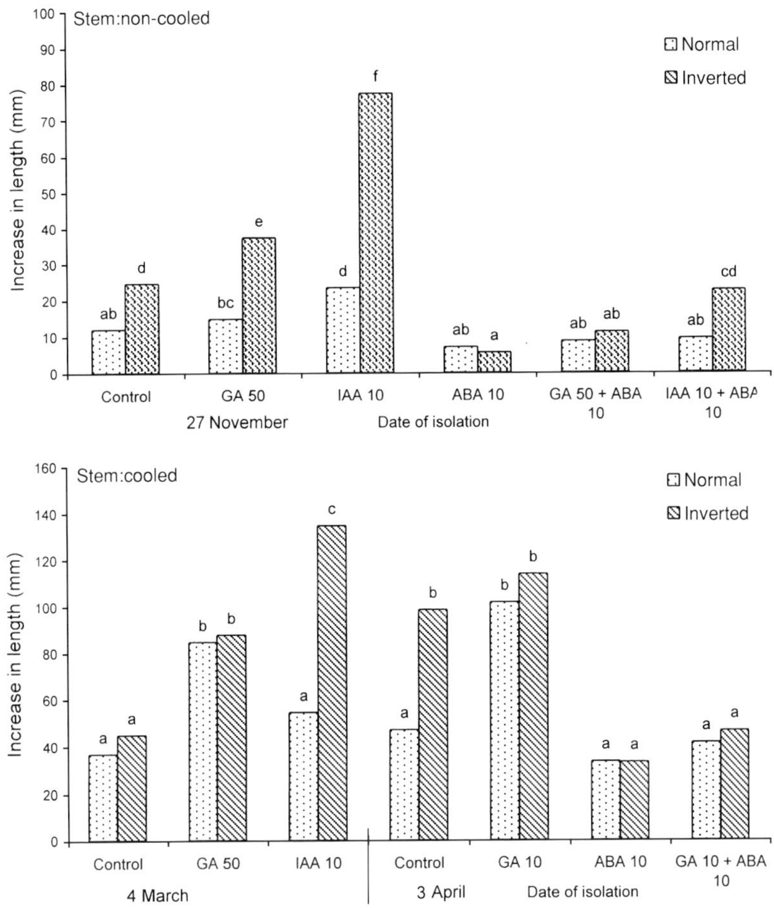

Fig. 16.6. The effect of plant growth regulators on the increase in length of stem explants isolated from non-cooled and cooled tulip bulbs of cv. 'Gudoshnik', cultured in the normal and in the inverted position in the dark. Different letters indicate significant differences according to Duncan's *t*-test ($P = 0.05$), values are calculated separately for each date of stem isolation (based on data in Kawa and Saniewski, 1990c).

stimulatory effect of GA was stronger in the light (Kawa and Saniewski, 1990b). On the later dates (November–December) the growth of control explants cultured in the dark was greater than in the previous experiment, but the stimulatory effect of GA was also observed (Fig. 16.7). The growth of leaves cultured in the light was large even on the control medium, and addition of GA to the medium did not affect leaf growth (Gabryszewska and Saniewski, 1984; Kawa and Saniewski, 1990b). The data indicate that gibberellins play an important role in leaf growth. Light had a stimulatory effect on leaf growth, possibly through the stimulatory action on gibberellin biosynthesis.

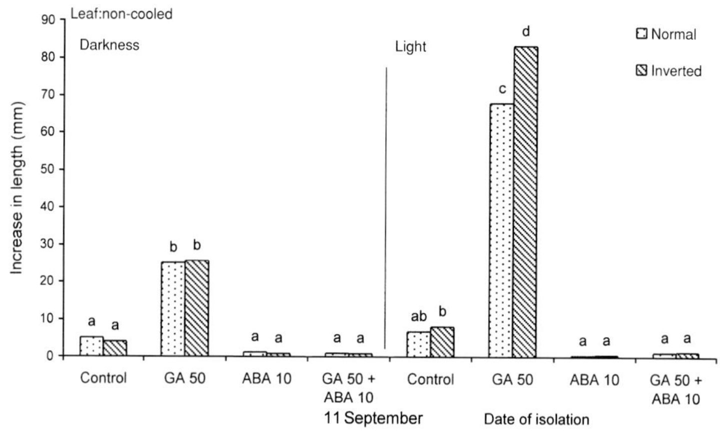
Leaf:non-cooled
Darkness
Light
Normal
Inverted
Increase in length (mm)
Control
GA 50
ABA 10
GA 50 + ABA 10
11 September
Date of isolation

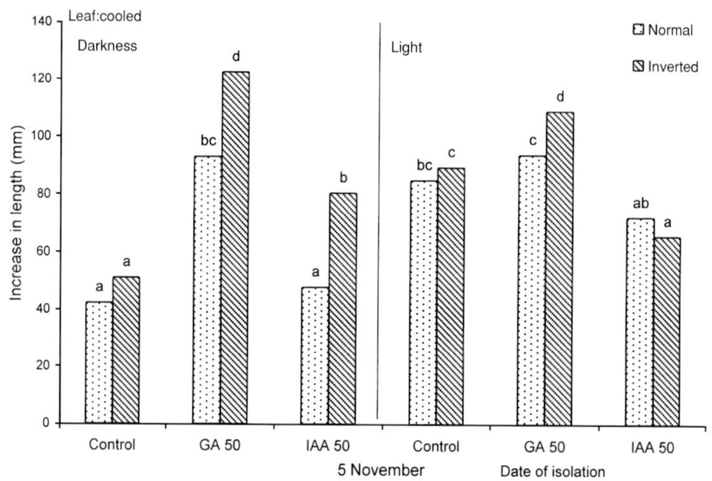
Leaf:cooled
Darkness
Light
Normal
Inverted
Increase in length (mm)
Control
GA 50
IAA 50
5 November
Date of isolation

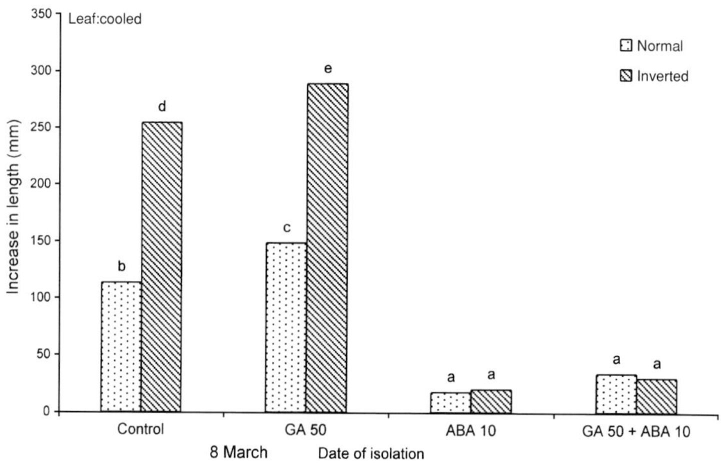
Leaf:cooled
Normal
Inverted
Increase in length (mm)
Control
GA 50
ABA 10
GA 50 + ABA 10
8 March
Date of isolation

The growth of explants isolated from cooled bulbs was much greater than in the case of explants isolated from non-cooled bulbs (Fig. 16.7). The increase in length and weight was very large even on the medium without GA, and the stimulatory effect of GA was weak; the endogenous level of gibberellins was probably sufficient for leaf elongation. The growth of leaf explants isolated from cooled bulbs and cultured in the inverted position was stronger than in those cultured in the normal position, both with and without GA treatments (Kawa and Saniewski, 1989, 1990b). It is possible that in the inverted position of leaf explants accumulation of basipetally transported substances, including auxin, takes place and auxin together with gibberellins stimulate leaf growth. Similar growth of the leaves isolated from cooled bulbs was observed in the dark and in the light.

It should be mentioned that the growth of leaf explants isolated from non-cooled and cooled bulbs is strongly inhibited by ABA exogenously applied to the medium (Fig. 16.7) (Gabryszewska and Saniewski, 1984; Kawa and Saniewski, 1990b). Thus, it seems that the presence of ABA in non-cooled tulip bulbs would inhibit the growth of leaves, and the cutting of leaves from the plant would cause leaf growth because the sources of growth inhibitors are eliminated.

Gibberellic acid greatly stimulated the growth of pistils isolated from non-cooled tulip bulbs and to a much lesser degree the growth of the organ isolated from cooled bulbs (Fig. 16.8) (Kawa and Saniewski, 1985, 1986, 1990a). However, the addition of GA together with IAA to the medium stimulated the growth of the pistil isolated from non-cooled bulbs, more than in the presence of only GA in the medium. The addition of abscisic acid to the medium alone or together with GA greatly inhibited the growth of the pistil explants from non-cooled and cooled bulbs (Fig. 16.8) (Kawa and Saniewski, 1990a).

Conclusions

High levels of free gibberellin and low content of abscisic acid as a result of cooling are the most important factors in the control of dormancy release of tulip bulbs. It is probable that gibberellins, produced during bulb cooling, stimulate flower bud development, mostly pistil and leaves, as a source of auxin biosynthesis. Another possibility is that gibberellins stimulate auxin release or auxin response systems in pistil and leaves. It is

Fig. 16.7. The effect of plant growth regulators on the increase in length of leaf explants isolated from non-cooled and cooled tulip bulbs of cv. 'Gudoshnik', cultured in the normal and in the inverted position, in the light and in the dark. Different letters indicate significant differences according to Duncan's *t*-test ($P = 0.05$), values are calculated separately for each date of leaf isolation (based on data in Kawa and Saniewski, 1990b).

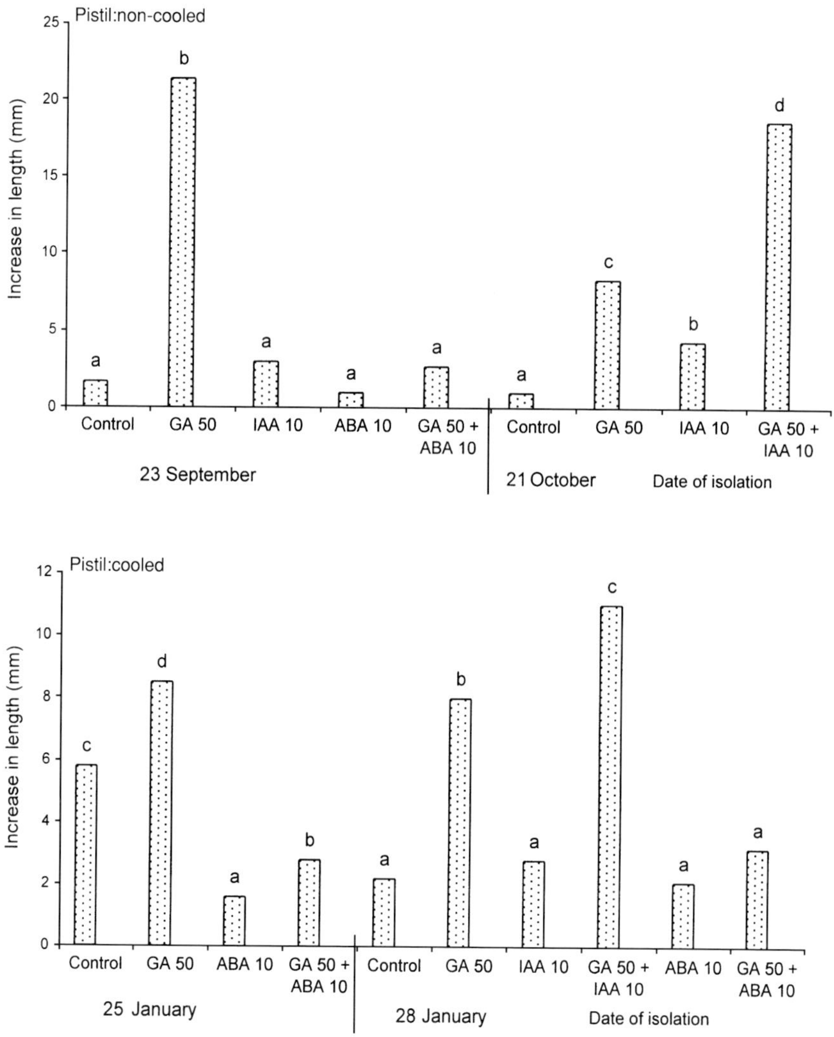

Fig. 16.8. The effect of plant growth regulators on the increase in length of pistil explants isolated from non-cooled and cooled tulip bulbs of cv. 'Gudoshnik', cultured in the dark. Different letters indicate significant differences according to Duncan's *t*-test ($P = 0.05$), values are calculated separately for each date of pistil isolation (based on data in Kawa and Saniewski, 1990a).

also well known that gibberellins increase auxin transport and enhance auxin biosynthesis (see Saniewski, 1989). During tulip sprouting, the leaves play an important role in the flower bud development (Saniewski, 1986). Excision of all leaves, when the length of stem was about 4.0 cm fully inhibited stem elongation and flower bud blasting occurred. It is possible

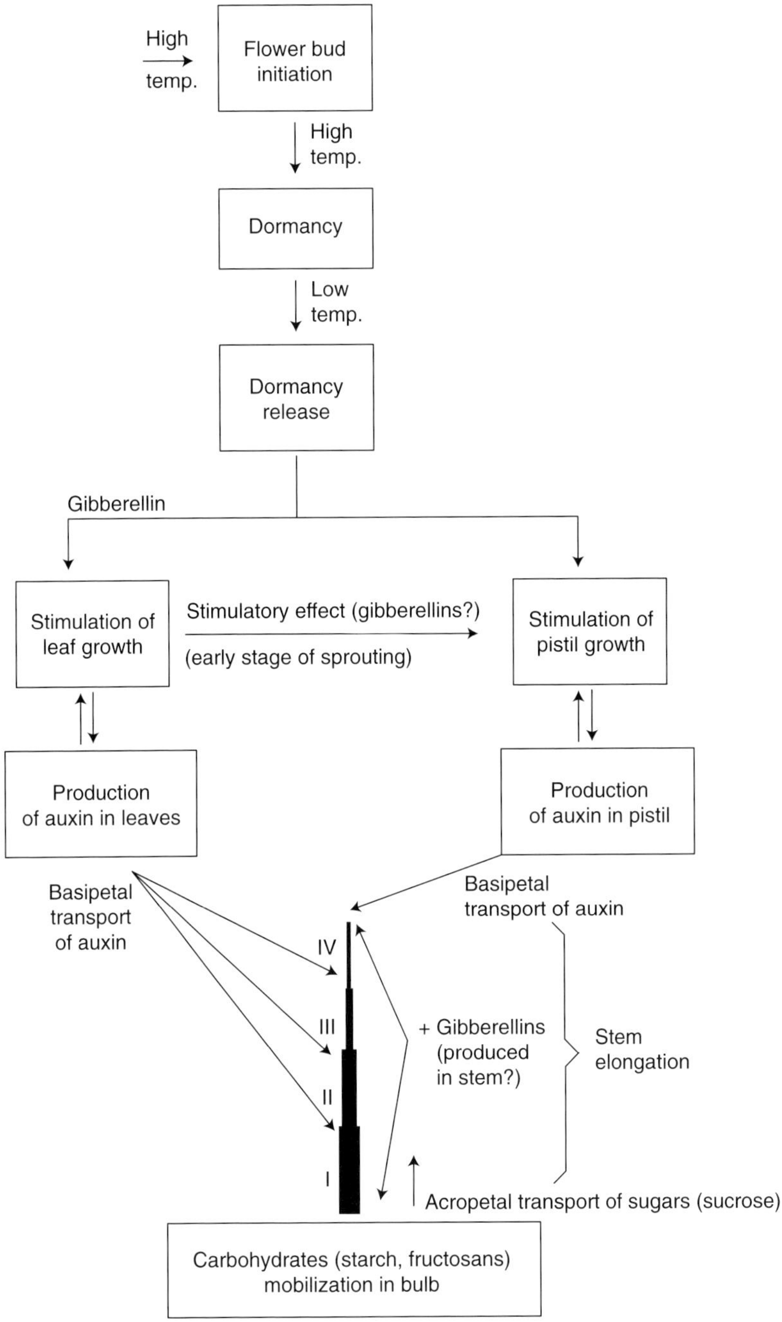

Fig. 16.9. Simple model scheme of hormonal control of tulip growth and development.

that the leaves also provide gibberellins, which stimulate the pistil growth (Kawa and Saniewski, 1986), and/or the young leaves, as a sink and producer of hormones, intensify the flow of substrates from the mother bulb to the upper part. These substrates can also reach the flower bud located close to the leaves (Saniewski, 1986). Excision of leaves, when the stem length was about 8.0 cm, caused normal elongation of the stem, natural perianth development, and the pistil growth was higher than in intact plants. It seems probable that competition between leaves and a flower bud, as sinks for substrate from mother bulbs, is eliminated and more substrate is supplied to the flower bud causing a stronger growth of the pistil. In conclusion it seems that gibberellins produced during the cooling of bulbs play an important role in flower bud development and leaves, and another pool of gibberellins are synthesized during shoot growth which, together with auxin, produced in pistil and leaves, control stem growth in tulips (Fig. 16.9).

References

Aung, L.H. and De Hertogh, A.A. (1967) The occurrence of gibberellin-like substances in tulip bulbs (*Tulipa* sp.). *Plant Cell Physiology* 8, 201–205.

Aung, L.H. and De Hertogh, A.A. (1968) Gibberellin-like substances in non-cold and cold treated tulip bulbs (*Tulipa* sp.). In: Wightman, F. and Setterfield, G. (eds) *Biochemistry and Physiology of Plant Growth Substances.* Runge Press, Ottawa, pp. 943–956.

Aung, L.H. and De Hertogh, A.A. (1979) Temperature regulation of growth and endogenous abscisic acid-like content of *Tulipa gesneriana* L. *Plant Physiology* 63, 1111–1116.

Aung, L.H. and Rees, A.R. (1974) Changes in endogenous gibberellin levels in *Tulipa* bulblets during ontogeny. *Journal of Experimental Botany* 25, 745–751.

Aung, L.H., De Hertogh, A.A. and Staby, G.L. (1971) The alteration of bulb hormones content by environmental stimuli. *Acta Horticulturae* 23, 156–161.

Banasik, L. and Saniewski, M. (1985) The effect of different auxins on tulip stalk elongation. *Acta Horticulturae* 167, 193–204.

Boonekamp, P.M., Beijersbergen, J.C.M. and Franssen, J.M. (1990) The development of flowering assays for cold-treated tulip bulbs. *Acta Horticulturae* 266, 177–181.

Bylov, W.N. and Smirnova, Z.I. (1979) Possibility of gibberellin application for early forcing *Tulipa gesneriana* cv. London. *Acta Horticulturae* 91, 173–177.

Cocozza Talia, M. and Stellacci, P. (1979) A research on the effect of gibberellin upon tulip flowering. *Acta Horticulturae* 91, 167–172.

Franssen, J.M. and Voskens, P.G.J.M. (1992) Methods to determine abscisic acid and indole-3-acetic acid, and determinations of these hormones in tulip cv. Apeldoorn bulbs as related to the cold treatment. *Acta Horticulturae* 325, 267–276.

Gabryszewska, E. and Saniewski, M. (1983) Auxin control of tulip stalk elongation *in vitro*. *Scientia Horticulturae* 19, 153–159.

Gabryszewska, E. and Saniewski, M. (1984) Hormonal control of tulip leaf elongation *in vitro*. *Acta Physiologiae Plantarum* 6, 19–25.

Gilford, J.McD. and Rees, A.R. (1973) Growth of the tulip shoot. *Scientia Horticulturae* 1, 143–156.

Hanks, G.R. (1984) Factors affecting the response of tulip to gibberellins. *Scientia Horticulturae* 23, 379–390.

Hanks, G.R. (1985) The response of 9°C-tulips to gibberellins. *Scientia Horticulturae* 27, 153–161.

Hanks, G.R. and Rees, A.R. (1977) Stem elongation in tulip and narcissus: the influence of floral organs and growth regulators. *The New Phytology* 78, 579–591.

Hanks, G.R. and Rees, A.R. (1980) Growth substances of tulip: the activity of gibberellin-like substances in field-grown tulips from planting until flowering. *Zeitschrift für Pflanzenphysiologie* 98, 213–223.

Jones, S.K. and Hanks, G.R. (1984) Treatment of tulips with gibberellic acid by vacuum infiltration. *Journal of Horticultural Science* 59, 241–252.

Kamerbeek, G.A., Beijersbergen, J.C.M. and Schenk, P.K. (1972) Dormancy in bulbs and corms. In: Goren, N. and Mehdel, K. (eds) *Proceedings of XVIIIth International Horticultural Congress, Tel Aviv*, vol. V, pp. 233–239.

Kawa, L. and Saniewski, M. (1985) Stimulatory effect of gibberellic acid on tulip pistil growth *in vitro*. *Acta Horticulturae* 167, 205–210.

Kawa, L. and Saniewski, M. (1986) The effect of gibberellic acid and abscisic acid on tulip pistil growth *in vitro*. *Acta Horticulturae* 177, 129–133.

Kawa, L. and Saniewski, M. (1989) The effect of gibberellic acid on tulip leaf growth *in vitro*. *Acta Horticulturae* 251, 93–100.

Kawa, L. and Saniewski, M. (1990a) The effect of plant growth regulators on tulip pistil growth *in vitro* (in Polish). *Prace Instytutu Sadownictwa i Kwiaciarstwa, Seria B – Rosliny Ozdobne* 15, 65–72.

Kawa, L. and Saniewski, M. (1990b) The effect of plant growth regulators on tulip leaf growth *in vitro* (in Polish). *Prace Instytutu Sadownictwa i Kwiaciarstwa, Seria B – Rosliny Ozdobne* 15, 73–82.

Kawa, L. and Saniewski, M. (1990c) The effect of plant growth regulators on tulip stalk elongation *in vitro* (in Polish). *Prace Instytutu Sadownictwa i Kwiaciarstwa, Seria B – Rosliny Ozdobne* 15, 83–94.

Kawa-Miszczak, L., Wegrzynowicz-Lesiak, E. and Saniewski, M. (1992) The effect of removal of roots and application of plant growth regulators on tulip shoot growth. *Acta Horticulturae* 325, 71–76.

Okubo, H. and Uemoto, S. (1985) Changes in endogenous gibberellin and auxin activities during first internode elongation in tulip flower stalk. *Plant Cell Physiology* 26, 709–719.

Okubo, H. and Uemoto, S. (1986) Changes in endogenous gibberellin during internode elongation of tulip flower stalk as affected by light conditions. *Journal of the Japanese Society for Horticultural Science* 55, 215–220.

Op den Kelder, M., Benschop, M. and De Hertogh, A.A. (1971) Factors affecting floral stalk elongation of flowering tulips. *Journal of the American Society for Horticultural Science* 96, 603–605.

Rakhimbayev, I.R., Syrtanova, G.A. and Solomina, V.F. (1978) The effect of cold treatment on the level of biological activity of endogenous growth regulators of tulip bulbs (in Russian). *Plant Physiology (Russian)* 25, 249–253.

Rebers, M. (1992) The role of gibberellins in the cold requirement of tulip. *Acta Horticulturae* 325, 253–258.

Rebers, M., Romeijn, G., Knegt, E. and Van der Plas, L.H.W. (1994a) Effects of exogenous gibberellins and paclobutrazol on floral stalk growth of tulip sprouts isolates from cooled and non-cooled tulip bulbs. *Physiologia Plantarum* 92, 661–667.

Rebers, M., Vermeer, E., Knegt, E., Shelton, C.J. and Van der Plas, L.H.W. (1994b) Gibberellins in tulip bulb sprouts during storage. *Phytochemistry* 36, 269–272.

Rebers, M., Vermeer, E., Knegt, E., Shelton, C.J. and Van der Plas, L.H.W. (1995) Gibberellins level and cold-induced floral stalk elongation in tulip. *Physiologia Plantarum* 94, 687–691.

Rudnicki, R.M., Nowak, J. and Saniewski, M. (1976) The effect of gibberellic acid on sprouting and flowering of some tulip cultivars. *Scientia Horticulturae* 4, 387–397.

Saniewski, M. (1986) Differential action of IAA and NAA, dependent on the transport direction, on perianth and pistil growth in tulips. *Acta Horticulturae* 177, 135–140.

Saniewski, M. (1989) The use of paclobutrazol, an inhibitor of gibberellin biosynthesis, for study of hormonal control of tulip stem elongation. *Bulletin of the Polish Academy of Sciences, Biological Sciences* 37, 55–64.

Saniewski, M. and De Munk, W.J. (1981) Hormonal control of shoot elongation in tulips. *Scientia Horticulturae* 15, 363–372.

Saniewski, M. and Gabryszewska, E. (1983) Hormonal control of tulip stalk elongation *in vitro*. *Acta Horticulturae* 131, 275–282.

Saniewski, M. and Kawa-Miszczak, L. (1992) Hormonal control of growth and development of tulips. *Acta Horticulturae* 325, 43–54.

Saniewski, M. and Okubo, H. (1997) Auxin induces stem elongation in non-precooled and precooled derooted and rooted tulip bulbs. *Journal of the Faculty of Agriculture, Kyushu University* 42, 53–61.

Saniewski, M. and Okubo, H. (1998a) Inhibitory effect of naphthylphthalamic acid (NPA) on stem growth induced by auxin in precooled tulip bulbs. *Journal of the Faculty of Agriculture, Kyushu University* 43 (1–2), 59–66.

Saniewski, M. and Okubo, H. (1998b) Effects of 2,3,5-triiodobenzoic acid (TIBA) on stem growth induced by indole-3-acetic acid (IAA) and naphthylacetic acid (NAA) in precooled rooted tulip bulbs. *Journal of the Faculty of Agriculture, Kyushu University* 43 (1–2), 11–23.

Saniewski, M. and Wegrzynowicz-Lesiak, E. (1993) Continuous supply of auxin is necessary for tulip stem growth. *Journal of Fruit and Ornamental Plant Research* 12, 59–66.

Saniewski, M., Kawa, L. and Wegrzynowicz, E. (1990a) The effect of abscisic acid on pistil and stem growth in tulips. *Prace Instytutu Sadownictwa i Kwiaciarstwa, Seria B – Rosliny Ozdobne* 15, 95–104.

Saniewski, M., Kawa, L. and Wegrzynowicz, E. (1990b) Stimulatory effect of gibberellins on tulip stem elongation. *Prace Instytutu Sadownictwa i Kwiaciarstwa, Seria B – Rosliny Ozdobne* 15, 104–112.

Saniewski, M., Kawa-Miszczak, L., Wegrzynowicz-Lesiak, E. and Okubo, H. (1999a) Gibberellin induces shoot growth and flowering in nonprecooled derooted bulbs of tulip (*Tulipa gesneriana* L.). *Journal of Faculty of Agriculture, Kyushu University* 43, 411–418.

Saniewski, M., Kawa-Miszczak, L., Wegrzynowicz-Lesiak, E. and Okubo, H. (1999b) Inhibitory effect of abscisic acid on shoot growth and flowering induced by gibberellic acid in non-cooled derooted bulbs of tulip (*Tulipa gesneriana* L.). *Journal of Faculty of Agriculture, Kyushu University* 44, 25–32.

Saniewski, M., Kawa-Miszczak, L., Wegrzynowicz-Lesiak, E. and Okubo, H. (1999c) Abscisic acid inhibited shoot growth induced by gibberellic acid in cooled derooted bulbs of tulip (*Tulipa gesneriana* L.). *Journal of Fruit and Ornamental Plant Research* (in press).

Singh, B.N., Galston, E., Dashek, W. and Walton, D.C. (1979) Abscisic acid levels and metabolism in the leaf epidermal tissue of *Tulipa gesneriana* L. and *Commelina communis* L. *Planta* 146, 135–238.

Syrtanova, G.A. and Rakhimbayev, I.R. (1973) Endogenous gibberellins during dormancy and growth of tulip bulbs (in Russian). *Plant Physiology (Russian)* 20, 721–724.

Syrtanova, G.A., Turetzkaja R. and Rakhimbayev, I.R. (1973) Natural auxins and inhibitors in dormant and growing tulip bulbs (in Russian). *Plant Physiology (Russian)* 20, 1133–1136.

Syrtanova, G.A., Rakhimbayev, I.R. and Kefeli, W.I. (1975) Native growth inhibitor of tulip (in Russian). *Plant Physiology (Russian)* 22, 165–169.

Terry, P.H., Aung, L.H. and De Hertogh, A.A. (1982) Identification of abscisic acid in *Tulipa gesneriana* L. by gas–liquid chromatography and mass spectroscopy. *Plant Physiology* 70, 1574–1576.

Van Bragt, J. (1971) Endogenous gibberellin levels and floral stalk elongation in tulip cv. Apeldoorn. *Mededelingen Faculteit Landbouwwetenschappen Gent* 36, 1301–1307.

Van Bragt, J. and Van Ast, K.J. (1976) Substitution of the cold requirement of tulip cv. Apeldoorn by GA_3. *Scientia Horticulturae* 4, 117–122.

Van Bragt, J. and Zijlstra, F.A. (1971) Effects of gibberellins on flowering of tulip cv. Apeldoorn. *Zeitschrift für Pflanzenphysiologie* 64, 139–144.

17

ATP, ADP and NTP Contents in Vegetative and Floral Peach Buds During Winter: Are They Useful for Characterizing the Type of Dormancy?

Marc Bonhomme,[1] Rémy Rageau[1] and Michel Gendraud[2]

[1]INRA, UA Bioclimatologie-PIAF, Domaine de Crouelle, F-63039 Clermont-Ferrand cedex 2, France; [2] UA Bioclimatologie-PIAF, Université Blaise Pascal, 24 Avenue des Landais, F-63177 Aubière cedex, France

Introduction

Energy metabolism and especially purine metabolism have been mentioned many times in discussions on endodormancy control (Gendraud, 1977; Le Floc'h and Lafleuriel, 1981; Le Floc'h, 1984; Gendraud and Petel, 1990; Le Floc'h and Faye, 1995; Lecomte *et al.*, 1998). The unique test of endodormancy available today, the 'nucleotides' test (NT), is based on the functioning of this metabolism. It compares the capacity of synthesis of ATP and non-adenylic tri-phosphate nucleotides (NTP) in the primordia contained inside the bud with or without a supply of adenosine, a precursor of this metabolism (Gendraud, 1975). A significant increase of the NTP pools obtained after incubation on adenosine compared with incubation on water characterizes an absence of endodormancy. This test has been applied, with pertinence, to the study of vegetative bud dormancy of various woody species (oak, Barnola *et al.*, 1986; ash, Lavarenne *et al.*, 1982; peach tree, Balandier, 1992; Balandier *et al.*, 1993). Unfortunately, it is not operative on floral primordia (Bonhomme, 1998; Bonhomme *et al.*, 1999) and, as in all tests, the test design generates some constraints.

First, this test needs an incubation of buds over a relatively long time (16 h) at 10°C. So, the observed data may not exactly reflect the initial dormancy state of the primordia at the beginning of its incubation: the state of the primordia may have changed during the incubation.

(eds J.-D. Viémont and J. Crabbé)

Second, to allow comparisons, quantities of nucleotides measured have to be expressed on the same basis for the different buds (most often protein, sometimes DNA content). This requires supplementary assays, increasing duration, costs and risks of errors in obtaining results. It would therefore be interesting to define a method of characterization, relevant for both vegetative and floral primordia that avoids or reduces these disadvantages. We have tested the hypothesis of a close correlation between the state of endodormancy of the primordia, on one hand, and their contents in adenylic (ATP) and non-adenylic tri-phosphate nucleotides (NTP), which could be considered as instantaneous markers of their energy metabolism activity and, on the other hand, ADP content, the ATP/ADP ratio able to be a marker of the degree of phosphorylation coming from the oxidative phosphorylations.

The results were compared, for both vegetative and floral primordia, with typical results obtained for several years by different tests and measures allowing us to characterize the different dormancies defined by Lang *et al.* (1987): endo-, para- and ecodormancy.

Plant Material and Treatments

The ATP, ADP and NTP characterization was obtained on 'Redhaven' peach trees, grafted on GF305, planted in 1990 and grown thereafter in an orchard at the laboratory for the natural conditions treatment, during the winters of 1994–1995 and 1995–1996. The complete cold deprivation treatment used to prevent dormancy release was applied to 15 peach trees, issued from identical scions but planted in 220-L containers, transferred on 13 October 1994 in a heated greenhouse with temperature maintained above 15°C.

Meteorological and Phenological Characterization of the 2 Years of ATP, ADP and NTP Measurements

In 1994–1995, the temperatures slowly and regularly decreased in autumn (Fig. 17.1). This period was followed by a colder spell from mid-December to mid-January. Terminal and axillary vegetative buds broke, respectively, on 10 and 17 February. The average dates for appearance of the pistil between petals in floral buds and full blossom were, respectively, 6 and 20 March.

In the 1995–1996 winter, cold arrived earlier and went on until the end of December. After a warmer spell, another cold period lasted from 16 January to 13 March, interrupted only by one warmer week. Terminal vegetative buds broke on 18 March and axillary buds on 24 March, i.e. about 5 weeks later than in 1995. The pistil appeared between petals on

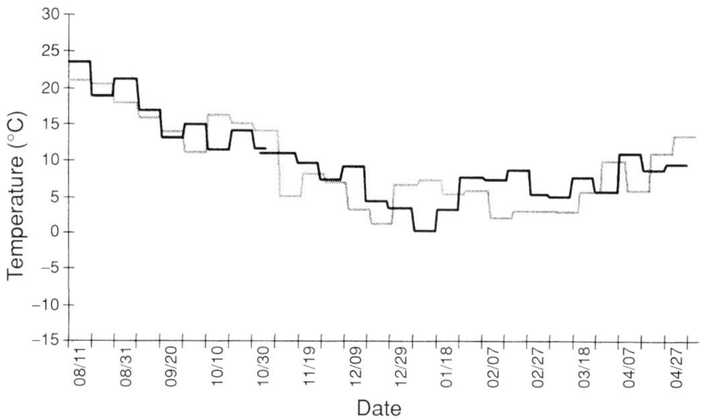

Fig. 17.1. Dynamics of mean temperature by decade during autumn and winter 1994–1995 (black line) and 1995–1996 (grey line).

18 March and the full blossom appeared on 30 March, i.e. 2 weeks later than in the previous year. A run of the 'dynamic model' of Fishman *et al.* (1987a,b) for the estimation of satisfaction of chilling requirements of the buds, situates the end of the endodormancy (45 'portions' for the variety 'Redhaven') at 15 December in 1994 and 20 December in 1995.

Nucleotide Content

At the beginning of October 1994, in vegetative primordia, ATP content values were around 4 pmol μg^{-1} of protein (Fig. 17.2A). This content slowly decreased until mid-January (decline of 30% in 3 months), then it increased rapidly and strongly (the content tripled in 1 month). In 1995–1996 (Fig. 17.2B), a first increase occurred in mid-December, followed by a transitory diminution beginning in January. The increase resumed until mid-March.

In floral primordia, in 1994–1995 (Fig. 17.2C), ATP contents were about 1 pmol μg^{-1} of protein at the beginning of October, and twice as high in 1995–1996 (Fig. 17.2D). We observed then a moderate increase of ATP contents from mid-December followed by a rapid and strong increase in February (content multiplied by 3.5 in 15 days). In 1995–1996 an increase occurred at the beginning of January. After a stabilization until mid-February, a strong increase occurred at the beginning of March, a short time before bud break. In vegetative primordia, ADP content was equivalent (in 1994–1995) or slightly inferior (in 1995–1996) to that of ATP until mid-January. Then it did not increase in proportion to ATP content. In floral primordia, in 1994–1995, the ADP content decreased from the beginning of October to the end of January, when it became lower than

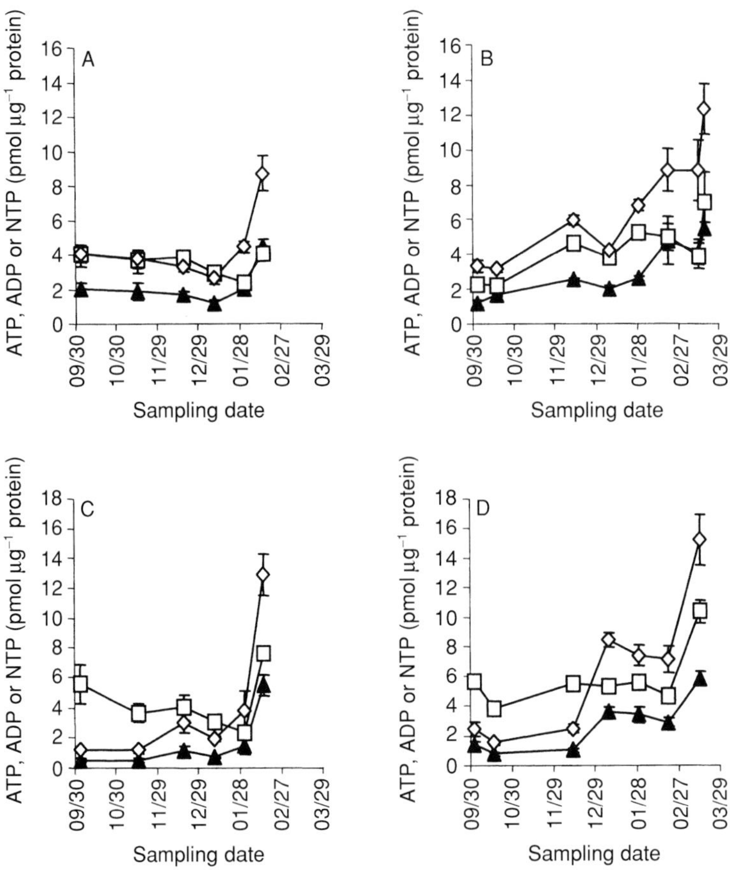

Fig. 17.2. Dynamics, in the orchard, of ADP (□), ATP (◇) and non-adenylic tri-phosphate (NTP ▲) nucleotide contents, on a protein basis, in vegetative (A and B) and floral primordia (C and D) during the 1994–1995 (A and C) and 1995–1996 (B and D) winters. Bars represent standard error.

ATP content; then it increased strongly (3.2-fold in 15 days). In 1995–1996, from mid-December to mid-February ADP content was lower than ATP content and was stable. Then, it increased strongly at the beginning of March but not to the same extent as ATP content.

NTP contents represented between 30 and 50% of ATP, in both 1994–1995 and 1995–1996, in vegetative as well as in floral primordia. They changed in proportion to those of ATP but were particularly low in floral primordia, about one-quarter of the values measured in vegetative primordia (0.5 pmol μg^{-1} protein in 1994–1995, 1 pmol μg^{-1} in 1995–1996).

Under chilling deprivation, for both vegetative and floral primordia (Fig. 17.3), from October to January, ATP content was low, ADP was slightly or much higher than ATP (respectively in the case of vegetative primordia

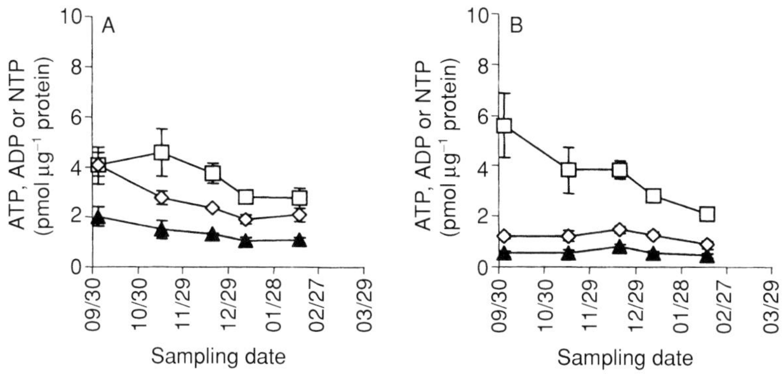

Fig. 17.3. Dynamics, under cold deprivation, of ADP (□), ATP (◇) and non-adenylic tri-phosphate (NTP ▲) nucleotide contents, on a protein basis, in vegetative (A) and floral primordia (B) during autumn and winter 1994–1995. Bars represent standard error.

and floral primordia), and NTP represented half of the ATP content. ATP content in vegetative primordia became slightly lower under chilling deprivation than under natural conditions; ADP content remained equivalent under the two climatic conditions. Nucleotide contents did not increase under this treatment until the last date of sampling while they increased very strongly in February in the buds of trees in the orchard.

ATP/ADP Ratio in Buds

For vegetative primordia, the ATP/ADP ratio remained constant (close to 1 in 1994–1995 and 1.4 in 1995–1996) from the beginning of October until mid-January in 1994–1995 and the end of January in 1995–1996 (Fig. 17.4A). It increased rapidly thereafter, approaching 2. For floral primordia (Fig. 17.4B), the ratio was first very low(< 0.5) then increased dramatically. This change occurred at the end of January in 1995, earlier (end of December) in 1996. The ratio became close to 1.5 and stabilized. Under cold deprivation, for vegetative primordia, the ATP/ADP ratio became slightly lower than that in the orchard and stabilized at a very low level. In floral primordia, the ATP/ADP ratio remained at the very low initial level until their complete necrosis.

How these Results Occur for Vegetative Primordia in Comparison with 'Classical' Dormancy Results

Consequent to the increase of ATP content in January, the ATP/ADP ratio seems to increase depending on the year, but the same maximal values are

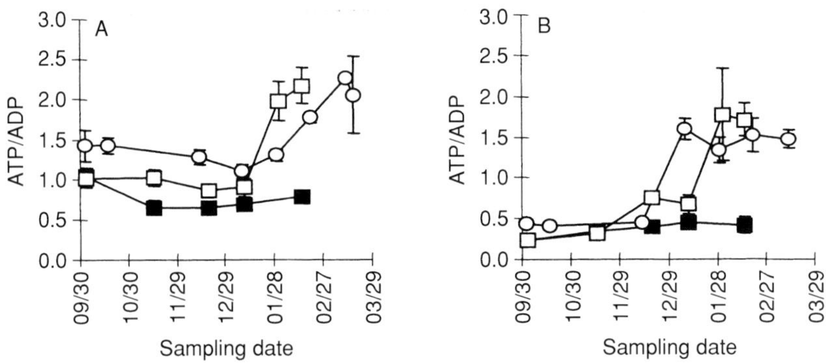

Fig. 17.4. Dynamics of the ATP/ADP ratio in vegetative (A) and floral (B) primordia, in the orchard during 1994–1995 (□) and 1995–1996 (○) winters or under cold deprivation (■). Bars represent standard error.

reached before bud break. Robert (1996) correlated a strong and rapid increase in ATP content with the end of endodormancy in vegetative buds of strawberry. We cannot establish such a link between the increase of ATP/ADP ratio and the end of endodormancy if we refer to usual results of the 'nucleotides' test in peach buds. This test, the only test of endodormancy available today (Gendraud, 1977, 1981; Rageau *et al.*, 1995), indicates, in the climatic context of Clermont-Ferrand, that the end of endodormancy is reached before the end of December (Balandier *et al.*, 1993; Bonhomme *et al.*, 1997; Bonhomme, 1998; Fig. 17.5). The increase in ATP/ADP ratio consecutive to the strong increase in ATP appears too late (not before the end of January) to reveal a direct link with endodormancy.

On the other hand, the changes in the ATP/ADP ratio seem to be consistent with the results of the 'one node cuttings' test obtained (but unpublished) during many years at Clermont-Ferrand. The 'one node cuttings' test indicates the end of endo- and/or short distance paradormancy (Balandier *et al.*, 1993; Rageau *et al.*, 1995); it corresponds to the crossing of a critical value of MTB (mean time to bud break): 12 days at 25°C. For 'Redhaven' peach buds, endo- and/or short distance paradormancy usually ends at the beginning of January in the orchard while under chilling deprivation MTB never crosses the critical value (Laroche, 1992; Rageau *et al.*, 1995; Bonhomme, 1998; Fig. 17.6). The increase in the ATP/ADP ratio seems to begin, in the orchard, very shortly after the positive response of the 'one node cuttings' test, i.e. the MTB value is below the critical value and endo- and/or short distance paradormancy is over. The change in the energy state of the primordia characterized by the ATP/ADP ratio increase seemed to be slowed by the cooler spell in February 1996 but did not turn backwards. So, we can consider that at this period, the ecodormancy is not able to change strongly the energy state of the buds. The measure of nucleotides contents of buds reveals neither that they are under endo-

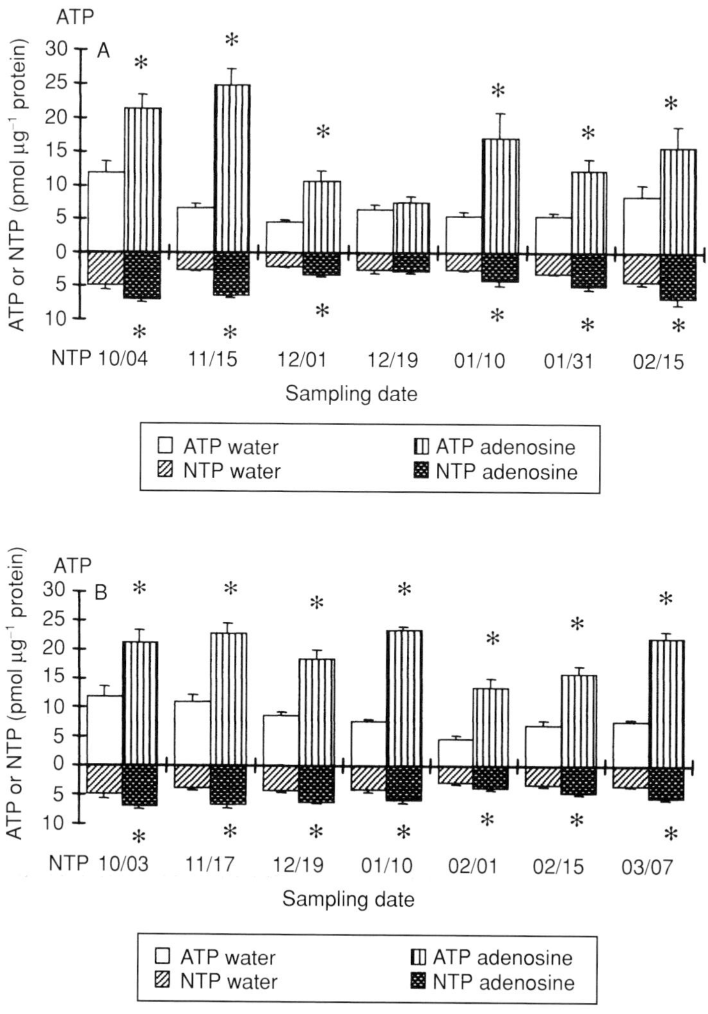

Fig. 17.5. Typical nucleotides test results on vegetative buds collected from the orchard (A) or under cold deprivation (B). A significant increase (*) of NTP pools after incubation on adenosine compared with incubation on distilled water indicates that buds were free of endodormancy. Bars represent standard error (from Bonhomme, 1998).

and/or short distance paradormancy nor that they are free of these dormancies. Consistently, under chilling deprivation, vegetative primordia are maintained strictly in the same energy state unfavourable to growth as they were at the beginning of October.

Therefore, we propose 1.5 as a threshold value for the ATP/ADP ratio, above which vegetative primordia no longer undergo endo- and/or paradormancy. It would be better to characterize the end of these dormancies

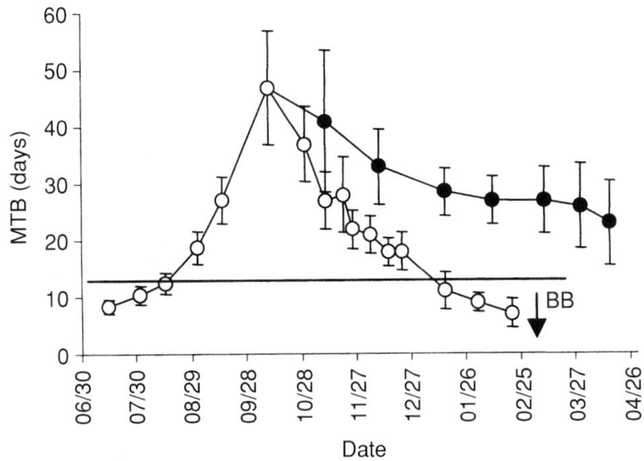

Fig. 17.6. Typical dynamics of MTB of vegetative buds collected from the orchard (○) and under cold deprivation (●). Bars represent standard error. For 'Redhaven', under the limit value of 12 days, we considered buds free of endo- and short-distance paradormancy (from Rageau, 1987; Laroche, 1992; Balandier *et al.*, 1993; Bonhomme, 1998).

by the beginning of the increase of the ATP/ADP ratio, but the crossing of this critical value is easier to locate than the real beginning of the increase of the ratio.

How these Results Occur for Floral Primordia in Comparison with 'Classical' Dormancy Results

During the period October-mid-December, the weakness of ATP contents, by themselves and comparing with those of ADP, constitutes an important difference with what is observed in vegetative primordia. On the other hand, as in vegetative primordia, the NTP content always represents between 30 and 50% of the ATP content.

The 'nucleotides' test is inoperative on floral primordia (Bonhomme, 1998), so we do not have any real test of endodormancy. The fresh and the dry weight of the floral primordia slowly increase in the orchard from October to January. However, it is only after this period and the impact of chilling that the floral primordia recover full growth competence in a large range of temperatures. The period of reduced growth competence is, in the literature, identified with a period of endodormancy.

It seems that a good coincidence exists between the date when the ATP/ADP ratio increases and the date when full growth competence recovery of the primordia in a large range of temperatures is observed, for example, through the 'Tabuenca' test (Tabuenca, 1964). This test consists of a comparison of the primordia weights at the sampling date and after

7 days in favourable conditions for growth (25°C, long days). We consider the test to be positive and the primordia free of endo- and/or para-dormancy when the weight of primordia grows threefold or more in 7 days. This acquisition of rapid growth capacity occurs generally in the first 10 days of January in the climatic conditions of Clermont-Ferrand (Fig. 17.7). The cooling in January–February 1996 slowed both the real growth of primordia in the orchard (data not shown) and the increase of nucleotide contents but, apparently, did not affect the ATP/ADP ratio. This stayed close to the 0.5 value (in any case always inferior to 1.0) during the period October–January, and close to the 1.5 value (in any case always superior to 1.0) after the resumption of rapid growth. After that, the law of the relative growth of the primordia, according to the temperature, no longer varies (Rageau, 1982) and the real growth curve obtained in natural conditions depends only on the action of temperatures through this law.

The weakness of ATP and NTP contents during the first phase induces a physiological situation unfavourable to strong growth. This also provides, partly, an explanation for the fact that the 'nucleotides' test is ineffectual on floral primordia (Bonhomme, 1998). We believe that the lack of ATP (and/or the weakness of its regeneration) does not allow provision of the energy needed for phosphorylation of the non-adenylic di-phosphate nucleotides in sufficient quantity to allow an increase in the NTP content. The immediate use of the ATP synthesized to provide the energy needed for the different stages of the biosynthesis of NTP would also explain the

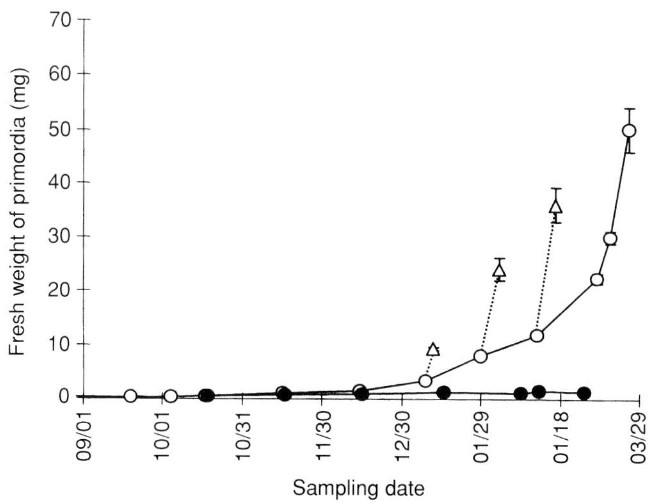

Fig. 17.7. Typical dynamics of fresh weight of floral primordia in the orchard (○), under cold deprivation (●), and after 7 days at 25°C ('Tabuenca' test Δ). Bars represent standard error (from Maerere, 1990; Bonhomme *et al.*, 1997; Bonhomme, 1998).

absence of an increase in ATP contents which is sometimes observed in the primordia incubated on adenosine.

Therefore, the change in the ATP/ADP ratio seems, in vegetative as well as in floral buds, to depend only on the passage from a state of endo- and/or paradormancy to a free state. The measure could, therefore, have the same discriminating value between ecodormancy and other dormancies as a 'whole tree' test, which consists of submitting a whole tree to thermal conditions favourable to growth (25°C) (Rageau *et al.*, 1995). The threshold value of 1.0 could therefore be considered as a good marker of this change of state *in situ*.

According to results concerning the dynamics of fresh (and dry) weight of primordia under chilling deprivation obtained during several years at the laboratory (Maerere, 1990; Bonhomme *et al.*, 1997; Bonhomme, 1998), it appears that the incapacity of growth at mild temperatures that characterizes primordia from October onwards persists as long as cold deprivation is applied. After approximately 6 months, this treatment leads to death of the primordia. Measurement of nucleotide contents and the ATP/ADP ratio indicates that the initial energy status of floral primordia does not improve and even shows a tendency to degrade in these conditions. This differs from what happens in vegetative primordia: the level of the energy status of these remains higher, which could explain, partly, the fact that only vegetative primordia can survive in such conditions. Figure 17.8 summarizes the train

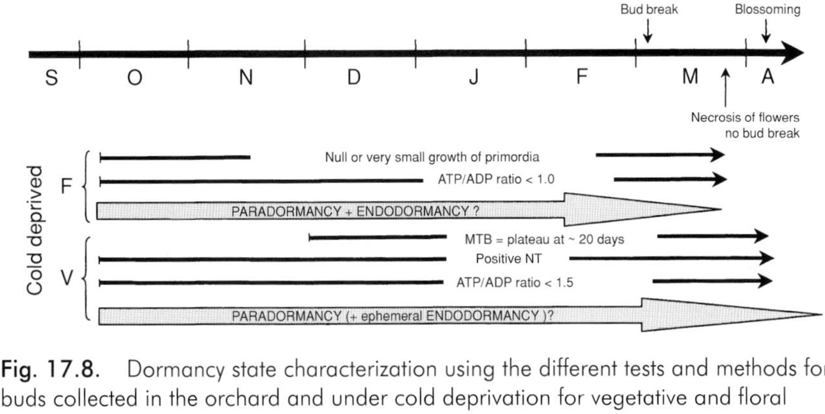

Fig. 17.8. Dormancy state characterization using the different tests and methods for buds collected in the orchard and under cold deprivation for vegetative and floral primordia. The dotted line represents the interannual variability. MTB, mean time to bud break; NT, 'nucleotide' test.

of dormancy states in vegetative and floral primordia obtained by the different tests and the measurement of the ATP/ADP ratio in the orchard and under cold deprivation.

Conclusions

The ATP/ADP ratio did not prove to be the marker of endodormancy that we were looking for. However, it reveals the state of primordia submitted to endodormancy and/or short- or long-distance paradormancy inhibitions. The ratio is not sensitive to the ecodormancy liable to block real growth after the end of endodormancy. It seems, actually, to be a good marker for the primordia of their ability to respond at favourable temperatures and start strong growth *in situ*. The ATP/ADP ratio presents the advantages of being easy to use, without bud incubation and without any proteins or DNA assays as references, on both floral and vegetative primordia.

Acknowledgement

The authors thank ATT for checking the English.

References

Balandier, P. (1992) Etude dynamique de la croissance et du développement des bourgeons de quelques cultivars de pêcher, cultivés à diverses altitudes sous climat tropical de l'île de la Réunion. Thèse Dr, Université Blaise Pascal, Clermont-Ferrand II, France.

Balandier, P., Gendraud, M., Rageau, R., Bonhomme, M., Richard, J.P. and Parisot, E. (1993) Bud break delay on single node cuttings and bud capacity for nucleotide accumulation as parameters for endo and paradormancy in peach trees in tropical climate. *Scientia Horticulturae* 55, 249–261.

Barnola, P., Crochet, A., Payan, E., Gendraud, M. and Lavarenne, S. (1986) Modifications du métabolisme énergétique et de la perméabilité dans le bourgeon apical et l'axe sous-jacent au cours de l'arrêt de croissance momentané de jeunes plants de chêne. *Physiologie Végétale* 24, 307–314.

Bonhomme, M. (1998) Physiologie des bourgeons végétatifs et floraux de pêcher dans deux situations thermiques contrastées pendant la dormance: capacité de croissance, force de puits et répartition des glucides. Thèse Dr, Université Blaise Pascal, Clermont-Ferrand II, France.

Bonhomme, M., Rageau, R., Richard, J.P. and Gendraud, M. (1997) Dormancy of peach floral buds: biological and tentative biochemical approaches. *Acta Horticulturae* 441, 167–174.

Bonhomme, M., Rageau, R., Richard, J.P., Erez, A. and Gendraud, M. (1999) Influence of three contrasted climatic conditions on endodormant vegetative and floral peach buds: analyses of their intrinsic growth capacity and their potential sink strength compared with adjacent tissues. *Scientia Horticulturae* 80, 157–171.

Fishman, S., Erez, A. and Couvillon, G.A. (1987a) The temperature dependence of dormancy breaking in plants: mathematical analysis of a two-step model involving a cooperative transition. *Journal of Theoretical Biology* 124, 473–483.

Fishman, S., Erez, A. and Couvillon, G.A. (1987b) The temperature dependence of dormancy breaking in plants: computer simulation of processes studied under controlled temperatures. *Journal of Theoretical Biology* 126, 309–321.

Gendraud, M. (1975) Contribution à l'étude du métabolisme des nucléotides di- et tri-phosphates de tubercules de Topinambour cultivés *in vitro* en rapport avec leurs potentialités morphogénétiques. *Plant Science Letters* 4, 53–59.

Gendraud, M. (1977) Etude de quelques aspects du métabolisme des nucléotides de pousses de Topinambour en relation avec leurs potentialités morphogénétiques. *Physiologie Végétale* 15, 121–132.

Gendraud, M. and Petel, G. (1990) Modifications in intercellular communications, cellular characteristics and change in morphogenetic potentialities of Jerusalem artichoke tubers (*Helianthus tuberosus* L.). In: Millet, B. and Greppin, H. (eds) *Intra- and Extracellular Communications in Plants: Reception, Transmission, Storage and Expression of Messages.* INRA, Paris, pp. 171–175.

Lang, G. A., Early, J.D., Martin, G.C. and Darnell, R.L. (1987) Endo-, para- and ecodormancy: physiological terminology and classification for dormancy research. *Hortscience* 22, 371–377.

Laroche, A. (1992) Etude physiologique et biochimique de la dormance des bourgeons végétatifs du pêcher privé ou non de froid hivernal. DEA, Unité Associée INRA-Université Blaise Pascal, Clermont-Ferrand II, France.

Lavarenne, S., Champciaux, M., Barnola, P. and Gendraud, M. (1982) Métabolisme des nucléotides et dormance des bourgeons chez le frêne. *Physiologie Végétale* 20, 371–376.

Le Floc'h, F. (1984) La regulation du métabolisme des nucléotides puriques chez le topinambour, *Hélianthus tuberosus* L., en relation avec la dormance. Thèse Dr, Université de Clermont-Ferrand II, France.

Le Floc'h, F. and Faye, F. (1995) Metabolic fate of adenosine and purine metabolism enzymes in young plants of peach tree. *Journal of Plant Physiology* 145, 398–404.

Le Floc'h, F. and Lafleuriel, J. (1981) The purine nucleosidases of Jerusalem Artichoke shoots. *Phytochemistry* 20, 2127–2129.

Lecomte, I., Faye, F. and Le Floc'h, F. (1998) Purine nucleotide metabolism in peach tree buds: characterization of dormancy release and growth ability 'marker enzymes'. *Acta Horticulturae* 465, 521–531.

Maerere, A. (1990) Privation de froid pendant la dormance et developpement des bourgeons floraux chez le pêcher. DEA, Université Blaise Pascal, Clermont-Ferrand II; INRA, Laboratoire de Bioclimatologie, Clermont-Ferrand, France.

Rageau, R. (1982) Etude expérimentale des lois d'action de la température sur la croissance des bourgeons floraux du pêcher (*Prunus persica* L. Batsch) pendant la post-dormance. *Comptes Rendus de l'Académie d'Agriculture de France* 68, 709–718.

Rageau, R. (1987) Le pêcher – références et techniques. In: *L'Arbre et Son Milieu,* Vol. 4, *Exigences Climatiques.* CTIFL, Paris, pp. 74–93.

Rageau, R., Julien, J.L. and Ollat, N. (1995) Approche du contrôle de la croissance des bourgeons dans le contexte de l'arbre entier. In: Cruiziat, P. et Lagouarde J.P. (eds) *Actes de l'école chercheur INRA en Bioclimatologie, T1, de la plante au couvert végétal.* Département de Bioclimatologie, INRA, Thiverval-Grignon, France, pp. 107–120.

Robert, F. (1996) Recherche de marqueurs morphologiques et biochimiques de la dormance du fraisier (*Fragaria* × *ananassa* Duch.). Thèse Dr, Université Blaise Pascal, Clermont-Ferrand II, France.

Tabuenca, M.C. (1964) Necesidades de frio invernal de variedades de Albaricoquero, Melocotonero y Peral. *Anales de la Estacion Experimentale de Aula Dei* 7, 113–132.

18

Alcohols that Break Seed Dormancy: the Anaesthetic Hypothesis, Dead or Alive?

Marc Alan Cohn[1] and Henk W.M. Hilhorst[2]

[1]Department of Plant Pathology and Crop Physiology, Louisiana Agricultural Experiment Station, LSU Agricultural Center, Baton Rouge, LA 70803, USA; [2]Department of Plant Physiology, Wageningen Agricultural University, Arboretumlaan 4, 6703 BD Wageningen, The Netherlands

Introduction

The mechanisms that underlie seed vigour, viability and dormancy continue to elude our understanding. These 'vivid' properties of seeds represent a major challenge to our collective ingenuity and represent a barrier to our quest for a stable world food supply and sound stewardship of genetic resources and the environment. The agricultural limitations imposed by seed dormancy have been recognized for centuries (Evenari, 1984). Employing a combination of environmental stimuli (light and temperature) and chemical treatments, most seeds can be empirically manipulated to perform as desired in a laboratory setting. However, these procedures are difficult, if not impossible, to apply in a field environment. In no case do we understand the mechanisms involved, although parts of the puzzle have been identified. For example, there is little to no controversy that phytochrome is the receptor for light-stimulated germination (Shinomura, 1997; Casal and Sánchez, 1998). Receptors for some plant hormones (Fluhr, 1998) and genes that contain plant hormone response elements (reviewed in Ritchie and Gilroy, 1998) have also been identified. Great advances via molecular genetics have offered substance to the role of abscisic acid-regulated genes in the development of seed dormancy (Koornneef *et al.*, 1998). Based upon the vast signal transduction literature, it is no surprise that evidence has begun to accumulate that protein phosphorylation/dephosphorylation events have a regulatory role in seed processes (Walker-Simmons, 1998). Yet with much diligent work, we have

(eds J.-D. Viémont and J. Crabbé)

yet to establish the biochemical sequences of events from the dormant seed to visible germination.

A large number of non-hormonal organic and inorganic chemicals that break seed dormancy (Bewley and Black, 1982) and generally terminate developmental arrest (Footitt and Cohn, unpublished) have been identified. Various hypotheses have been introduced to explain what these substances are doing in seeds. While these ideas have targeted hormonal balance, the pentose phosphate shunt, alternative respiration, catalase, gene regulation and membranes (reviewed in Bewley and Black, 1982, 1994; Cohn, 1987; Hilhorst, 1995, 1998; Bewley, 1997), the supporting data remain controversial. Given the wide variety of active substances, it is not clear how many different biological targets or activation pathways may exist within the seed. However, recent structure–activity studies and analysis have suggested that a majority of these individual chemicals may have a common mode of action (Cohn, 1997). Implicit in this view is the concept that applied chemicals are not necessarily active themselves, but require metabolism to a common active form. Direct and indirect supporting evidence has begun to emerge in support of this view (Cohn *et al.*, 1989; Corbineau *et al.*, 1991; Footitt *et al.*, 1995; Lin, 1997; Lin and Cohn, 1997).

One chemical class involved in these recent studies is the simple *n*-alkanols. Short-chain alcohols have been identified as stimulators of plant growth (e.g. Mer, 1958, 1961; Gudjónsdóttir and Burström, 1962; Morré *et al.*, 1965) and as dormancy-breaking chemicals for some time. The initial hypothesis for their mode of action postulated that alcohols physically interacted with membranes to facilitate changes of state that would activate dormant seeds (Taylorson and Hendricks, 1979). What emerges from the current state of the literature, therefore, is an important paradox. Do alcohols directly act as dormancy-breaking chemicals or are they only the precursors that must be metabolized for dormancy-breaking action? Therefore, it is timely to revisit the classic papers of Taylorson and others on this subject and examine them in light of recent findings.

Birth of the Anaesthetic Hypothesis

Pecket and Al-Charchafchi (1978) provided initial evidence that an alcohol could promote germination of a light-sensitive lettuce cultivar (*Lactuca sativa* L. var. Cannington Forcing). A brief 0.3–6 h ethanol (0.1–0.9 M) treatment increased germination by 17–19%; propanol elicited a similar response. The authors proposed that enhanced membrane permeability could be associated with the observed response given that calcium, viewed as a membrane stabilizer, prevented alcohol-stimulated germination. During the course of experiments on dormancy in *Panicum dichotomiflorum*, Taylorson and Hendricks (1980/81) independently discovered and extended this idea, supported by their complementary observations suggesting that membranes were involved in the dormancy-breaking process (reviewed in Taylorson and Hendricks, 1977) as well as reports that anaesthetics can break bud dormancy (Vegis, 1965). In a key report (Taylorson and Hendricks, 1979), dormancy in a number of grass species was broken by chemicals such as ethanol, methanol, acetone, chloroform and ethyl ether. Response to a chemical could be further mediated by appropriate incubation temperatures and the presence of light. However, some grasses were not responsive to chemical treatment, and dicot germination was generally inhibited. In addition, some chemicals elicited only a small stimulation, at best (isopropanol, acetaldehyde, propanol and butanol). To explain the dormancy-breaking activity of chemicals, Taylorson and Hendricks (1979) suggested 'that this action depends on modifying the properties of a membrane(s) in a manner related to the actions of other anesthetics'.

Pressure Reversibility

The key evidence further supporting the anaesthetic model was provided by a pressure-reversal experiment (Hendricks and Taylorson, 1980). Pressures of > 0.8 MPa, during the exposure of dormant *Panicum capillare* L. seeds to aqueous ethanol or ethyl ether, negated the stimulation by the chemical. That a positive pressure would overcome the dormancy-breaking action of these chemicals seemed to be consistent with hypotheses concerning anaesthetic action in animal systems, particularly the critical membrane volume hypothesis (Miller, 1975); the chemical partitions into the membrane due to its lipophilic nature and expands the membrane, altering its properties. Pressure would reverse this membrane expansion. Both the lipid and protein components of membranes could be chemical targets.

Pressure treatments inhibited dormancy-breaking when applied together with: (i) ethanol or ethyl ether in *Panicum capillare* (Hendricks and Taylorson, 1980); (ii) propanol in *Echinochloa crus-galli* (Taylorson, 1988); and (iii) 2,2,2,-trifluoroethanol in *Rumex crispus* (Taylorson, 1987). However, in *Amaranthus retroflexus*, pressure applied together with ethanol actually increased the dormancy-breaking response (Taylorson, 1989), and pressure could not reverse the activity of butanol in *E. crus-galli* (Taylorson, 1991). If non-lethal pressure was applied to *E. crus-galli* seeds for 24–48 hours *prior* to propanol treatment under ambient conditions, the action of the alcohol was also reduced, and Taylorson (1991) recognized that the pressure effects were not uniformly preventing the dormancy-breaking action of alcohols. Such a differential response to pressure with different anaesthetics has not been observed in animals (Seeman, 1972). The diversity of observations in seeds has not as yet been experimentally addressed, but they weaken the anaesthetic hypothesis as originally formulated.

Leakage of Cellular Components

As an additional indicator of alterations in cell membrane properties, the effects of applied chemicals upon leakage of cellular components have been studied. If a substance is perturbing a membrane structure such as the plasmalemma as part of its mode of action, one would expect to observe increased leakage to the incubation medium bathing the seeds. The degree of leakage was estimated by measuring the conductivity, absorbance at 280 nm, amino acid levels and pH of the medium after chemical treatment (Taylorson, 1988, 1989; Taylorson and Di Nola, 1990). While differences between chemically treated and control seeds were occasionally measured, no consistent correlations were observed between leakage and the effectiveness of dormancy-breaking treatments. The dose–response for leakage, when it was observed, was greatly different from the dose–response for dormancy-breaking.

Changes in Membranes and Proteins

Di Nola and Taylorson (1990) performed a series of ultrastructural and biochemical studies (Di Nola *et al.*, 1990, 1991) on dormancy-breaking in *E. crus-galli.* Propanol broke dormancy while isopropanol was inactive. Propanol treatment elicited elongation of mitochondria, protein body fusion, and the disappearance of lipid bodies in proximity to the plasmalemma. These alterations occurred prior to visible germination and did not occur in water-imbibed or isopropanol-treated seeds.

Di Nola *et al.* (1991) measured changes in the thermotropic transition properties of both microsomal and plasmalemma vesicles prepared from *E. crus-galli* seeds treated with water, 300 mM propanol, 10 mM benzyl alcohol (breaking dormancy), 300 mM isopropanol or water. Steady-state fluorescence anisotropy data, with 1,6-diphenyl-1,3,5-hexatriene (DPH) as a probe, were analysed on Arrhenius plots to detect membrane phase transitions as a function of temperature. The general trends of the results showed that membranes from seeds exposed to a dormancy-breaking chemical exhibited no discontinuities in the Arrhenius plots after a 24 h treatment or during subsequent germination, while discontinuities remained in membrane preparations from isopropanol- or water-treated seeds. In previous work (Di Nola and Taylorson, 1990), it was shown that there was no evidence of visible germination immediately after the 24 h dormancy-breaking treatments. However, Di Nola *et al.* (1991) also treated membranes from water-imbibed seeds with the various chemicals *in vitro*. The trends were the same as in chemically treated seeds, but they were independent of alcohol concentration from 5 to 300 mM. According to Taylorson (1988), a minimum of 100 mM propanol was required to elicit a germination response. In addition, Taylorson and Di Nola (1990) examined phase transitions in *R. crispus* as influenced by seed incubation temperature, ethanol and 2,2,2-trifluoroethanol (TFE). The membrane phase transitions were not entirely consistent with seed behaviour particularly for chemically treated seeds. Membranes from secondarily dormant *Cucumis sativus* seeds treated with alcohols also displayed fluidity patterns that were either consistent or inconsistent with a role for membrane phase transitions in the dormancy-breaking process (Sreenivasulu and Amritphale, 1998). Therefore, in order to clearly link the changes in membrane properties to dormancy-breaking, additional work would be required.

1. Membrane preparations from treated seeds exposed to sub-threshold concentrations of dormancy-breaking chemicals would help to clarify whether or not phase transitions are associated with chemical exposure *per se* or with the termination of development arrest.
2. Analysis of a fully dry-afterripened non-dormant control would help to determine whether or not the phase transitions are associated with pre-visible germination vs. dormancy-breaking processes.
3. Analysis of phase transitions during the chemical dormancy-breaking phase (0–24 h) would also help to pinpoint the timing of membrane-related events.
4. Other dormancy-breaking substances from different chemical classes should be studied to determine if the altered phase transitions are associated specifically with alcohols or not.
5. Seeds should be treated with a combination of dormancy-breaking alcohol and 4-methyl pyrazole (see below) to determine if the membrane transitions are a primary, secondary or coincidental event.
6. Do the trends reported also occur in other species?

Therefore, while there is little question that the phase transitions do occur, their relevance to the dormancy-breaking process requires much further investigation.

Di Nola *et al.* (1990) investigated the changes in membrane proteins in *E. crus-galli* using similar dormancy-breaking treatments and controls. A 23 kDa protein increased prior to visible germination in association with propanol treatment but not with isopropanol or water treatments. Caveats similar to those raised for the phase transition studies are also relevant here. In addition, many changes in the one-dimensional protein profiles were associated with the various pre-treatments and controls employed. It will be highly challenging to sift through this maze of protein changes to rule out artefacts associated with changes in incubation conditions. Clearly, there is much work still to be done.

The Bioassays

A substantial volume of experimentation shows that alcohols and related substances break dormancy of seeds (Table 18.1). The hints of difficulties with the anaesthetic hypothesis can be detected by examination of this bioassay data. Alternatives to the anaesthetic hypothesis were initially raised in Taylorson and Hendricks' own papers but were not experimentally addressed until recently.

Taylorson and Hendricks (1980/81) recognized that: (a) there were differences in relative activity between chemicals in a single species; (b) that the dormancy-breaking response of a species was contact time and concentration dependent; and (c) that seeds of various species responded differently to the same chemicals. In addition, (d) different genotypes within the same species had different sensitivities to the same chemical (Adkins *et al.*, 1984a), and (e) there was an interaction between alcohol response and the extent of dry-afterripening (Taylorson, 1988). Two general questions emerge from this diversity of responses: (i) what is the explanation

Table 18.1. Activity of alcohols and other small organic molecules as dormancy-breaking chemicals.

Species	Reference	Methanol	Ethanol	Pro-panol	Iso-propanol	Butanol	Pentanol
Amaranthus albus	10, 11	—	+(R)	+++(R)	—		
Amaranthus retroflexus	13	++	+++	—	++	—	—
Avena fatua	7, 8	– – –	+++	+		+	
Avena sativa	16	—	+++	+++	—	+++	
Cucumis sativus	22, 23	+++	+++	+++			
Digitaria ischaemum	1		++(R)				
Digitaria sanguinalis	1, 6	—	+++(R)		—		
Echinochloa crus-galli	1, 6, 10, 12, 14, 15	+	+++(R)	+++	+/–	++	+++
Eleusine indica	1		—				
Eragrostis cilianensis	1		—				
Lactuca sativa	4, 21		+++(R)				
Oryza sativa (red rice)	17, 18, 19	+++	+++	+++	+++	++	+++
Panicum capillare	1, 3, 13	—	+++	++/–	—	+/–	++
Panicum dichotomiflorum	1, 4, 6	++(R)	+++	—	+/–	—	
Poa annua	1		—				
Rumex crispus	6, 9, 10, 20	+++(R)	—	—	++(R)		
Setaria faberi	1, 5	++	++(R)	+	+		
Setaria glauca	1		—				
Sorghum halepense	1		—				

+, Positive effects; (R), red light interaction; —, negative or no effect.
(1) Taylorson and Hendricks (1979). (2) Taylorson (1979). (3) Hendricks and Taylorson (1980). (4) Taylorson and Hendricks (1980/81). (5) Taylorson (1982a). (6) Taylorson (1982b). (7) Adkins *et al.* (1984a). (8) Adkins *et al.* (1984b). (9) Taylorson (1984). (10) Taylorson (1987).

for these observations; and (ii) do these observations also compromise the membrane as the target?

Differing lipid solubilities (octanol/water partition coefficients, log $K_{o/w}$) of organic chemicals could readily account for (a) and (b). A correlation between log $K_{o/w}$ and relative activity among dormancy-breaking chemicals has been established for *Echinochloa crus-galli* (Taylorson, 1988), *Panicum capillare* (Taylorson, 1989) and *Oryza sativa* (Cohn *et al.*, 1989). However, there was no significant relationship between chemical lipophilicity and alcohol activity in *Amaranthus retroflexus* (Taylorson, 1989). On balance, however, these studies clearly indicate a relationship between log $K_{o/w}$ vs. relative activity of dormancy-breaking alcohols and related chemicals. These data appear to be consistent with the anaesthetic hypothesis. However, secondary alcohols, such as isopropanol (Table 18.1) and 2-pentanol (Cohn *et al.*, 1991), elicit little to no dormancy-breaking action despite having log $K_{o/w}$ values that should permit activity. If the target was the lipid bilayer, a modifying effect of functional group position within the range of active log $K_{o/w}$ values would not be expected, e.g. secondary

Table 18.1. *Continued.*

Benzyl alcohol	Acetone	Chloroform	Ether	Ethylene (2)	TFE	Acetaldehyde	Propionaldehyde	Ethyl acetate	Halothane	Procaine
					—	++				
—		—	+						—	
	—	—	—			+++		—		—
	+++					+++				
	—	+	+	—					—	
+++	—	++/−	+/−	—	—				—	
				—						
		+++	++							
	+++					+++	+++	+++		
++	—	+	++/−	—						
	+++(R)	+++(R)	+++(R)	—		+			++	++
				—						
	+	—	—		+++	—			—	
	− − −	− − −	− − −	—	− − −	++	++	++		
				—						
				—						

(11) Chadoeuf-Hannel and Taylorson (1985). (12) Taylorson (1988). (13) Taylorson (1989). (14) Di Nola *et al.* (1990). (15) Di Nola *et al.* (1991). (16) Corbineau *et al.* (1991). (17) Cohn *et al.* (1989). (18) Cohn *et al.* (1991). (19) Lin (1997). (20) Taylorson and Di Nola (1990). (21) VanDerWoude (1985). (22) Amritphale *et al.* (1993). (23) Sreenivasulu and Amritphale (1998).

alcohols have anaesthetic activity in tadpoles and other organisms (Alifimoff *et al.*, 1987). In addition, dependence of activity upon log $K_{o/w}$ does not always indicate a lipid target. Correlations between alcohol–protein interactions and log $K_{o/w}$ have been reported for purified enzymes, e.g. alcohol dehydrogenase (Hansch *et al.*, 1972), firefly luciferase (Moss *et al.*, 1991), and cytochrome c oxidase (Hasinoff and Davey, 1989).

Dealing with why seeds of various species responded differently to the same chemicals is more problematic since there are numerous ways to rationalize this response. There may be methodological problems in construction of the response curves. In the course of generating response curves, one will encounter concentrations of a chemical that do not elicit a germination response. Four problems arise: (i) is the chemical inactive because it does not penetrate the seed; (ii) is the chemical inactive because the tissue concentration is suboptimal; (iii) is the chemical physiologically inactive; or (iv) has the seed been killed by an overdose? The simplest first step to diagnose these problems is the performance of a viability test on non-responsive seeds after the chemical treatment (e.g. Adkins *et al.*, 1984a; Corbineau *et al.*, 1991; Cohn, 1996a). The viability test is particularly critical when evaluating alcohols, since they are known preservatives of plant tissue. In the absence of a viability test, a dead seed could be mistaken for an ungerminated healthy one. If seeds were still viable, one would need to continue dose–response experiments until toxicity was observed. Unfortunately, this test was not routinely employed during the first decade of work on the anaesthetic hypothesis. Therefore, when chemicals were found to be inactive, the reason was not always clear.

The same concentration of one chemical is not most active in all species and for all other chemicals. This means that new dose–response curves must be constructed for a given set of chemicals for each species. Log $K_{o/w}$ values can be an excellent guide to predicting active concentration ranges for a series of dormancy-breaking chemicals (e.g. Cohn, 1997). However, not only is the active chemical concentration a function of lipophilicity, but so is the range of chemical concentrations between bioassay activity and toxicity: the greater the lipophilicity, the more narrow the range between activity and toxicity (Fig. 18.1). So, in the construction of a dose–response curve, appropriate concentration steps for ethanol might miss the entire activity window for pentanol. As a consequence of these operational problems, it is challenging to draw definite conclusions from Table 18.1 without experimental reinvestigation.

For example, most applied chemicals are active as dormancy-breaking agents in *E. crus-galli* and *O. sativa*. The same cannot be said for many other species (Table 18.1). Yet *A. retroflexus* dormancy can be broken only by methanol, ethanol, isopropanol and ether, while propanol, butanol, pentanol, and chloroform are inactive (Taylorson, 1989). Considering the log $K_{o/w}$ values for the inactive substances, their chemical structures and biological activity in other species, such results are surprising.

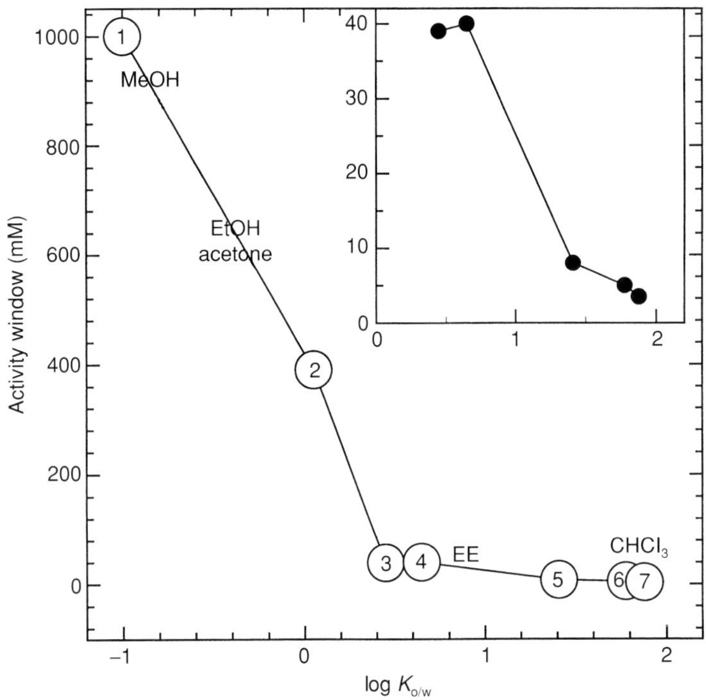

Fig. 18.1. The size of the chemical bioactivity window (the concentration range required for 50% stimulation of dormancy-breaking to 50% inhibition of germination of dehulled red rice) as a function of the octanol/water partition coefficients (log $K_{o/w}$) for (1) 1,5-pentanediol, (2) isopropanol, (3) acetaldehyde, (4) 2-butanol, (5) pentanol, (6) cyclohexylmethanol, and (7) cyclohexanecarboxaldehyde. Predicted activity ranges based upon log $K_{o/w}$ for methanol (MeOH), ethanol (EtOH), acetone, ethyl ether (EE), and chloroform ($CHCl_3$). Inset: expanded scale for compounds 3–7. Dehulled red rice was incubated at various chemical concentrations for 1 day, then rinsed and transferred to water for 7 days at 30°C.

The dependency of alcohol activity upon genetic background (d) and dry-afterripening (e) have yet to be investigated in detail. At this time, these phenomena must simply serve as cautionary observations that might require consideration when interpreting new data. It is hoped that the power of genetics will eventually be utilized to understand aspects of alcohol action.

A Metabolic Alternative: the 'Dynamic Duo' of ADH1 and Cytochrome P-4502E1

The anaesthetic model suggests that alcohols and related substances directly interact with membranes. A role for metabolic conversion of the applied chemical was ruled out because some dormancy-breaking alcohols, e.g. methanol and isopropanol, are not usually substrates for seed alcohol dehydrogenase (ADH1) (Duffus, 1968; Eriksson, 1968; Leblová and Perglerova, 1976; Corbineau *et al.*, 1991; Lin, 1997). In addition, 'The similar effect on germination of ethanol and of the much more slowly metabolized chloroform and ethyl ether is strong evidence against a metabolic function of ethanol' (Taylorson and Hendricks, 1979). None the less, Taylorson and Hendricks (1980/81) cite their unpublished data showing rapid uptake of ethanol by *Panicum* seeds during a 2 h exposure and a further rapid exponential loss of ethanol ($t_{½}$ = 1.5 h) after transfer to water.

What is the fate of the alcohol? Clearly, it might simply diffuse out of the seeds. On the other hand, alcohol metabolism may also play a significant role (Adkins *et al.*, 1984a,b; Cohn *et al.*, 1989; Corbineau *et al.*, 1991). Many dormancy-breaking alcohols are substrates for ADH1 in dormant seeds (Corbineau *et al.*, 1991; Lin, 1997), and Footitt *et al.* (1995) showed rapid propanol uptake (also saturating at 2 h) and metabolism of propanol to 3-hydroxypropionic acid during the dormancy-breaking process, more than 20 h prior to visible germination. An inhibitor of ADH1, 4-methyl pyrazole (4-MP), prevented the dormancy-breaking action of ethanol (Corbineau *et al.*, 1991) and other *n*-alkanols (Lin and Cohn, 1997). However, 4-MP did not inhibit the dormancy-breaking action of the products of alcohol metabolism, monocarboxylic acids (Lin, 1997). This rules out a general inhibitory effect of 4-MP and implies that metabolism of alcohols by ADH1 is absolutely required to elicit dormancy-breaking activity.

Cytochrome P-4502E1

These results do not account for the activity of other small organic substances such as diethyl ether, chloroform, methanol, acetone and isopropanol (Table 18.1). Recently, the discovery of cytochrome P-4502E1 has provided a metabolic alternative to account for these activities as well (reviewed in Lieber, 1997). All of the anaesthetic-like dormancy-breaking compounds that are not ADH1 substrates (and many that are) can be metabolized readily by cytochrome P-4502E1.

The cytochrome P-4502E1 mono-oxygenation reaction requires NADPH and consumes oxygen (Lieber, 1997). Therefore, increased oxygen uptake typically observed after treating dormant seeds with alcohols (e.g. Adkins *et al.*, 1984b; Corbineau *et al.*, 1991) may not reflect normal respiration but rather the activity of cytochrome P-450. With ethanol as a substrate, the K_m (Michaelis binding affinity constant) for the P-450-mediated reaction is 10 mM ethanol (Lieber, 1997), while the apparent K_m

for ADH1 in red rice seeds is 19 mM ethanol (Lin, 1997). Furthermore, the pH optimum for P-450-mediated reactions is 6.8–7.4, in contrast with the very basic pH optimum (typically pH 9.6) for ADH1. If these properties hold for P-450 in seeds, it is clear that P-4502E1 would play a significant role in the dormancy-breaking process.

Bioassay data add strength to this speculation. Pyrazole and 4-methyl pyrazole are inhibitors of *both* ADH1 and P-4502E1. Methanol, isopropanol, isobutanol and cyclohexylmethanol all break dormancy in red rice and are not ADH1 substrates. Yet, they are prevented from breaking dormancy by 4-methyl pyrazole without subsequent loss of viability (Lin, 1997). Therefore, the physiological data suggest the presence of P-4502E1, although this form of P-450 has not been characterized in plants.

A minor technical, yet important, factor must also be considered. Reagent grade ethyl ether and chloroform are not pure. They contain stabilizers (Technical Services, Aldrich Chemical Co., Missouri, 1998, personal communication), usually ethanol, at concentrations as high as 1%. Therefore, seed dormancy broken by low concentrations of ethanol and high concentrations of ether or chloroform may represent an artefact.

The light requirement for activity of dormancy-breaking alcohols

One of the striking features reported in several Taylorson papers and in VanDerWoude (1985) was the red light requirement for the dormancy-breaking activity of many tested compounds (Table 18.1). While it was initially proposed that this response was consistent with a role for membranes interacting with phytochrome, recent data suggest a 'metabolic alternative'. In dormant photoblastic lettuce seeds, Toyomasu *et al.* (1995) observed gibberellic acid- or red-light-mediated upregulation of a gene whose sequence had 90% similarity to alcohol dehydrogenase. Its message was not detected in dry seeds, was not expressed in red/far-red reversal experiments, and was expressed in GA- or red-light-treated seeds far earlier than the visible germination event. While the presence of the protein product or ADH1 activity was not measured, these data suggest that light would induce one enzyme capable of metabolizing alcohols. Other data (e.g. Taylorson and Hendricks, 1979) appear to show that non-ADH1 substrates such as methanol, acetone, chloroform and ethyl ether also display light-dependent dormancy-breaking activity. However, the narrow concentration-activity band separating light-dependent versus light-independent activity of ethanol in comparison with the broad range of tested concentrations for the non-ADH1 substrates also suggests that light-independent dormancy-breaking activities might have been overlooked. A possible alternative is that the P-450-mediated activity is also light-induced.

Trifluoroethanol (TFE)

TFE is a competitive inhibitor of ADH1 and has a reverse bimodal effect on the germination of *Rumex crispus.* At low concentrations it partially inhibits germination, while at higher concentrations dormancy is broken (Taylorson, 1987; Taylorson and Di Nola, 1990). How can this be explained? In animals, competitive inhibitors of ADH1 can induce cytochrome P-4502E1, although TFE has not been evaluated (Koop *et al.*, 1985; Lieber, 1997). Reduced seed germination could be due to the inhibitory presence of TFE, but there is no apparent ADH1 even in the untreated seeds (Taylorson, 1984, 1987). As TFE concentration increased, P-450 would be induced. One or more cytochrome P-450s can metabolize TFE to trifluoroacetic acid in rats (Kaminsky *et al.*, 1992). Therefore, TFE may simply be a precursor for the generation of another active dormancy-breaking acid, similar to results obtained with other alcohols (Cohn, 1996b). In passing, it is also interesting that *R. crispus* does not germinate in the presence of alcohols that would be classified as ADH1 substrates (Table 18.1). Only inducers of cytochrome P-450 such as methanol, isopropanol, acetone and TFE break dormancy in this species. This response profile is consistent with the apparent absence of detectable seed ADH1 in *R. crispus.*

Implications for Additional Research

Recent findings strongly justify experimental pursuit of alternatives to the anaesthetic hypothesis to explain the dormancy-breaking action of small organic molecules. Based upon these results, some of the experimentally testable predictions could be:

1. A cytochrome P-450 with the ability to metabolize dormancy-breaking chemicals should be present or inducible in seeds that respond to P-450 substrates.
2. For seeds that require light for the action of dormancy-breaking substances, ADH1/P-450 should be absent from dark-imbibed seeds and should be induced by light treatments. Seeds that respond to small molecular weight organics in the dark should have the ADH1 and/or P-450 system.
3. Pyrazoles should not only inhibit dormancy-breaking, but should block ^{13}C-alcohol metabolism.
4. Metabolism of ether and chloroform should be detectable with the appropriate kinetics during the dormancy-breaking process.
5. The dormancy-breaking activity of ether, chloroform, trifluoroethanol and halothane should be inhibited by pyrazoles.

Recent data from structure–activity bioassays, metabolism of an alcohol measured via ^{13}C NMR, as well as enzyme substrate and inhibitor studies (summarized above and in Cohn, 1996b, 1997) are mutually consistent:

alcohol metabolism is required for elicitation of its dormancy-breaking activity. The anaesthetic hypothesis as originally formulated must be modified or discarded.

Acknowledgements

We thank James Moroney, Alberto Gianinetti, Lucia Doherty, and John Larkin for critical reviews of the manuscript prior to its submission. MAC expresses his deepest appreciation to Ray Taylorson for his friendship, sound advice and creative seed science.

References

Adkins, S.W., Naylor, J.M. and Simpson, G.M. (1984a) The physiological basis of seed dormancy in *Avena fatua.* V. Action of ethanol and other organic compounds. *Physiologia Plantarum* 62, 18–24.

Adkins, S.W., Simpson, G.M. and Naylor, J.M. (1984b) The physiological basis of seed dormancy in *Avena fatua.* VI. Respiration and the stimulation of germination by ethanol. *Physiologia Plantarum* 62, 148–152.

Alifimoff, J.K., Firestone, L.L. and Miller, K.W. (1987) Anesthetic potencies of secondary alcohol enantiomers. *Anesthesiology* 66, 55–59.

Amritphale, D., Dixit, S. and Singh, B. (1993) Effect of acetone on the induction and breakage of secondary dormancy in seeds of cucumber. *Journal of Experimental Botany* 44, 1621–1626.

Bewley, J.D. (1997) Seed germination and dormancy. *The Plant Cell* 9, 1055–1066.

Bewley, J.D. and Black, M. (1982) *Physiology and Biochemistry of Seeds in Relation to Germination. 2. Viability, Dormancy and Environmental Control.* Springer-Verlag, Berlin.

Bewley, J.D. and Black, M. (1994) *Seeds. Physiology of Development and Germination,* 2nd edn. Plenum Press, New York.

Casal, J.J. and Sánchez, R.A. (1998) Phytochromes and seed germination. *Seed Science Research* 8, 317–329.

Chadoeuf-Hannel, R. and Taylorson, R.B. (1985) Anaesthetic stimulation of *Amaranthus albus* seed germination: interaction with phytochrome. *Physiologia Plantarum* 65, 451–454.

Cohn, M.A. (1987) Mechanisms of physiological seed dormancy. In: Frazier, G.W. and Evans, R.A. (eds) *Seed and Seedbed Ecology of Rangeland Plants.* USDA-ARS, Washington, DC, pp. 14–20.

Cohn, M.A. (1996a) Operational and philosophical decisions in seed dormancy research. *Seed Science Research* 6, 147–153.

Cohn, M.A. (1996b) Chemical mechanisms of breaking seed dormancy. *Seed Science Research* 6, 95–99.

Cohn, M.A. (1997) QSAR modelling of dormancy-breaking chemicals. In: Ellis, R.H., Black, M., Murdoch, A.J. and Hong. T.D. (eds) *Basic and Applied Aspects of Seeds.* Kluwer Academic, Dordrecht, pp. 289–295.

Cohn, M.A., Jones, K.L., Chiles, L.A. and Church, D.F. (1989) Seed dormancy in red rice. VII. Structure–activity studies of germination stimulants. *Plant Physiology* 89, 879–882.

Cohn, M.A., Church, D.F., Ranken, J. and Sanchez, V. (1991) Hydroxyl group position governs activity of dormancy-breaking chemicals. *Plant Physiology* 96, S-63.

Corbineau, F., Gouble, B., Lecat, S. and Come, D. (1991) Stimulation of germination of dormant oat (*Avena sativa* L.) seeds by ethanol and other alcohols. *Seed Science Research* 1, 21–28.

Di Nola, L. and Taylorson, R.B. (1990) Ultrastructural changes in radicles of barnyardgrass (*Echinochloa crus-galli* [L.] Beauv.) seeds during transition from dormancy to germination. *Israel Journal of Botany* 39, 303–312.

Di Nola, L., Mischke, C.F. and Taylorson, R.B. (1990) Changes in the composition and synthesis of proteins in cellular membranes of *Echinochloa crus-galli* (L.) Beauv. seeds during the transition from dormancy to germination. *Plant Physiology* 92, 427–433.

Di Nola, L., Taylorson, R.B. and Berlin, E. (1991) Thermotropic properties of cellular membranes in dormant and non-dormant *Echinochloa crus-galli* (L.) Beauv. seeds. *Journal of Experimental Botany* 42, 113–121.

Duffus, J.H. (1968) Alcohol dehydrogenase in the barley embryo. *Phytochemistry* 7, 1135–1137.

Eriksson, C.E. (1968) Alcohol:NAD oxidoreductase (E.C. 1.1.1.1.) from peas. *Journal of Food Science* 33, 525–532.

Evenari, M. (1984) Seed physiology: its history from antiquity to the beginning of the 20th century. *The Botanical Review* 50, 119–142.

Fluhr, R. (1998) Ethylene perception: from two-component signal transducers to gene induction. *Trends in Plant Science* 3, 141–146.

Footitt, S., Vargas, D. and Cohn, M.A. (1995) Seed dormancy in red rice. X. A ^{13}C NMR study of the metabolism of dormancy-breaking chemicals. *Physiologia Plantarum* 94, 667–671.

Gudjónsdóttir, S. and Burström, H. (1962) Growth-promoting effects of alcohols on excised wheat roots. *Physiologia Plantarum* 15, 498–504.

Hansch, C., Schaeffer, J. and Kerley, R. (1972) Alcohol dehydrogenase structure–activity relationships. *Journal of Biological Chemistry* 247, 4703–4710.

Hasinoff, B.B. and Davey, J.P. (1989) The inhibition of a membrane-bound enzyme as a model for anaesthetic action and drug toxicity. *Biochemical Journal* 258, 101–107.

Hendricks, S.B. and Taylorson, R.B. (1980) Reversal by pressure of seed germination promoted by anesthetics. *Planta* 149, 108–111.

Hilhorst, H.W.M. (1995) A critical update on seed dormancy. I. Primary dormancy. *Seed Science Research* 5, 61–73.

Hilhorst, H.W.M. (1998) The regulation of secondary dormancy. The membrane hypothesis revisited. *Seed Science Research* 8, 77–90.

Kaminsky, L.S., Fraser, J.M., Seaman, M. and Dunbar, D. (1992) Rat liver metabolism and toxicity of 2,2,2-trifluoroethanol. *Biochemical Pharmacology* 44, 1829–1837.

Koop, D.R., Crump, B.L., Nordblom, G.D. and Coon, M.J. (1985) Immunochemical evidence for induction of the alcohol-oxidizing cytochrome P-450 of rabbit liver microsomes by diverse agents: ethanol, imidazole, trichloroethylene, acetone, pyrazole, and isoniazid. *Proceedings of the National Academy of Sciences USA* 82, 4065–4069.

Koornneef, M., Leon-Kloosterziel, K.M., Schwartz, S.H. and Zeevaart, J.A.D. (1998) The genetic and molecular dissection of abscisic acid biosynthesis and signal transduction in *Arabidopsis*. *Plant Physiology and Biochemistry* 36, 83–89.

Leblová, S. and Perglerova, E. (1976) Soybean alcohol dehydrogenase. *Phytochemistry* 15, 813–815.

Lieber, C.S. (1997) Cytochrome P-4502E1: its physiological and pathological role. *Physiological Reviews* 77, 517–544.

Lin, T.-Y. (1997) The metabolism of dormancy-breaking compounds and its role in dormancy-breaking processes. MSc Thesis, Louisiana State University, Baton Rouge, Louisiana.

Lin, T.-Y. and Cohn, M.A. (1997) The involvement of alcohol oxidation via ADH in the seed dormancy-breaking process. *Plant Physiology* 114, S-294.

Mer, C.L. (1958) Growth-promoting effect of ethanol on oat seedlings. *Nature* 182, 1812–1813.

Mer, C.L. (1961) Carbon dioxide and ethanol as factors controlling the growth of etiolated oat seedlings. *Nature* 191, 260–261.

Miller, K.W. (1975) The pressure reversal of anesthesia and the critical membrane hypothesis. In: Fink, B.R. (ed.) *Progress in Anesthesiology*, vol. 1. *Molecular Mechanisms of Anesthesia.* Raven Press, New York, pp. 341–351.

Morré, D.J., Rogers, B.J. and Gamble, R. (1965) Promotion of plant growth by long chain alcohols and organic solvents. *Phyton* 22, 7–12.

Moss, G.W.J., Franks, N.P. and Lieb, W.R. (1991) Modulation of the general anesthetic sensitivity of a protein: a transition between two forms of firefly luciferase. *Proceedings of the National Academy of Sciences, USA* 88, 134–138.

Pecket, R.C. and Al-Charchafchi, F. (1978) Dormancy in light-sensitive lettuce seeds. *Journal of Experimental Botany* 29, 167–173.

Ritchie, S. and Gilroy, S. (1998) Gibberellins: regulating genes and germination. *New Phytologist* 140, 363–383.

Seeman, P. (1972) The membrane actions of anesthetics and tranquilizers. *Pharmacological Reviews* 24, 583–665.

Shinomura, T. (1997) Phytochrome regulation of seed germination. *Journal of Plant Research* 110, 151–161.

Sreenivasulu, Y. and Amritphale, D. (1998) Chemical stimulation of germination and membrane fluidity change in secondarily dormant cucumber seeds. *Current Science* 75, 1396–1399.

Taylorson, R.B. (1979) Response of weed seeds to ethylene and related hydrocarbons. *Weed Science* 27, 7–10.

Taylorson, R.B. (1982a) Anesthetic effects on secondary dormancy and phytochrome responses in *Setaria faberi* seeds. *Plant Physiology* 70, 882–886.

Taylorson, R.B. (1982b) Interaction of phytochrome and other factors in seed germination. In: Khan, A.A. (ed.) *The Physiology and Biochemistry of Seed Development, Dormancy and Germination.* Elsevier Biomedical Press, Amsterdam, pp. 323–346.

Taylorson, R.B. (1984) Prevention of action of far-red-absorbing phytochrome in *Rumex crispus* L. seeds by ethanol. *Plant Physiology* 74, 223–226.

Taylorson, R.B. (1987) Reverse bimodal action of 2,2,2-trifluoroethanol on *Rumex crispus* seed germination. *Physiologia Plantarum* 69, 716–720.

Taylorson, R.B. (1988) Anaesthetic enhancement of *Echinochloa crus-galli* (L.) Beauv. seed germination: possible membrane involvement. *Journal of Experimental Botany* 39, 50–58.

Taylorson, R.B. (1989) Responses of redroot pigweed (*Amaranthus retroflexus*) and witchgrass (*Panicum capillare*) seeds to anaesthetics. *Weed Science* 37, 93–97.

Taylorson, R.B. (1991) Interactions of alcohols and increased air pressure on *Echinochloa crus-galli* (L.) Beauv. seed germination. *Annals of Botany* 68, 337–340.

Taylorson, R.B. and Di Nola, L. (1990) Phytochrome and membrane behaviour as influenced by temperature and anaesthetic treatment of *R. crispus* seeds. *Seed Science and Technology* 18, 549–558.

Taylorson, R.B. and Hendricks, S.B. (1977) Dormancy in seeds. *Annual Review of Plant Physiology* 28, 331–354.

Taylorson, R.B. and Hendricks, S.B. (1979) Overcoming dormancy in seeds with ethanol and other anesthetics. *Planta* 145, 507–510.

Taylorson, R.B. and Hendricks, S.B. (1980/81) Anesthetic release of seed dormancy – an overview. *Israel Journal of Botany* 29, 273–280.

Toyomasu, T., Yamauchi, T., Yamane, H., Murofushi, N. and Inoue, Y. (1995) cDNA cloning and characterization of gibberellin-responsive genes in photoblastic lettuce seeds. *Bioscience Biotechnology and Biochemistry* 59, 1846–1849.

VanDerWoude, W.J. (1985) A dimeric mechanism for the action of phytochrome: evidence from photothermal interactions in lettuce seed germination. *Photochemistry and Photobiology* 42, 655–661.

Vegis, A. (1965) Die bedeutung von physikalischen und chemischen außenfaktoren bei der induktion und beendigung von ruhezuständen bei organen und geweben höherer pflanzen. *Encyclopedia of Plant Physiology* 15 (2), 534–668.

Walker-Simmons, M.K. (1998) Protein kinases in seeds. *Seed Science Research* 8, 193–200.

Are Cellular Membranes Involved in the Control of Seed Dormancy?

19

Henk W.M. Hilhorst[1] and Marc A. Cohn[2]

[1]*Department of Plant Physiology, Wageningen Agricultural University, Arboretumlaan 4, 6703 BD Wageningen, The Netherlands;* [2]*Department of Plant Pathology and Crop Physiology, 302 Life Sciences Building, Louisiana State University, Baton Rouge, LA 70803, USA*

Introduction

Seed dormancy can be reversible, i.e. in the same seed dormancy may repeatedly be relieved and imposed. This characteristic has been described for seeds in the soil seed bank and is termed 'dormancy cycling'. To our knowledge reversibility of coat-imposed dormancy or dormancy due to embryo immaturity has never been described. Thus we can safely assume that dormancy cycling is located in the embryo. Physiological dormancy has often been used as a term for embryo-located dormancy (Baskin and Baskin, 1998). It comprises an internal block to germination that can be eliminated or imposed by the action of environmental factors.

Dormancy cycling in the field is regulated by long-term fluctuations of soil temperature in concert with transient signals from other environmental factors, such as light, nitrate and soil water potential. Cycling is the basis for emergence of seedlings at defined periods of the year (Karssen, 1982; Baskin and Baskin, 1985; Bouwmeester and Karssen, 1993; Hilhorst *et al.*, 1996), and is an important mechanism of survival. During unfavourable seasons for growth and reproduction, dormancy cycling prevents seed germination (Vleeshouwers *et al.*, 1995). Suicidal seedling growth is thus avoided. It is by no means clear how these long-term temperature fluctuations are perceived and stored by the seeds. How can it be explained that summer annuals require low, and winter annuals require elevated temperatures to break their dormancy? To add to the enigma, dormancy cycling can be simulated under laboratory conditions by short-term temperature treatments. Dormancy of many weedy species can be broken

(eds J.-D. Viémont and J. Crabbé)

and induced by a few days at a constant temperature whereas in the field this may take weeks or months (Totterdell and Roberts, 1979; Hilhorst, 1990a).

An obvious difference between laboratory and field conditions is the moisture content of the seeds. In the field, seed moisture content will show long- and short-term fluctuations, whereas under laboratory conditions the seeds remain fully imbibed. Evidently, under natural conditions, temperature is never as constant as under laboratory conditions. Assuming that only the seeds in the top soil layer of 0–5 cm contribute to seedling emergence at a given time, temperature fluctuations of the atmosphere will be dampened to a lesser extent than at greater soil depths (Forcella *et al.*, 1997). Also in this top layer, fluctuations in the soil water potential will be large compared with those at greater depths. This implies that moisture content of seeds in this layer will also vary considerably. Thus, for a large part of their life time seeds will be only partially hydrated. It therefore appears safe to assume that at low hydration levels, many processes in seeds occur at much slower rates than when fully imbibed. In addition, induction and breakage of dormancy may directly depend on de- and rehydration cycles. For example, spring field emergence of giant foxtail depended on the first day that the soil water potential at 5 cm depth reached a value of −2 MPa (Forcella *et al.*, 1997). If this occurred too early in spring, emergence was low due to the induction of secondary dormancy.

Bouwmeester and Karssen (1993) have shown that dehydration of *Sisymbrium officinale* seeds prior to the germination test had a strong dormancy breaking effect during annual dormancy cycling. These examples indicate that soil water potential may influence dormancy. However, in a burial experiment with seeds of *Arabidopsis thaliana*, a parallel experiment, in which the seeds were kept fully imbibed in Petri dishes under outside temperature conditions, showed that there were no principal differences in dormancy cycling between the buried seeds and the seeds in the dishes (Derkx and Karssen, 1994). These examples suggest an intimate interplay between time, temperature and water potential of the environment in the triggering of germination. Effectively, this means that the 'summation' of temperature–time–water units determines the dormancy level of seeds and/or the fraction of a seed population that is dormant. Insufficient numbers of species have been studied to ascertain the contribution of water potential variation, but the evidence to date clearly emphasizes the importance of seasonal temperature fluctuations in dormancy cycling.

To explain the regulation of dormancy cycling of seeds in the soil seed bank the following assumptions have to be made: (i) seeds must be able to 'sense' the temperature; (ii) seeds must be able to 'store' temperature information; and (iii) seeds must be able to respond to similar temperatures in opposite directions: winter vs. summer annuals.

The Concept of the Involvement of Cellular Membranes in the Regulation of Dormancy

Over the past 25 years a role of membranes in the regulation of dormancy and germination has been frequently suggested (for review see Hilhorst, 1998). Some examples are listed below.

Changes in dormancy are accompanied by changes in membrane permeability

Hendricks and Taylorson (1976, 1978, 1979) have provided evidence that transferring seeds to temperatures above their germination optimum reduced germinability and increased leakage of amino acids. In the ten species tested, germinability was reduced at temperatures above 30°C. This coincided with a sharp increase in leakage of amino acids. It is not likely that the seeds were killed at these temperatures, but this cannot be excluded since controls were lacking in these studies. In two species this leakage did not increase and germination appeared to be optimal at the elevated temperatures. It was argued that at between 30 and 35°C a phase transition occurred in the plasma membranes, resulting in increased leakage. Apparently, these elevated temperatures were outside the 'germination temperature window'. The width of this window may be regarded as a measure of the extent of dormancy (Hilhorst, 1998).

Changes in membrane proteins are associated with changes in dormancy

Based on studies with light-requiring lettuce seeds, VanDerWoude (1985) formulated a hypothesis for the involvement of membrane-associated phytochrome receptors in the regulation of dormancy. Treatment for 4 h at 4°C, 1 h at 28°C or 5–28 h in 1% ethanol sensitized the seeds for red light (660 nm) induced germination, expressed as a very low fluence response. The different treatments resulted in very similar (fluence) responses, indicating the possible operation of a common underlying mechanism. Because ethanol, anaesthetics and other agents that modify membrane properties had been suggested to influence phytochrome-controlled germination (Taylorson and Hendricks, 1979), it was concluded that the underlying mechanism was associated with membranes. It was argued that membrane fluidity was increased under the applied conditions, thereby increasing the probability of phytochrome binding to its membrane-bound receptor.

Dormancy of seeds of barnyardgrass (*Echinochloa crus-galli*) can be easily manipulated by both temperature and application of propanol. Dormancy is broken by a 24 h treatment with *n*-propanol but not by *iso*-propanol at the same concentration. For germination, alternating temperatures are required (Di Nola *et al.*, 1989). The transition from the dormant to the germinating state was marked by changes in the composition of membrane-associated proteins, mainly in the plasma membrane-enriched fractions. A 23 kDa protein increased after treatment with *n*-propanol but decreased upon treatment with *iso*-propanol. In addition, synthesis of 15 and 18 kDa proteins was enhanced in dormant seeds. Thus, changes in dormancy may be related to changes at the level of membrane proteins, although this is probably not mediated by a direct effect of the alcohols on membranes (Cohn and Hilhorst, Chapter 18, this volume).

The receptor occupancy theory

Changes in dormancy of seeds of *Sisymbrium officinale* and *Arabidopsis thaliana* coincided with changes in sensitivity or responsiveness to light and nitrate (Hilhorst, 1990a,b; Hilhorst *et al.*, 1996). The pattern of seasonally increasing and decreasing responsiveness calculated from dose–response curves could best be explained by assuming that the number of active receptors for phytochrome and nitrate increased and decreased with the relief or onset of dormancy, respectively. It was envisaged that the availability of the receptor sites depended on their synthesis, as well as on their accessibility for their ligands phytochrome and nitrate. Accessibility may be a function of (temperature controlled) membrane fluidity, in that the degree of fluidity determines the magnitude of movements of the receptor protein within the membrane.

From the above it is clear that the evidence for a role of membranes in the regulation of seed dormancy is at best circumstantial. Yet, a role for membranes cannot be neglected, for the simple reason that temperature is the principal regulatory factor in dormancy cycling and cellular membranes are considered to be very sensitive temperature sensors in all living organisms (Murata and Los, 1997). In addition, Derkx *et al.* (1993) have shown that relief and induction of dormancy in seeds of *Sisymbrium officinale* are not accompanied by changes in activity of the respiratory pathways. Respiratory activity did not increase until the non-dormant seeds received a germination trigger. Therefore, temperature regulation of dormancy may not proceed through a simple effect on kinetics of (general) metabolic reactions.

Influence of Temperature on Properties of Membranes and Membrane-bound Proteins

Homeoviscous and homeophasic adaptation

The phase behaviour and physical properties of lipids in biological membranes are exquisitely sensitive to changes in temperature. Because membranes have a multitude of functions, temperature-induced perturbations in membrane organization will have a long-range impact on cellular functioning (Hazel, 1995). Thus, membranes may be the primary target for temperature perception at the cellular level (Minorsky, 1989; Vigh *et al.*, 1995; Murata and Los, 1997). However, this high sensitivity of membranes to temperature also poses a serious challenge to poikilotherms to maintain physiological function under altered thermal circumstances. In all living organisms membranes adapt to changes in the environmental temperature through homeoviscous adaptation (HVA) and homeophasic adaptation (HPA) (Vigh *et al.*, 1998). Effectively, cells compensate for temperature-induced cellular disturbances by changing their membrane lipid composition during adaptation to temperature shift (HVA) or by preventing transition to the gel phase (HPA). Changes may occur in lipid composition, fatty acid unsaturation, proportions of lipid classes, and in the lipid : protein ratio, in order to maintain homeostatic equilibrium. Each of these changes results in alterations in dynamic properties of the lipid fraction of membranes. Also, the activities of membrane-associated proteins, such as ion channels, receptors, transporter proteins and enzymes, may change dramatically. For instance, phagocytotic activity of the amoeba *Acanthamoeba castellanii* was highly dependent on temperature-induced changes in plasma-membrane lipid order, an effect mediated by the enzyme *n*-6-desaturase, which increased fatty acid unsaturation at low temperature (Avery *et al.*, 1995).

In thermal adaptation, generally, two temporally differing responses to the temperature shift can be distinguished: (i) an instant response, during which the transition temperatures between lipid phases may change dramatically; and (ii) a slower response during which the difference between transition temperatures and the (new) ambient temperature is restored (acclimation).

Gene expression controlled by the membrane's physical state

Evidence is accumulating that the membrane's physical state may control gene expression. Most notable is the induction of transcription of genes coding for heat-shock proteins (hsps) (reviewed in Vigh *et al.*, 1998). It has been demonstrated that genetic manipulation of the ratios of saturated : unsaturated fatty acids had a significant influence on transcriptional activity of the *hsp70* and *hsp82* genes (Carratu *et al.*, 1996). Also physical manipulation of the membrane's physical state, e.g. by temperature acclimation, by catalytic lipid hydrogenation or by adding fluidizers such as benzyl-alcohol, may affect activity of heat-shock genes. In this way it was shown that *hsp17* of *Synechocystis* functions as a 'fluidity gene', maintaining fluidity of the chloroplast thylakoid membranes (Horvath *et al.*, 1998). Apparently, some of the heat-shock proteins are able to bind to the membrane, prevent denaturation of membrane-associated proteins and restore membrane fluidity. In *Saccharomyces cerevisiae* hsps induced by temperature shock prevented impairment of the mitochondrial ATPase activity (Patriarca and Maresca, 1990). Also membrane properties of other organisms may affect the activity of hsp or other chaperone-like proteins, e.g. the GroESL chaperonins in *Escherichia coli* which bind to the membrane (Bochkareva *et al.*, 1998). To our knowledge, gene expression controlled by the physical state of the membrane has not been demonstrated in plants.

Influence of Organics on Properties of Membranes and Membrane-bound Proteins

Interactions of organics with membranes and membrane-bound proteins

There is a wealth of literature on the interaction between chemical compounds, including anaesthetics, and membranes. From these reports it is clear that there are several types of interaction. For instance, short-chain alcohols partition preferentially at the membrane–water interface and would thereby stabilize a lamellar membrane structure. The more hydrophobic longer-chain alcohols may partition deeper into the hydrophobic acyl chain region and may perturb the acyl chain packing and, thus, promote hexagonal phase formation (Tilcock and Cullis, 1987). The state of the membrane, e.g. liquid crystalline phase vs. gel phase, also influences the effectiveness of anaesthetic action (Thewalt *et al.*, 1985). A notable result of the effect of anaesthetics on membranes is an increased permeability, which appears not to be specific for any type of ion (Barchfeld and Deamer, 1985, 1988).

In the more recent literature the direct action of anaesthetic compounds on membranes has largely been abandoned. There is now good evidence that the principal target site of anaesthetics are membrane-associated proteins, in particular, hydrophobic pockets of the protein that are exposed to water (Franks and Lieb, 1994). Several examples make clear that lipid partitioning coefficients of the anaesthetics can be mimicked by anaesthetic-binding sites on proteins. Thus, the current hypothesis is that anaesthetics bind directly to membrane-bound proteins, which can be receptors.

Alcohols can elicit and suppress the heat-shock response

Recent studies have demonstrated that a heat-shock expression vector in the yeast *S. cerevisiae* can be activated by incubation with alcohols in the 1–1000 mM range. There was a log-linear correlation between the number of carbons in the alcohols and the alcohol concentration required, to induce a maximal heat-shock response. The effectiveness of methanol was 100 times lower than that of *n*-pentanol. However, no evidence was provided that the alcohols interacted with membranes nor was it shown that alcohol dehydrogenase was induced during the treatment with the alcohols (Curran and Khalawan, 1994). In contrast, hsp synthesis in *Drosophila* cultured cells is inhibited by short-chain alcohols (Munks and Turner, 1994), and the cell lines used in this study had no detectable levels of alcohol dehydrogenase activity. However, the direct action of alcohols, as opposed to their metabolic products, was not conclusively established in either of these studies (cf. Cohn and Hilhorst, Chapter 18, this volume).

Do targets of organics and temperature overlap in the regulation of dormancy?

There is now good evidence that alcohols and other organics may break dormancy through their metabolization (Cohn and Hilhorst, Chapter 18, this volume). However, there is no reason to reject the possibility that the organics interact with membranes. In order to be metabolized, the organics must pass the plasma membrane and possibly also organellar membranes. Because of their amphiphilic nature these compounds will partition in membranes and affect the membrane dynamic properties (Golovina *et al.*, 1998). The products of metabolism of organics, such as alcohols, may include compounds that may be involved in dormancy breaking through their perturbation of intracellular pH (Cohn and Hilhorst, Chapter 18, this volume) as has been observed upon breaking of dormancy in red rice embryos (Footitt and Cohn, 1992). Both metabolism and increased membrane permeability (or activation of proton pumps) may result in a change of the intracellular pH. There is a large body of evidence that intracellular pH plays an important role in changing developmental patterns in many biological systems (Kurkdjian and Guern, 1989). However, it is by no means clear what the metabolic steps are following the changes in cytoplasmic pH. So far, convincing evidence is lacking that alcohols, or their metabolites interact with membranes or membrane proteins in seeds to alter their state of dormancy.

Can organics play a role in the regulation of dormancy under natural conditions?

Although soil may contain a vast array of organic substances there is no evidence that they play a role in the regulation of dormancy or germination (Baskin and Baskin, 1998). Yet, many seeds are exposed to limited oxygen availability during the germination process due to seed coverings that may limit oxygen diffusion, or to a low partial oxygen pressure, e.g. in wet soils (Bewley and Black, 1994). The resulting fermentation processes may yield ethanol at substantial concentrations. However, this seems to occur in germinating seeds only and, thus, cannot be involved in dormancy relief. It is difficult to envisage why seeds have the machinery to have their dormancy relieved by the action of organic compounds. The ability to metabolize organics through the (concerted) action of alcohol dehydrogenase (ADH) and the cytochrome P-450 system (Cohn and Hilhorst, Chapter 18, this volume) may serve two functions: (i) to generate energy; and (ii) to detoxify the potentially toxic substances. However, it appears that 'the cure may be worse than the disease'. The active organics have an optimal concentration in the relief of dormancy (Cohn *et al.*, 1989; Cohn, 1997). Supra-optimal concentrations will lessen the response and, ultimately, seeds may lose their viability. Seeds from *Nicotiana plumbaginifolia* that are deficient in the expression of seed ADH are considerably more resistant to ethanol than the wild type. The dose that inhibited 50% germination of the wild type was 50 mM whereas the mutants were inhibited to 50% in the range 125–175 mM ethanol (Rousselin *et al.*, 1990). As the ethanol is metabolized via ADH, it must be assumed that one or more of the metabolites are more toxic than the ethanol itself.

If breaking of dormancy by organics is not relevant under natural conditions we may hypothesize that the breaking of dormancy by organics is 'accidental'. Metabolization of the organic substances may yield products that trigger signal transduction pathways that bypass the natural requirements for the relief of dormancy. The family of cytochrome P-450-dependent monooxygenases includes several enzymes that mediate the conversion of kaurene to kaurenoic acid in the gibberellins synthetic pathway (Chapple, 1998). It has also been shown that (+)-abscisic acid 8′-hydroxylase is a cytochrome P-450 monooxygenase (Krochko *et al.*, 1998). A scenario can be envisaged in which the organic induces or activates cytochrome P-450 monooxygenases that catalyse steps in gibberellin biosynthesis or abscisic acid degradation, which may have been suppressed in the dormant state. The effect of the ethanol would be comparable with administering exogenous GAs to dormant seeds, which may trigger germination.

Can the Action of Temperature on Cellular Membranes Explain Dormancy Cycling?

After discussing the interaction between temperature (and organics) and cellular membranes and membrane proteins, we can now address the question of whether membranes have the properties that would fulfil the requirements formulated at the beginning of this chapter.

Seeds must be able to sense the temperature

It is clear that membranes change their dynamic state in response to temperature changes. There is direct evidence for this in many organisms, including cyanobacteria, yeast, fish and mammals (Vigh *et al.*, 1998). In plants most research efforts have aimed at elucidating the mechanism of chilling sensitivity. Most of the evidence suggests that the unsaturation of lipids is associated with the chilling resistance. Genetic manipulation of the level of unsaturation of chloroplast membranes may alter photosynthetic activity and growth patterns (Nishida and Murata, 1996). In sunflower (*Helianthus annuus* L.) seeds the activity of one of the fatty acid desaturating enzymes, oleoyl phosphatidyl choline desaturase (ODS), is increased by low temperatures. Treatment of the seeds for only 2 h at 10°C resulted in an increase of over 30% of the relative content of the 18 : 2 fatty acids in the microsomal fractions at the expense of 18 : 1 fatty acids (Sarmiento *et al.*, 1997). As there were no data on performance of these seeds prior to and after the cold treatment, it remains unclear whether the treatment altered the state of dormancy.

Seeds must be able to 'store' temperature information

As pointed out above, organisms may respond instantaneously to temperature shock by producing heat-shock proteins that may serve to protect membrane proteins and maintain membrane fluidity. However, this response is transient. On a longer term the organisms will acclimate to the new thermal conditions by adjusting the composition of their membrane lipids in such a way that the initial fluidity is largely restored. For example, the ratios of the saturated fatty acids, and mono-, di- and tri-unsaturated fatty acids will change (Hazel, 1995). It is obvious that the altered lipid composition contains information about the thermal history of the organism. The biochemical and physical restructuring of cell membranes within their lipid components is well-known in cold acclimation of whole plants and plant organs, such as buds and bulbs (Guy, 1990).

Similar temperatures can act in opposite directions: activation/inactivation

In a study of the modulation of membrane-bound protein kinase C (PKC) activity by the lipid bilayer structure, Slater *et al.* (1994) found that PKC activity was either attenuated or potentiated depending on the ratios of phosphatidylcholine, phosphatidylethanolamine and cholesterol. The PKC activity was not so much dependent on membrane fluidity as on spacing of the phospholipid head groups. Although the study was performed with a model system (vesicles) it clearly shows that membrane structure, as determined by its lipid composition, can have a dramatic impact on the function of membrane-bound proteins.

It is clear that we cannot (yet) claim that cellular membranes are the key regulatory unit in temperature-regulated dormancy. However, the potential of cellular membranes to exert this function cannot be neglected, particularly now that it is becoming evident that membrane dynamic properties may influence gene expression. As yet, only a relatively small number of studies have been undertaken to identify genes that are differentially expressed in dormant and non-dormant seeds. However, it is surprising that many of the genes that are expressed in dormant seeds are stress-related genes, encoding for heat-shock proteins and LEA (late embryogenesis abundant) proteins. In *Arabidopsis thaliana* levels of the small heat-shock protein were determined in a range of mutants that were reduced in abscisic acid (ABA) or in their sensitivity to ABA. It appeared that the *abi3-1* and *abi3-6* mutants had severely reduced levels whereas levels in the other mutants were undistinguishable from wild type. It was suggested that the synthesis of seed-specific heat-shock protein was correlated with the ABI3 response pathway. This pathway is involved in desiccation tolerance and dormancy. The *abi3* mutants are essentially non-dormant (Wehmeyer *et al.*, 1996). Li and Foley (1995) have reported the expression of a group 1 LEA protein in seeds from the dormant M73 line of *Avena fatua*. Both the transcription activity and transcript level remained high in the imbibing dormant seeds but declined in afterripened seed during imbibition (Li and Foley, 1996). However, upregulation of stress-related proteins has also been reported. During dormancy breakage in seeds of the Douglas fir (*Pseudotsuga menziesii*) three genes that belonged to three different classes of LEA proteins were expressed at a high level during the 4 weeks of cold treatment whereas at 20°C expression decreased after 1 week (Jarvis *et al.*, 1996).

Parallels between Heat and Cold Responses and Changes in Dormancy

One of the most striking parallels between heat and cold responses and changes in dormancy is the resulting change in thermotolerance. In many organisms heat or cold shock results in an adaptation to the new temperature circumstances, i.e. functioning is restored to normal at formerly non-permissive temperatures (Hazel, 1995; Vigh *et al.*, 1998). In seeds, a chilling treatment may break dormancy. One of the results of such a treatment is the widening of the germination temperature 'window' (Hilhorst, 1998). In other words, non-dormant seeds have a greater thermotolerance (for germination) than dormant seeds, and vice versa. Secondly, organics, such as alcohols, break dormancy (Cohn and Hilhorst, Chapter 18, this volume) but they may also induce events usually associated with the response to heat shock. For example, in the yeast *S. cerevisiae* alcohols lower the threshold for the maximal activation of a heat-shock transcription factor (Curran and Khalawan, 1994), and in *Synechocystis* benzylalcohol could induce heat-shock protein expression at temperatures at which the cells were acclimatized (Horvath *et al.*, 1998). Thus, it appears possible that the expression of a number of stress proteins would be associated with the regulation of dormancy. Whether these proteins are components of the signal transduction paths that are involved in the regulation of dormancy remains to be shown.

Conclusions and Prospects

There is widespread consensus on the involvement of cellular membranes in the response to alterations in temperature and other environmental stresses in many organisms. Parallels can be drawn between the stress responses and changes in seed dormancy, which is largely regulated by temperature. So far only circumstantial evidence provides support for a role of membranes in the regulation of dormancy (Hilhorst, 1998). However, the current state of technology now allows us to study stress responses and dormancy in seeds in similar detail as in model (unicellular) organisms such as *Synechocystis*. It is feasible to influence expression of heat-shock proteins and desaturases in seeds by means of anti-sense techniques and observe the effects on dormancy. In addition, the availability of mutants of *Arabidopsis thaliana* with altered fatty acid composition or dormancy will assist in assessing the role of membranes. Also, studying membrane dynamic properties in seeds is coming within reach. With the use of specific spin labels that penetrate membranes to different depths, electron paramagnetic resonance spectrometry can provide information about the label's micro-environment in the membrane (E. Golovina, 1999, Wageningen, personal communication).

References

Avery, S.V., Lloyd, D. and Harwood, J.L. (1995) Temperature-dependent changes in plasma-membrane lipid order and the phagocytotic activity of the amoeba *Acanthamoeba castellani* are closely correlated. *Biochemical Journal* 312, 811–816.

Barchfeld, G.L. and Deamer, D.W. (1985) The effect of general anesthetics on the proton and potassium permeabilities of liposomes. *Biochimica et Biophysica Acta* 819, 161–169.

Barchfeld, G.L. and Deamer, D.W. (1988) Alcohol effects on lipid bilayer permeability to protons and potassium: relation to the action of general anesthetics. *Biochimica et Biophysica Acta* 944, 40–48.

Baskin, J.M. and Baskin, C.C. (1985) The annual dormancy cycle in buried weed seeds: a continuum. *BioScience* 35, 492–498.

Baskin, J.M. and Baskin, C.C. (1998) *Seeds. Ecology, Biogeography, and Evolution of Dormancy and Germination.* Academic Press, San Diego.

Bewley, J.D. and Black, M. (1994) *Seeds. Physiology of Development and Germination,* 2nd edn. Plenum Press, New York.

Bochkareva, E.S., Solovieva, M.E. and Girshovich, A.S. (1998) Targeting of GroEL to SecA on the cytoplasmic membrane of *Escherichia coli. Proceedings of the National Academy of Sciences, USA* 95, 478–483.

Bouwmeester, H.J. and Karssen, C.M. (1993) Annual changes in dormancy and germination in seeds of *Sisymbrium officinale* (L.) Scop. *New Phytologist* 124, 179–191.

Carratu, L., Franceschelli, S., Pardini, C.L., Kobayashi, G.S., Horvath, I., Vigh, L. and Maresca, B. (1996) Membrane lipid perturbation modifies the set point of the temperature of heat shock response in yeast. *Proceedings of the National Academy of Sciences, USA* 93, 3870–3875.

Chapple, C. (1998) Molecular-genetic analysis of plant cytochrome P450-dependent monooxygenases. *Annual Review of Plant Physiology and Plant Molecular Biology* 49, 311–343.

Cohn, M.A. (1997) QSAR modelling of dormancy-breaking chemicals. In: Ellis, R.H., Black, M., Murdoch, A.J. and Hong, T.D. (eds) *Basic and Applied Aspects of Seed Biology.* Kluwer Academic Publishers, Dordrecht, pp. 289–295.

Cohn, M.A., Jones, K.L., Chiles, L.A. and Church, D.F. (1989) Seed dormancy in red rice. VII. Structure–activity studies of germination stimulants. *Plant Physiology* 89, 879–882.

Curran, B.P.G. and Khalawan, S.A. (1994) Alcohols lower the threshold temperature for the maximal activation of a heat shock expression vector in the yeast *Saccharomyces cerevisiae. Microbiology* 140, 2225–2228.

Derkx, M.P.M. and Karssen, C.M. (1994) Are seasonal dormancy patterns in *Arabidopsis thaliana* regulated by changes in seed sensitivity to light, nitrate and gibberellin? *Annals of Botany* 73, 129–136.

Derkx, M.P.M., Smidt, W.J., VanDerPlas, L.H.W. and Karssen, C.M. (1993) Changes in dormancy of *Sisymbrium officinale* seeds do not depend on changes in respiratory activity. *Physiologia Plantarum* 89, 707–718.

Di Nola, L., Mischke, C.F. and Taylorson, R.B. (1989) Changes in the composition and synthesis of proteins in cellular membranes of *Echinochloa crus-galli* (L.) Beauv. seeds during the transition from dormancy to germination. *Plant Physiology* 92, 427–433.

Footitt, S. and Cohn, M.A. (1992) Seed dormancy in red rice. VIII. Embryo acidification during dormancy-breaking and subsequent germination. *Plant Physiology* 100, 1196–1202.

Forcella, F., Wilson, R.G., Dekker, J., Kremer, R.J., Cardina, J., Anderson, R.L., Alm, D., Renner, K.A., Harvey, R.G., Clay, S. and Buhler, D.D. (1997) Weed seed bank emergence across the Corn Belt. *Weed Science* 45, 67–76.

Franks, N.P. and Lieb, W.R. (1994) Molecular and cellular mechanisms of general anaesthesia. *Nature* 367, 607–614.

Golovina, E.A., Hoekstra, F.A. and Hemminga, M.A. (1998) Drying increases intracellular partitioning of amphiphilic substances into the lipid phase. *Plant Physiology* 118, 975–986.

Guy, C.L. (1990) Cold acclimation and freezing stress tolerance: role of protein metabolism. *Annual Review of Plant Physiology and Plant Molecular Biology* 41, 187–223.

Hazel, J.R. (1995) Thermal adaptation in biological membranes: is homeoviscous adaptation the explanation? *Annual Review of Physiology* 57, 19–42.

Hendricks, S.B. and Taylorson, R.B. (1976) Variation in germination and amino acid leakage of seeds with temperature related to membrane phase change. *Plant Physiology* 58, 7–11.

Hendricks, S.B. and Taylorson, R.B. (1978) Dependence of phytochrome action in seeds on membrane organization. *Plant Physiology* 61, 17–19.

Hendricks, S.B. and Taylorson, R.B. (1979) Dependence of thermal responses of seeds on membrane transitions. *Proceedings of the National Academy of Sciences, USA* 76, 778–781.

Hilhorst, H.W.M. (1990a) Dose–response analysis of factors involved in germination and secondary dormancy of seeds of *Sisymbrium officinale*. I. Phytochrome. *Plant Physiology* 94, 1090–1095.

Hilhorst, H.W.M. (1990b) Dose–response analysis of factors involved in germination and secondary dormancy of seeds of *Sisymbrium officinale*. II. Nitrate. *Plant Physiology* 94, 1096–1102.

Hilhorst, H.W.M. (1998) The regulation of secondary dormancy. The membrane hypothesis revisited. *Seed Science Research* 8, 77–90.

Hilhorst, H.W.M., Derkx, M.P.M. and Karssen, C.M. (1996) An integrating model for seed dormancy cycling: characterization of reversible sensitivity. In: Lang, G. (ed.) *Plant Dormancy: Physiology, Biochemistry and Molecular Biology*. CAB International, Wallingford, UK, pp. 341–360.

Horvath, I., Glatz, A., Varvaszovszki, V., Török, Z., Páli, T., Balogh, G., Kovács, E., Nádasdi, L., Benkö, S., Joó, F. and Vigh, L. (1998) Membrane physical state controls the signaling mechanism of the heat shock response in *Synechocystis* PCC 6803: identification of *hsp17* as a 'fluidity gene'. *Proceedings of the National Academy of Sciences, USA* 95, 3513–3518.

Jarvis, S.B., Taylor, M.A., Banco, J., Corbineau, F. and Davies, H.V. (1997) Dormancy-breakage in seeds of Douglas fir (*Pseudotsuga menziesii*) (Mirb.) (France). Support for the hypothesis that LEA gene expression is essential for this process. *Journal of Plant Physiology* 151, 457–464.

Karssen, C.M. (1982) Seasonal patterns of dormancy in weed seeds. In: Khan, A.A. (ed.) *The Physiology and Biochemistry of Seed Development, Dormancy and Germination*. Elsevier Biomedical Press, Amsterdam, pp. 243–270.

Krochko, J.E., Abrams, G.D., Loewen, M.K., Abrams, S.R. and Cutler, A.J. (1998) (+)-Abscisic acid 8′-hydroxylase is a cytochrome P450 monooxygenase. *Plant Physiology* 118, 849–860.

Kurkdjian, A. and Guern, J. (1989) Intracellular pH: measurement and importance in cell activity. *Annual Review of Plant Physiology and Plant Molecular Biology* 40, 271–303.

Li, B. and Foley, M.E. (1995) Cloning and characterization of differentially expressed genes in imbibed dormant and afterripened *Avena fatua* embryos. *Plant Molecular Biology* 29, 823–831.

Li, B. and Foley, M.E. (1996) Transcriptional and posttranscriptional regulation of dormancy-associated gene expression by afterripening in wild oat. *Plant Physiology* 110, 1267–1273.

Minorsky, P.V. (1989) Temperature sensing by plants: a review and hypothesis. *Plant, Cell and Environment* 12, 119–135.

Munks, R.J.L. and Turner, B.M. (1994) Suppression of heat-shock protein synthesis by short-chain fatty acids and alcohols. *Biochimica et Biophysica Acta* 1223, 23–28.

Murata, N. and Los, D.A. (1997) Membrane fluidity and temperature perception. *Plant Physiology* 115, 875–879.

Nishida, I. and Murata, N. (1996) Chilling sensitivity in plants and cyanobacteria: the crucial contribution of membrane lipids. *Annual Review of Plant Physiology and Plant Molecular Biology* 47, 541–568.

Patriarca, E.J. and Maresca, B. (1990) Acquired thermotolerance following heat shock protein synthesis prevents impairment of mitochondrial ATPase activity at elevated temperatures in *Saccharomyces cerevisiae*. *Experimental Cell Research* 190, 57–64.

Rousselin, P., Lepingle, A., Faure, J.-D., Bitoun, R. and Caboche, M. (1990) Ethanol-resistant mutants of *Nicotiana plumbaginifolia* are deficient in the expression of pollen and seed alcohol dehydrogenase. *Molecular and General Genetics* 222, 409–415.

Sarmiento, C., Garcés, R. and Mancha, M. (1998) Oleate desaturation and acyl turnover in sunflower (*Helianthus annuus* L.) seed lipids during rapid temperature adaptation. *Planta* 205, 595–600.

Slater, S.J., Kelly, M.B., Taddeo, F.J., Ho, C., Rubin, E. and Stubbs, C.D. (1994) The modulation of protein kinase C activity by membrane lipid bilayer structure. *Journal of Biological Chemistry* 269, 4866–4871.

Taylorson, R.B. and Hendricks, S.B. (1979) Overcoming dormancy in seed with ethanol and other anesthetics. *Planta* 145, 507–510.

Thewalt, J.L., Wassall, S.R., Gorissen, H. and Cushley, R.J. (1985) Deuterium NMR study of the effect of n-alkanol anesthetics on a model membrane system. *Biochimica et Biophysica Acta* 817, 355–365.

Tilcock, C.P. and Cullis, P.R. (1987) Lipid polymorphism. *Annals of the New York Academy of Sciences* 492, 88–102.

Totterdell, S. and Roberts, E.H. (1979) Effects of low temperatures on the loss of innate dormancy and the development of induced dormancy in seeds of *Rumex obtusifolius* L. and *Rumex crispus* L. *Plant, Cell and Environment* 2, 131–137.

VanDerWoude, W.J. (1985) A dimeric mechanism for the action of phytochrome: evidence from photothermal interactions in lettuce seed germination. *Photochemistry and Photobiology* 42, 655–661.

Vigh, L., Los, D.A., Murata, N., Glatz, A., Kovács, E. and Horváth, I. (1995) Is the membrane the primary target in the biological perception of temperature? Effect of membrane physical state on the expression of stress-defense genes. In: Kader, J.-C. and Mazliak, P. (eds) *Plant Lipid Metabolism.* Kluwer Academic Publishers, Dordrecht, pp. 369–371.

Vigh, L., Maresca, B. and Harwood, J.L. (1998) Does the membrane's physical state control the expression of heat shock and other genes? *Trends in Biochemical Sciences* 23, 369–374.

Vleeshouwers, L.M., Bouwmeester, H.J. and Karssen, C.M. (1995) Redefining seed dormancy: an attempt to integrate physiology and ecology. *Journal of Ecology* 83, 1031–1037.

Wehmeyer, N., Hernandez, L.D., Finkelstein, R.R. and Vierling, E. (1996) Synthesis of small heat-shock proteins is part of the developmental program of late seed maturation. *Plant Physiology* 112, 747–757.

Intercellular Communication Channels and Intracellular Calcium Levels Involved in the Dormancy Development of Poplar (*Populus deltoides*) Plants

20

Ling-Cheng Jian,[1] Ji-Hong Li,[1] Paul H. Li[1] and Tony H.H. Chen[2]

[1]Laboratory of Plant Hardiness, Department of Horticultural Science, University of Minnesota, St Paul, MN 55108, USA; [2]Department of Horticulture, Oregon State University, ALS 4017, Corvallis, OR 97331, USA

Introduction

The development of vegetative dormancy is an important adaptive strategy for the survival and regrowth of temperate perennials, because the dormant plants are better equipped to withstand the potentially damaging low temperatures in the winter months.

As the apical meristem (growing point) is the centre responsible for the primary growth and development of the shoots, dormancy can be viewed as the development of the growing point with a stationary phase (Rees, 1981). During dormancy, the entire tissue of the apical meristem loses its development vector, and settles into an aspatially heterogeneous state (Kauffman, 1993).

It has been known that all cells in the apical meristem of any plant species are connected by plasmodesmata, forming a symplast (Robards and Lucas, 1990; Gunning and Steer, 1996). The plasmodesmal intercellular communication channels play an important role in plant growth and development. Plasmodesmal alterations, such as functional symplasmic domains, are considered to be involved in cell division and differentiation, tissue formation, organogenesis and other specialized functions including the responses and adaptations of plants to changes in the environment

 Dormancy in Plants (eds J.-D. Viémont and J. Crabbé)

(Robards and Lucas, 1990; Lucas *et al.*, 1993). There is evidence suggesting that intercellular communication channels are associated with the plant's growth and dormancy development. In winter, *Chara* cells undergo a period of vegetative dormancy as characterized by restricted cell-to-cell communication, inactive branch dactyls and restricted growth. In spring, active growing branch dactyls were characterized by extensive intercellular communications (Shepherd and Goodwin, 1992). In 4°C-stored potato tubers, symplasmic domains were formed in bud meristems (Lucas *et al.*, 1993).

For the last several years, we have used electron microscopy (EM) combined with cytochemical technology to investigate the alterations of the plasmodesmal ultrastructure and the subcellular localization of calcium and Ca^{2+}-ATPase activity in the cells of poplar apical meristems during short-day (SD)-induced dormancy. Our findings suggest that plasmodesmal intercellular communication channels and the levels of intracellular calcium are involved in the dormancy development of poplar plants.

Materials and Methods

Preparation of plant materials and measurement of dormancy

Dormant poplar (*Populus deltoides* Bartr. Ex Marsh) plants were prepared and their dormancy was measured according to Jian *et al.* (1997).

Materials used for EM-cytochemical fixations were apical buds that were collected from plants grown under 16-h long day (LD); exposed to 8-h short day (SD) for 20, 28, 35, 42, 49, 70 or 77 days; or exposed to SD for 42 days to first induce the dormancy and then transferred to LD for 30 days, at which time the bud dormancy was just broken.

Under SD photoperiod in a warm temperature regime, plant growth rate decreases immediately upon transferring to SD from LD. Plant growth completely stopped after 20 days SD. There was a gradual increase in the degree of dormancy as evidenced by the delayed bud break under LD, which increased from 14 to 29 days after 28 to 49 days SD, respectively. After 70 days SD, it took more than 3 months of LD before regrowth became visible (Jian *et al.*, 1997).

Preparation for the ER-plasmodesma observation

Because the plasmodesma is a structurally continuous system of the plasma membrane (PM) and the endoplasmic reticulum (ER), we used a modified potassium permanganate fixative for the fixation of membrane lipids (Luft, 1956; Whaley *et al.*, 1960; Jian *et al.*, 1973).

It has been known that potassium permanganate fixative is less desirable for fixing the cellular ultrastructure. We believe, however, after modifications of the method, that it performs well for fixing the membranous configuration, especially the lipid part of the membrane.

After removing the bud scales and young leaves, buds were cut into 0.5 × 0.5 × 0.5 mm slices, and immediately immersed in a 1.2% potassium permanganate ($KMnO_4$) solution containing 0.5% sodium chloride for 4 h at 0–4°C. After fixation, samples were washed at 0–4°C with 0.5% NaCl solution, and then passed once through a 30% ethanol washing. Samples were further fixed in a 70% ethanol for 12 h at 0–4°C. After the ethanol fixation, samples were dehydrated in a series of 80, 90 and 100% ethanol at room temperature and then embedded in Embed 812. After sectioning, sections were stained with uranyl acetate and lead citrate and examined under transmission electron microscopy (TEM).

Preparation for the Ca^{2+} EM-cytochemical observation

Subcellular calcium localization was analysed according to the method of Slocum and Roux (1982), and adapted by Wang and Jian (1994).

Apical buds were cut into 0.5 × 0.5 × 0.5 mm slices, and immediately immersed in a fixative containing 4% glutaraldehyde and 2% potassium antimonate in 0.1 M potassium phosphate buffer, pH 7.6, for 4 h at 4°C. Slices were washed three times, 20 min each, with 0.1 M potassium phosphate buffer, pH 7.6, containing 2% potassium antimonate. Samples were further fixed in potassium phosphate buffer containing 1% osmium tetroxide and 2% potassium antimonate at 4°C overnight. Samples were bathed twice in phosphate buffer containing 2% potassium antimonate followed by twice with pH 10 distilled water. Samples were examined under TEM after dehydration, embedding, sectioning and staining.

Parallel with the above sample preparation, a series of samples were fixed with glutaraldehyde-osmium fixative without potassium antimonate. These samples served as the control for the Ca^{2+} cytochemistry.

In order to confirm that the antimonate precipitate contained Ca^{2+}, chelation of calcium ions was performed with EGTA. The grids, mounted with sample sections that had been examined by EM, were immersed in 100 mM EGTA, pH 8, and incubated at 60°C for 1 h. The grids were then rinsed with water, stained with uranyl acetate again and examined under EM.

Preparation for the Ca^{2+}-ATPase EM-cytochemical observation

Cytochemical samples for Ca^{2+}-ATPase activity were prepared by the one-step lead method of Ando *et al.* (1981) with minor modifications.

Slices of apical buds were fixed in 4% paraformaldehyde in 0.1 M sodium cacodylate buffer, pH 7.2, for 1 h at room temperature. After fixation, samples were bathed twice in 0.1 M sodium cacodylate buffer, followed by twice with Tris-maleate buffer, pH 7.5, 25 min each at 4°C. Finally, samples were incubated at room temperature for 1 h in a reaction medium consisting of 50 mM Tris-maleate buffer (pH 7.5), 2 mM ATP-Na, 1.5 mM $CaCl_2$ and 3 mM $CeCl_3$.

Four controls were conducted to confirm the specificity of the reaction product:

1. ATP was omitted from the reaction medium.
2. Samples were pre-treated in Tris-maleate buffer containing 5 mM EGTA for 30 min and then incubated in the reaction medium without $CaCl_3$ in which 5 mM EGTA was added.
3. Samples were pre-treated for 30 min at room temperature in Tris-maleate buffer containing 10 mM sodium ortho-vanadate (Na_3VO_4), an inhibitor of the PM Ca^{2+}-ATPase activity (Gioglio *et al.*, 1991) and then incubated in the complete reaction medium containing 10 mM Na_3VO_4.
4. Samples were pre-treated for 30 min at room temperature in Tris-maleate buffer containing 0.1 mM erythrosin B, a specific inhibitor for Ca^{2+}-ATPase activity (Rasi-Caldogno *et al.*, 1989; Williams *et al.*, 1990), and then incubated in the complete reaction medium containing 0.1 mM erythrosin B.

After the incubation, samples were further fixed in a cacodylate buffer, pH 7.2, containing 4% glutaraldehyde for 3 h at 4°C and then fixed in a cacodylate buffer, pH 7.2, containing 1% osmium tetroxide (OsO_4) at 4°C overnight. After dehydration, embedding and sectioning, samples were stained with uranyl acetate and examined under EM.

Results

Changes in plasmodesmata and endoplasmic reticuli in the poplar apical bud meristematic cells during SD-induced dormancy

Under LD/warm conditions, endoplasmic reticulum (ER), mitochondrion (M), plastid (P) and Golgi body (G) were abundantly distributed in the meristematic cells. A relatively large-sized nucleus (N) was usually observed (Fig. 20.1A, B, D and E). Many ERs paralleled the plasma membrane (PM) and the cell wall (W) (Fig. 20.1A and B). Plasmodesmata (PD) existed between cells, indicating a perfectly integrated symplasmic network in the poplar apical bud meristematic cells (Fig. 20.1A–E).

In addition, a double layer of membranous structure of the nuclear envelope (NE) was clearly seen with ER connected to the outer layer of the nuclear membrane (Fig. 20.1A, B, D and E). Other ERs, while connected to

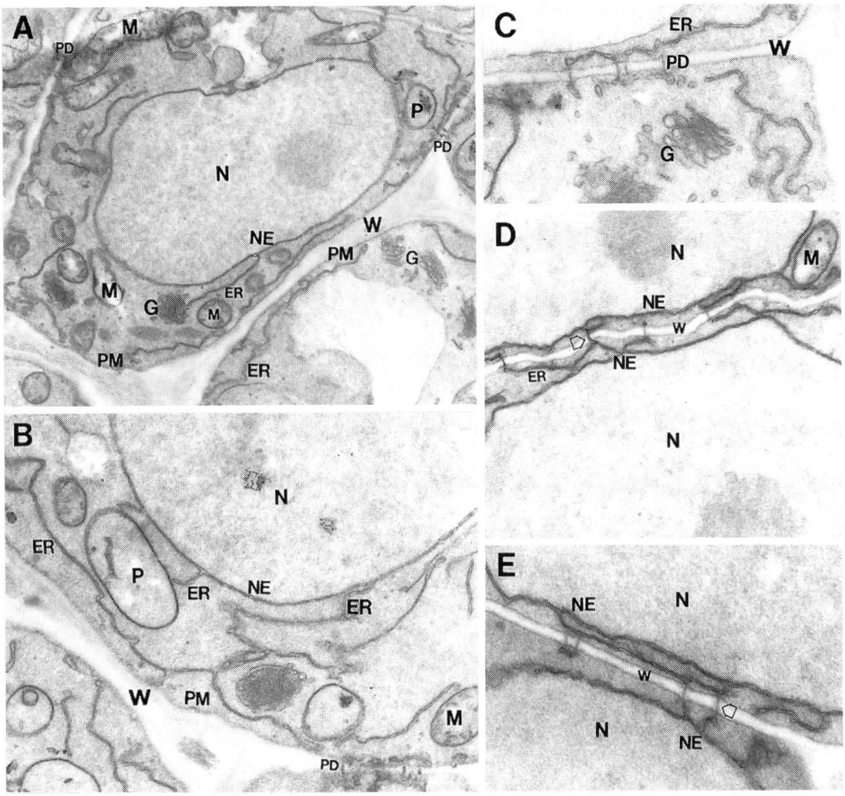

Fig. 20.1. The ultrastructure of poplar apical bud meristematic cells grown under 16-h long days. (LD). Cells contain relatively large-sized nuclei (N). Endoplasmic reticulum (ER), mitochondrion (M), plastid (P), Golgi body (G), plasma membrane (PM) and cell wall (W) are well defined (A) and (B). Many ERs run parallel to PM and W (A) and (B). A large number of plasmodesmata (PD) traverse the W of two adjacent cells (A–C). Some ERs connected to the outer layer of the nuclear envelope (NE) are also associated with M, P, G or PM (A) and (B). Also, some ERs, which connect to the NE of one cell, extend into an adjacent cell through PDs and then connect to the NE of the adjacent cell, forming an ER bridge-like connection between nuclei (D) and (E) (arrows). (A) × 10,400, (B) × 13,200, (C) × 15,290, (D) × 13,900, (E) × 13,400.

the NE, were extended into the interior cytoplasm or contacted the PM (Fig. 20.1A and B). Still others contacted M, P or G (Fig. 20.1A, B and D). The ER, while connected to the NE of one cell, extended into the adjacent cell through the PD and then connected to the NE of the second cell, forming a distinct bridge-like connection between nuclei (Fig. 20.1D and E). These structurally associated phenomena suggest that ERs may provide a network for coordinating various physiological processes among organelles and between cells.

During the SD-induced dormancy development, poplar apical bud cells showed several noticeable features about the ER distribution and its bridge-like connections between nuclei. First, the PD and ER-bridge between N-to-N still could be seen among some cells (Fig. 20.2A). Second, PDs were discontinued among many cells (Fig. 20.2B and C, arrows). Third, PD distribution among some cells still remained intact in a certain area (PD), but disconnected in another (arrows) (Fig. 20.2D). The third case implied that the symplasmic domain might be formed in the poplar apical meristematic cells during the development of the SD-induced dormancy.

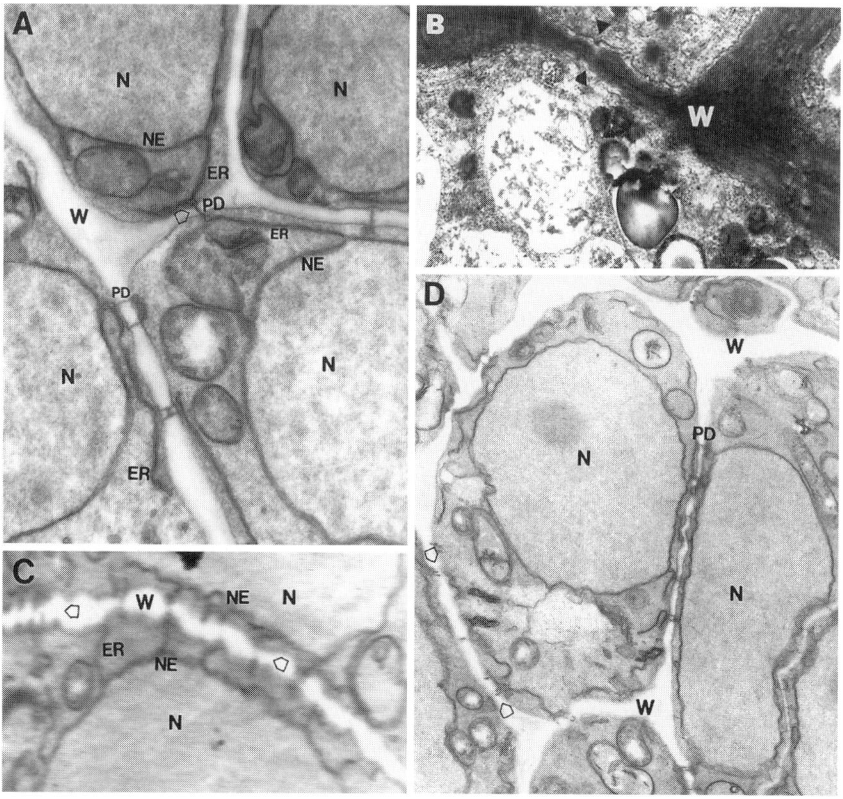

Fig. 20.2. The ultrastructure of poplar apical bud meristematic cells grown under 8-h short days (SD) when plants reached high degrees of dormancy. (A) After 42 days SD exposure, PDs between cells and bridge-like ER connections between nuclei are seen in the apical meristemic cells. (B) Sample sections have been prepared with glutaraldehyde-osmium procedure after 77 days SD. PDs are blocked and the PM at the entrance of the PD are fused with each other (arrow). (C) and (D) Samples after 70 days SD exposure. Many PDs are discontinued (C, arrows). Discontinued PDs (D, arrows) are seen in the peripheral cells of the symplasmic domain in which continued PDs exist. (A) × 15,400, (B) × 16,800, (C) × 14,000, (D) × 8400.

Another interesting phenomenon was also observed in poplar apical buds during dormancy development. After 28 days SD exposure when plants had just developed a measurable dormancy, the contraction of ERs occurred (Fig. 20.3A and B, at cER). However, some ERs were only partially contracted and the uncontracted portion of the ER was similar in appearance to that under LD (Fig. 20.3B). When the bud developed a deeper dormancy such as after 42 or 70 days SD exposure, more contracted ERs were seen in the apical meristematic cells (Fig. 20.3D, at cER). We believe that the contraction of ERs may cause the discontinuation of ER from the NE as shown in Fig. 20.3C and the disconnection of PD as shown in Fig. 20.3E (arrows).

Immediately after shoot growth resumed from the dormancy stage, contracted ERs were no longer observed in the apical bud cells. Associations of ERs with other organelles, ER-bridge connections between nuclei and the intercellular communication channels, the PD, were all

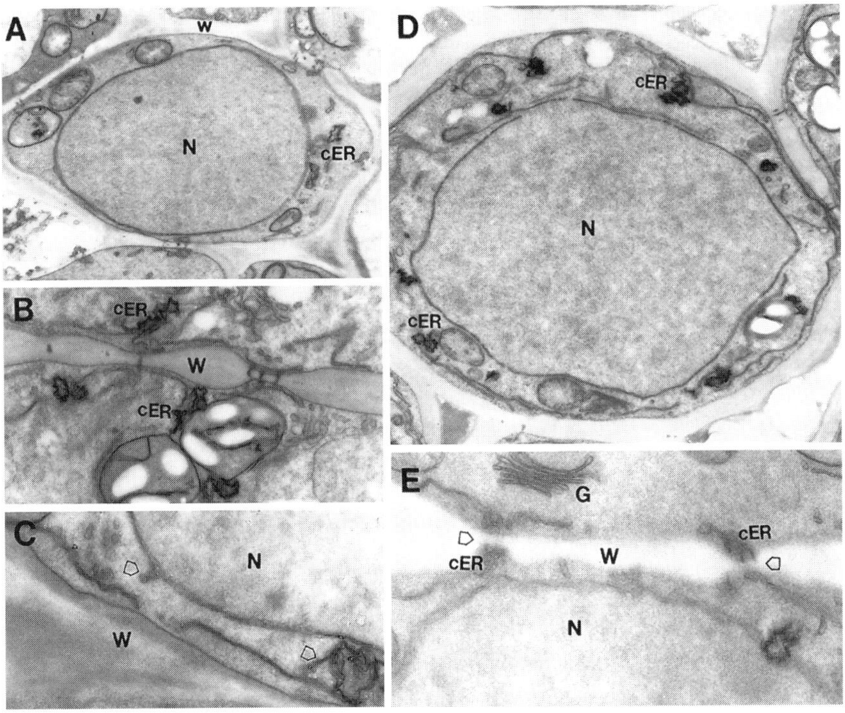

Fig. 20.3. The ultrastructure of poplar apical bud meristematic cells grown under various periods of SD exposure. (A–C) After 28 days SD exposure. (D) After 42 days SD exposure. (E) After 70 days SD exposure. Various forms of contracted ERs (cER) are observed in the cells after 28, 42, or 70 days SD exposure (A–E). Some ER, while still connected with NE, becomes disconnected at its other end (C, arrow). The contraction of ERs leads to the disconnection of PDs (E, arrows). (A) × 6400, (B) × 11,400, (C) × 15,700, (D) × 8550, (E) × 18,500.

re-established similar to the scenario before dormancy developed (Fig. 20.4A and B). This re-establishment further demonstrated that SD altered the features of the PD and the ER which seemed to bear a close relationship to the development of dormancy in poplar plants.

Alterations of the distribution and the concentration of calcium in poplar apical bud meristematic cells during SD-induced dormancy

EM revealed that poplar apical buds fixed with potassium antimonate showed electron-dense deposits, which were calcium deposits, and an indication where the calcium ions were distributed (Fig. 20.5A and B, arrows), as verified by the EGTA chelating control (Fig. 20.5C, arrows) and the control that was fixed without potassium antimonate fixative (Fig. 20.5D). Under LD, calcium deposits were mainly localized in the vacuole and the intercellular space (Fig. 20.5A and B, arrow), indicating that they are the major stores of Ca^{2+} in the cells of actively growing poplar plants. Rather large Ca^{2+} deposits were found in the plastids (Fig. 20.5A, arrow). Some deposits were localized in the cell wall and both ends of the PD, and there were few deposits showing in the cytosol and the nucleus.

After 20 days SD, changes in calcium distribution in bud cells had already occurred, even though the plant did not have any measurable dormancy. The calcium deposits decreased markedly in the intercellular

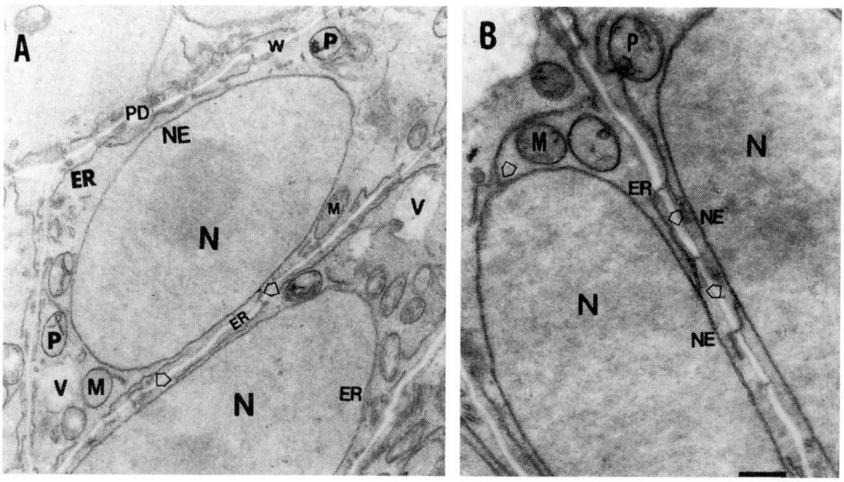

Fig. 20.4. The ultrastructure of poplar apical bud meristematic cells after breaking dormancy following 30 days LD exposure. (A) and (B) cERs are no longer observed. The scenario of endo-membrane organizations is similar to that before the dormancy developed as shown in Fig. 20.1, e.g. the associations of ER with PM or with organelles, PDs, and bridge-like ER connections between nuclei (A) × 7350, (B) × 12,500.

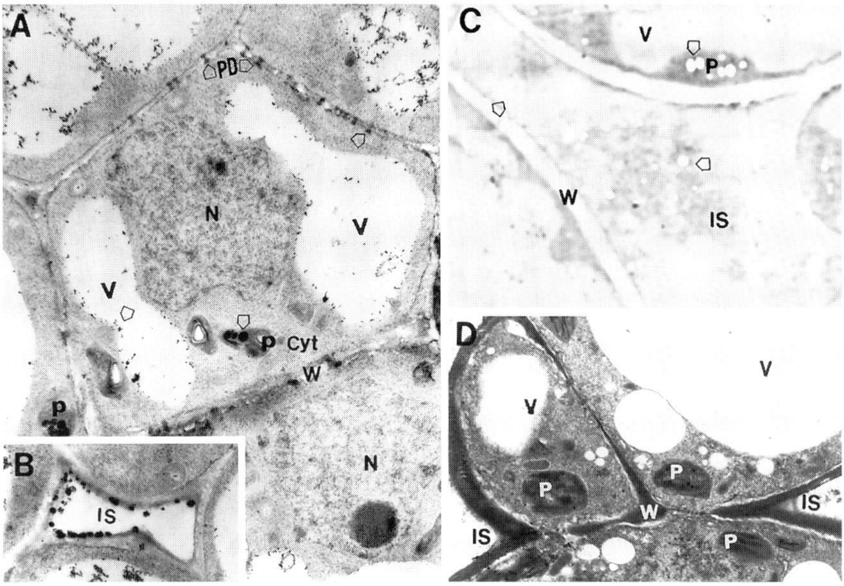

Fig. 20.5. Subcellular localization of Ca^{2+} antimonate deposits in poplar apical bud meristematic cells grown under LD. (A) and (B) The electron-dense calcium antimonate deposits (arrows) are mainly localized in the vacuole (V), intercellular spaces (IS), and the plastid (P). Some are observed at the entrances of PDs (PD/arrow) and in W. Few are seen in the cytosol (Cyt) and the nucleus (N). (C) Sample sections were treated with EGTA calcium-chelating agent. There are many transparent holes (arrows) shown in P, IS and V, where calcium deposits had been located before the EGTA treatment. (D) Tissue sample, being fixed without potassium antimonate, shows no calcium deposits. (A) × 6000, (B) × 6400, (C) × 3500, (D) × 3500.

space, some localized on the external side of the PM, and some were seen in the cytosol and the nucleus (Fig. 20.6A, arrows).

After 28 days SD exposure when the plant needed a 14 day LD to break its dormancy, few deposits were observed in the intercellular spaces (IS). Many deposits were distributed between the cell wall (W) and the PM. Small-sized deposits showed up in the cytosol and the nucleus (Fig. 20.6B and C). When plants were exposed to SD for 35 and 49 days, there were sizeable deposits clearly visible between CW and PM, and the deposits were significantly increased in the cytosol and the nucleus (Fig. 20.7A and B). Evidence suggested that SD is a driving force for the influx of Ca^{2+} in poplar cells. The high concentration of intracellular calcium scenario lasted more than 3 weeks which we do not believe has ever been reported for plants in the literature.

After 77 days SD when plants became deeply dormant, the calcium distribution was altered again. Cytosolic and nuclear Ca^{2+} deposits appeared to be decreased, and a large number of deposits were found in the IS and the CW (Fig. 20.8A–C).

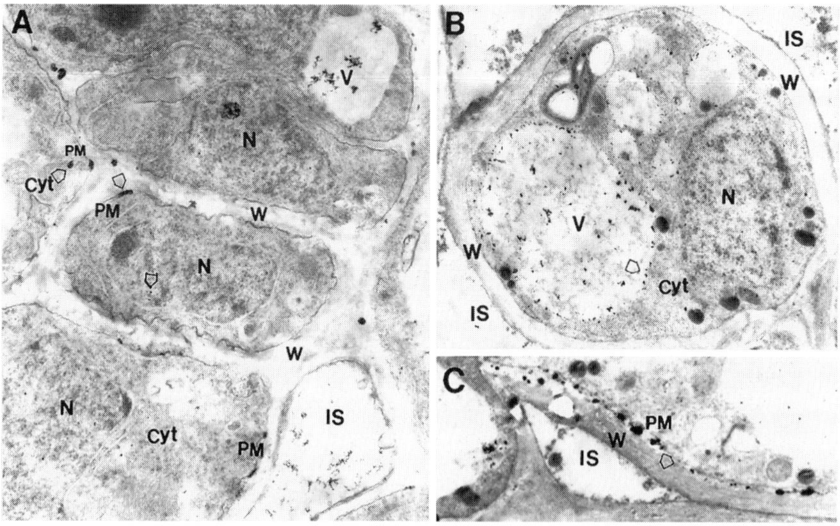

Fig. 20.6. Subcellular localization of Ca^{2+} antimonate deposits in poplar apical bud meristematic cells grown under SD. (A) After 20 days SD exposure, Ca^{2+} deposits are markedly reduced in IS; some are at the outer face of the PM (arrows); and some are in cytosol and nuclei (arrows). (B) and (C) After 28 days SD exposure, few Ca^{2+} deposits are found in IS, while a large number of the deposits are found between the W and the PM. Also, many small-sized deposits are observed in the Cyt, N, and V where the deposits are distributed near to the tonoplast. (A) × 9100, (B) × 6300, (C) × 7700.

Changes in the activity and localization of Ca^{2+}-ATPase in poplar apical bud meristematic cells during SD-induced dormancy

When tissue slices are incubated in the reaction medium containing ATP, $CaCl_2$ and $CeCl_3$, Ca^{2+}-ATPase hydrolyses the ATP to ADP and P_i. The inorganic P is then precipitated by cerium ions. The size of the fine-grained cerium phosphate deposits and their locations are indications of enzymatic activity and its distribution, respectively. We have used five types of controls to verify the truthfulness of the enzymatic scenario. They were that:(i) the reaction medium had no ATP (data not shown); (ii) $CaCl_2$ was omitted, but 5 mM EGTA was added (Fig. 20.9C); (iii) 10 mM Na_3VO_4 was added (Fig. 20.9D); (iv) 0.1 mM erythrosin B was added (Fig. 20.9E); and (v) the cells were fixed with routine glutaraldehyde-osmium tetroxide if any precipitous grains other than cerium precipitates exist (Fig. 20.9F). In all of the controls, we did not see any visible grains on the PM and/or NE.

In LD-grown plants, cerium phosphate-precipitated grains were localized on the cytoplasmic side of the PM and a small number of reaction products were also seen on the NE, indicating where the enzyme activity occurred (Fig. 20.9A and B, arrows).

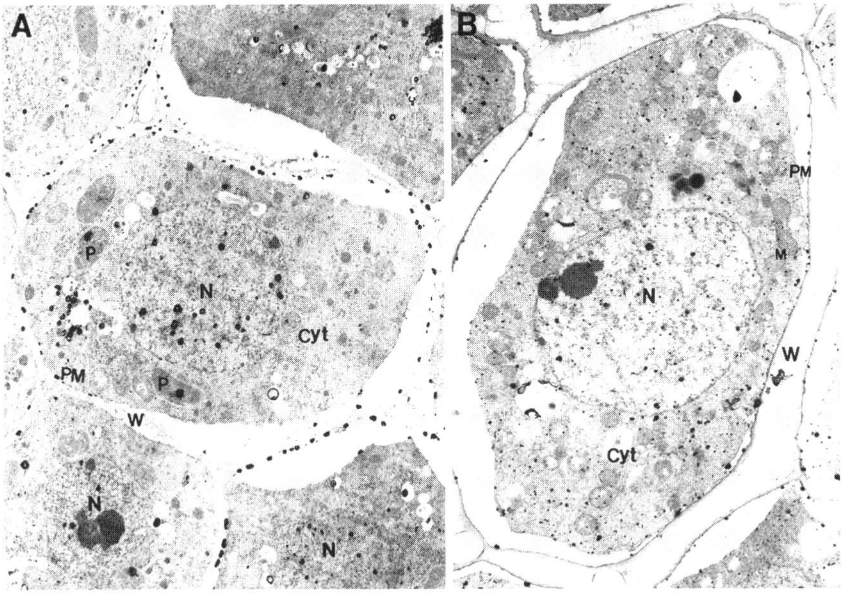

Fig. 20.7. Subcellular localization of Ca^{2+} antimonate deposits in poplar apical bud meristematic cells after plants exposed to 35 (A) and 49 days SD (B). Many large-sized Ca^{2+} deposits are widely distributed between the cell wall and the plasma membrane, and in cytosol and in nuclei. (A) and (B) × 6300.

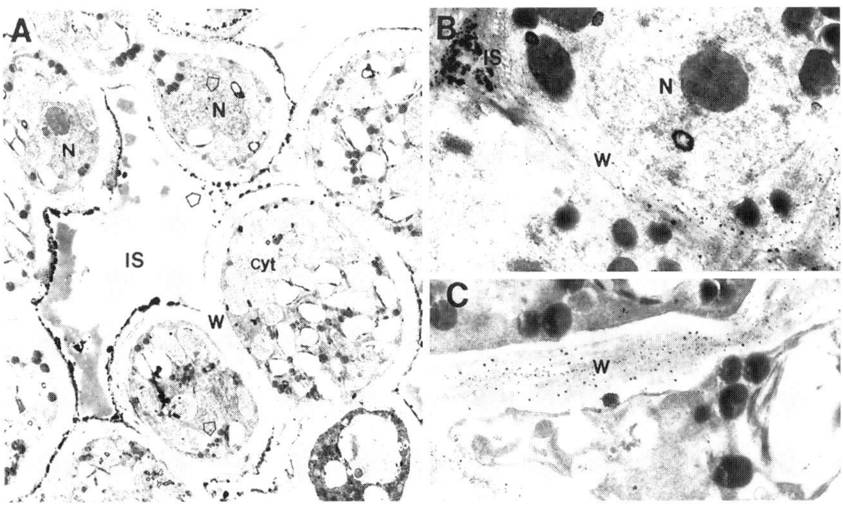

Fig. 20.8. Subcellular localization of Ca^{2+} antimonate deposits in poplar apical bud meristematic cells after 77 days SD exposure. (A–C) Ca^{2+} deposits are markedly reduced in the cytosol and in the nuclei and many deposits appear again in the IS and in the cell wall. (A) × 5600, (B) × 8400, (C) × 10,500.

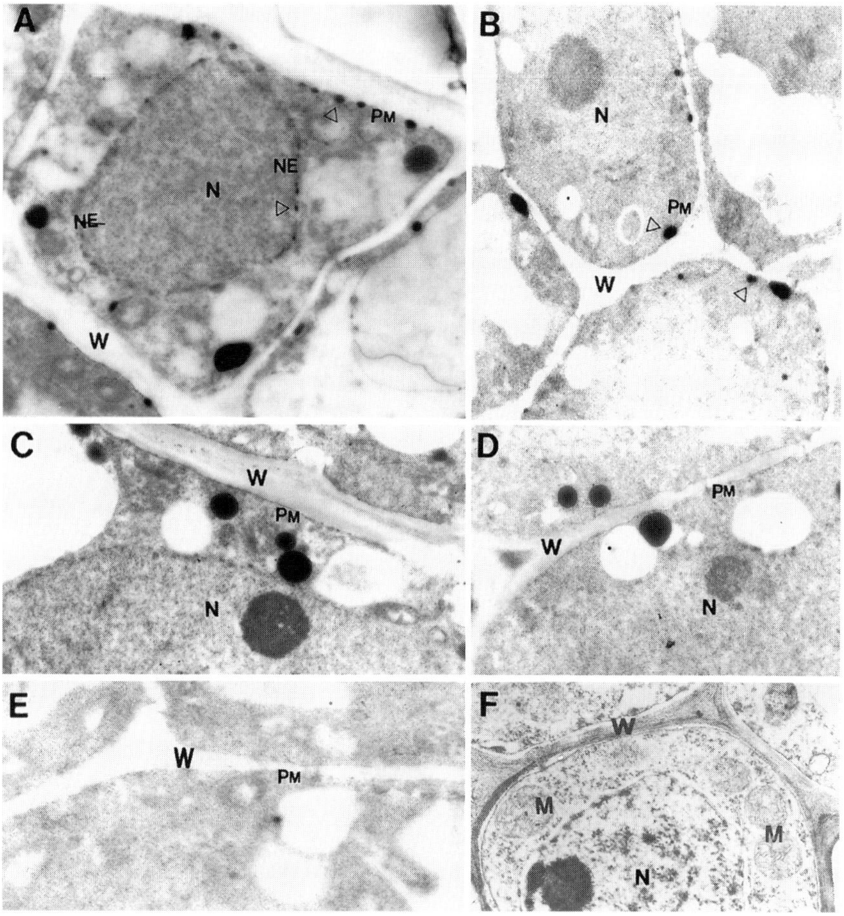

Fig. 20.9. EM-cytochemical localization of the Ca^{2+}-ATPase activity in poplar apical bud meristematic cells grown under LD exposure. (A) and (B). The reaction products – the electron-dense cerium phosphate precipitates (arrows), an indication of the Ca^{2+}-ATPase activity localization – are localized at the cytoplasmic side of the PM. A small number of the precipitates (arrow) are also found on the NE. (C–F) The controls with no cerium phosphate precipitates to be seen: (i) $CaCl_2$ was omitted but 5 mM EGTA was added to the reaction medium (C); (ii) 10 mM Na_3VO_4, an enzyme inhibitor, was added to the reaction medium (D); (iii) 0.1 mM erythrosin B, a special enzyme inhibitor, was added to the reaction medium (E); and (iv) the cells were fixed with glutaraldehyde-osmium tetroxide (F). (A) × 10,200, (B) × 7300, (C) × 8700, (D) × 8700, (E) × 9400, (F) × 14,500.

When poplar dormancy was rapidly developing during the period of 28 to 42 days SD exposure, the site of Ca^{2+}-ATPase activity was clearly altered. No cerium phosphate grains were present on the interior side of the PM. In contrast, many deposits were observed on the exterior face of the PM

(Fig. 20.10A–D). After 70 days SD exposure when buds were in deep dormancy, the reaction products of Ca^{2+}-ATPase activity reappeared on the interior side of the PM and some grains also appeared on the NE (Fig. 20.11A and B). When the dormancy was just broken, the enzyme activity was still on the interior side of the PM and on the NE (Fig. 20.11C and D, arrow).

Discussion

The methodology

For the ER-plasmodesma

Our structural understanding of the ER/plasmodesmal membrane systems is largely dependent on the methods used to preserve and contrast these elements for the EM. Unfortunately, there is no perfect method available. It has been shown that potassium permanganate fixation often causes membrane disruption and also destroys other inclusions such as ribosomes, microtubules and microfilaments (Hepler, 1981; Hepler *et al.*, 1990). Even the conventional fixation of glutaraldehyde-osmium tetroxide has its shortcomings. Owing to its relatively poor ability to stain membranes, membranous systems, especially the ER, have often gone undetected

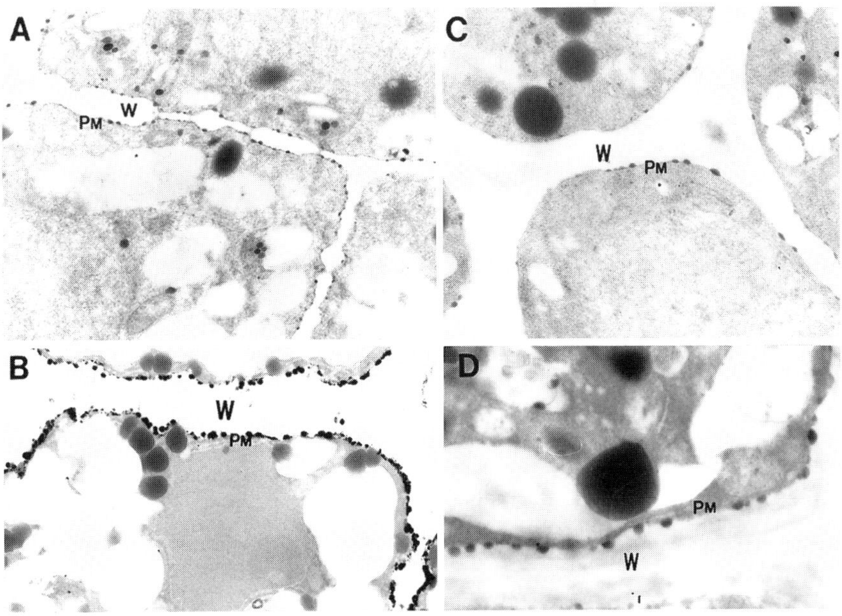

Fig. 20.10. When plants were exposed to SD for 28 days (A) and (B) and 42 days (C) and (D), no cerium reaction products were seen on the interior face of the PM, but on the exterior face. (A) × 8400, (B) × 13,300, (C) × 10,500, (D) × 13,300.

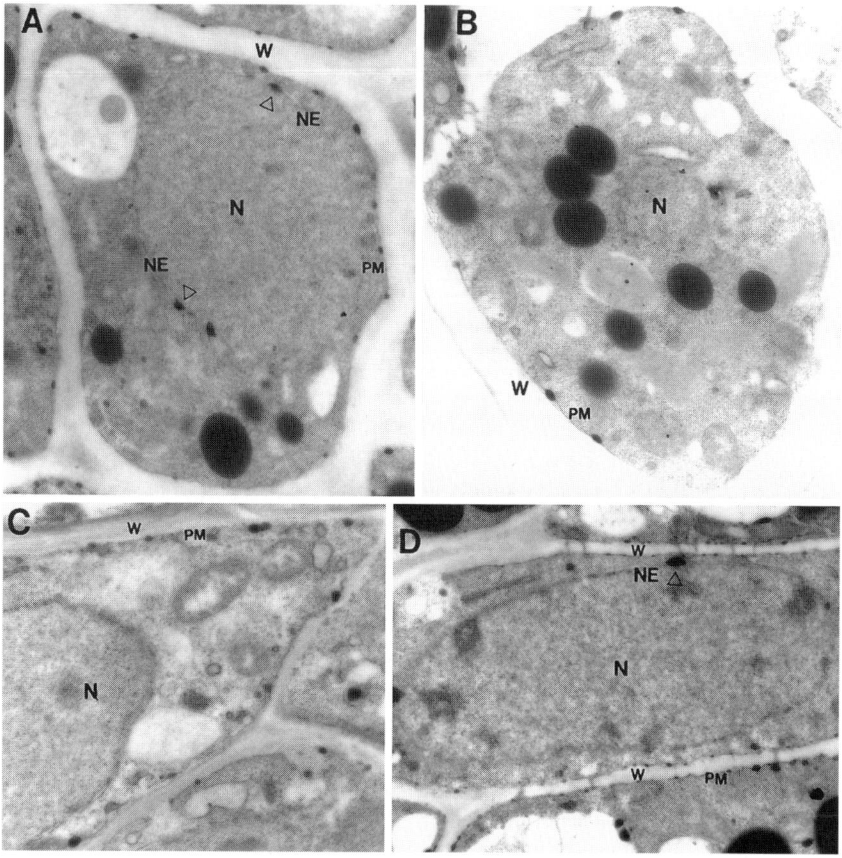

Fig. 20.11. EM-cytochemical localization of the Ca^{2+}-ATPase activity in deep dormant apical bud meristematic cells and in cells immediately after breaking dormancy. (A) and (B) Plants had been exposed to SD for 70 days. The reaction products reappear on the interior face of the PM. Some are also seen on the NE. (C) and (D) Plants had been broken from dormancy following a 30-day LD exposure. The same scenario as shown in (A) and (B) is repeated. (A) and (B) × 11,700, (C) and (D) × 9500.

(Hepler *et al.*, 1990). Using the high pressure freeze-fixation and freeze-substitution method, although the cytoskeleton components (the MTs and MFs) were well fixed, the membrane structure still could not be well preserved, because the acetone-OsO_4 freeze-substitution is less effective in fixing the lipid part of the membrane (Lichtscheidl *et al.*, 1990). Potassium permanganate has been recognized as a better fixative for membrane lipids. In addition, its osmotic ability to the PM is very high, so that it can rapidly fix the membrane structure within 1–2 min and thus is able to preserve the intrinsic features of the membrane organization. However, this method was generally abandoned after the 1960s (Luft 1956; Whaley *et al.*, 1960; Jian *et al.*, 1973).

The classic permanganate method has at least two unfavourable aspects. First is its low osmotic potential which may cause osmotic shock to the organelles and nuclei, resulting in swelling and sometimes even breakage of the envelope membrane. Second is its reduced effectiveness in fixing the proteins, thus membranous proteins could become dissociated from membrane bodies during the subsequent chemical treatments, resulting in membrane disruption. We have made several modifications from the classic method. First, we added NaCl to the fixative and also to the washing medium up to 0.5% concentration attempting to prevent or reduce the osmotic shock to the organelles and nuclei. Second, the fixation time was extended to 4 h from 2 and held at 4°C instead of room temperature. Third, after permanganate fixation, the samples were immediately immersed in 70% ethanol for 12 h at cold temperatures, allowing the samples to be further fixed, especially for the protein part of the membrane systems. We believe the ultrastructure of the membranous systems, especially the ER and ER-related PD, were well preserved without obvious membrane disruption.

For the localization of Ca^{2+}-ATPase activity

Evidence suggests that the activity of the PM Ca^{2+}-pump ATPase could be at least partially inhibited by the glutaraldehyde fixation (Ogawa *et al.*, 1986; Kortje *et al.*, 1990). In the present study, we, therefore, used 4% paraformaldehyde for sample fixation according to the method of Davis *et al.* (1987). Our results showed a higher activity reaction of the PM Ca^{2+}-ATPase (Figs 20.9–20.11). The reaction medium consisted of all elements but Mg^{2+} ions. It is known that in a biochemical assay with cell-free preparations, Mg^{2+} ions must be added to the reaction medium besides Ca^{2+} for the activation of Ca^{2+}-ATPase activity (Briskin, 1990; Rasi-Caldogno *et al.*, 1995; Bonza *et al.*, 1998). However, in dealing with the cytochemical assay with integral cells, no added Mg^{2+} ions are needed in the reaction medium for the Ca^{2+}-ATPase activity because endogenous Mg^{2+} ions exist in the cells (Pappas and Kriho, 1988). Nevertheless, we compared the poplar Ca^{2+}-ATPase activity in response to the reaction medium with and without Mg^{2+} ions. The results agreed with the conclusion from others that no Mg^{2+} is needed in the reaction medium for the cytochemical assay of poplar tissue

Possible role of the ER-plasmodesma system in dormancy development

The discovery of the plasmodesma generated the symplasmic concept, which converts colonies of independent cells into an interconnected symplast system (Gunning and Robards, 1976). The plasmodesmal complex comprises an outer plasma membrane, which continues with the neighbouring cells, and a centrally located/appressed form of ER which forms a continuum with the ER of the adjacent cells (Robards and Lucas, 1990). In our study, we observed that the ER, which connected to the outer layer of the NE of one cell, extended into the adjacent cell through the plasmodesma and then connected to the NE of the second cell, forming a distinct bridge-like connection from nucleus to nucleus (Fig. 20.1E and F; Fig. 20.2A). The ER-plasmodesma system, it is believed, forms a connection between cytoplasmas of two cells. We now provide evidence that the ER-plasmodesma system actually directly connects between the nuclei of two cells. Our findings further perfect the concept of plant symplast.

The plasmodesmal channel is a dynamic structure. When viral particles moved through the plasmodesmata or when plants were under anaerobic or osmotic stresses, the diameter of the plasmodesmal pores became widened (Cleland *et al.*, 1994; Wolf and Lucas, 1994; Schulz, 1995). In certain regions of the tissue, PD channels may be sealed or disconnected, forming symplasmic domains which are thought to be involved in cell differentiation, tissue formation, organogenesis and specialized physiological functions (Lucas *et al.*, 1993). In our study, we observed that the features of ER-plasmodesma altered during the development of SD-induced dormancy in poplar. The frequency of PDs in the cell wall between cells decreased and the diameter of the PD pores reduced (Jian *et al.*, 1997).

After a period of 28 to 70 days SD exposure, when poplar apical buds were developing dormancy, the contraction of ER (*c*ER) occurred. We believe the phenomenon of *c*ER, which is observed only under SD exposure, differs from the concentric configuration of ER (CER). *c*ER has an irregular shape as compared with the concentric whorl of CER. The contraction of ER under SD and the dispersal of *c*ER under LD are slow processes, occurring over a matter of weeks. CER was formed under anaerobic and/or ice-encased conditions (Andreeva and Grineva, 1970; Pomeroy and Andrews, 1978). The appearance of CER appeared to be a natural response to the unfavourable environmental conditions, which has been reported in many cases in animals and plants. Shih and Rappaport (1971) reported that CER dispersed in dormant potato cells when excised buds were treated with abscisic acid (ABA) and gibberellic acid (GA). Because both ABA, a growth inhibitor, and GA, a growth promoter, could disperse the CER from dormant cells of the potato tuber bud, they were unable to establish the relationship between the alteration of ER configuration into CER and the development of potato tuber dormancy.

We suggest that *c*ER, formed under SD, is somehow involved in the development of poplar dormancy. The contraction seemed to lead to the ER's disassociation between nucleus and organelles, the disconnection of the PD, and the breakage of ER-bridges between nuclei. It is possible that these alterations and the formation of the symplasmic domains may cause plant growth cessation, a prerequisite for dormancy development.

Alterations of subcellular calcium distribution in relation to SD signalling and to the dynamic changes in ER-plasmodesma system

Environmental stimuli such as light, high/low temperatures, etc., can increase cytosolic Ca^{2+} concentration (Bush, 1995), which acts as a second messenger (Gilroy *et al.*, 1991; Poovaiah and Reddy, 1993). During SD exposure, Ca^{2+} ions increased in both cytosol and nucleus in poplar apical bud cells. The initiation of the calcium increase occurred prior to the dormancy that can be measured and the high level of the intracellular calcium lasted for several weeks while dormancy was developing. It appears that the SD signal directs the intrinsic increase in intracellular calcium concentration.

There is evidence indicating that cytosolic Ca^{2+} concentration may be involved in the regulation of PD permeability (Erwee and Goodwin, 1983). Micro-injection of the buffered carboxyfluorescein dye containing IP_3 into the cells of the staminal hair of *Setcreasea purpurea* revealed that the elevated cytosolic Ca^{2+} concentration inhibited the dye's cell-to-cell diffusion via PD, suggesting that calcium may regulate the opening and closing of the PD pores by modifying the conformation of the structural proteins surrounding the PD pores (Tucker, 1988, 1990). Artificially increasing Ca^{2+} ion concentration in spring cells (actively growing cells) by applying ionophore A23187 or direct micro-injection of Ca^{2+} significantly restricted the intercellular communication to a level similar to that found in the winter cells (Shepherd and Goodwin, 1992). Eklund and Eliasson (1990) and Eklund (1991) reported that high cytosolic Ca^{2+} concentrations stimulated the non-cellulosic polysaccharide and lignin deposition in cell walls. The SD-induced increase in intracellular Ca^{2+} concentration may, therefore, be involved in the thickening of the cell walls which in turn leads to the constriction and closing of the PD pores.

Cytoskeleton combines with ER forming a cytoskeleton–ER complex (Lichtscheidl *et al.*, 1990; Reuzeau *et al.*, 1997). It is known that a high concentration of cytosolic Ca^{2+} may depolymerize one of the cytoskeleton components, the microtubule (Kiehart, 1981). When poplar plants were exposed to SD for 28 days or longer, increased cytosolic Ca^{2+} content was observed (Jian *et al.*, 1997). The timing that the ER initiated its contraction under SD coincided with the timing of the increased cytosolic Ca^{2+} concentration. It is thus plausible that SD triggers an influx of Ca^{2+} ions which may result in the depolymerization of the microtubule from the cytoskeleton–ER complex, which in turn causes the ER contraction.

Changes in sites of the PM Ca^{2+}-ATPase activity in relation to the alteration of intracellular Ca^{2+} levels during SD-induced dormancy development

Studies of the EM-cytochemical localization of PM Ca^{2+} pump ATPase activity have revealed that the reaction site is localized on the cytoplasmic side of the PM (Penniston, 1983; Pappas and Kriho, 1988; Mughal *et al.*, 1989; Kortje *et al.*, 1990; Gioglio *et al.*, 1991; Barry, 1992). Our findings agreed with the literature, but only when poplar plants were grown under LD, or at a late stage of dormancy development, e.g. after 70 days SD exposure. However, when the dormancy was developing during the period of 28–42 days SD exposure, no reaction products of the enzymatic activity were seen on the interior side of the PM. The dynamic of enzymatic activity seems to have an adaptive nature with altered intracellular Ca^{2+} levels during the SD exposure. Owing to the high affinity of the enzyme to calcium, it can reduce the cytoplasmic calcium concentration to an extremely low level (Penniston, 1983; Evans *et al.*, 1991). Active calcium pumping on the PM was necessary in order to maintain a calcium homeostasis when poplar plants were actively growing or reached deep dormancy, e.g. after 70 days SD exposure. During the period of 28–42 days SD exposure, while dormancy was developing, calcium influx occurred through the PM calcium channels, resulting in a high concentration of cytosolic calcium (Jian *et al.*, 1997). During this period, the activity of PM calcium pumping should theoretically be at a minimum.

During the period of 28–42 days SD exposure, we did, however, observe many cerium phosphate deposits on the exterior face of the PM in poplar (Fig. 20.10). We did not attempt to identify them. In animal cells, an ecto-Ca^{2+}-ATPase has been reported and its reaction products were localized on the exterior face of the PM (Kortje *et al.*, 1990; Barry, 1992; Maxwell *et al.*, 1995; Plesner, 1995; Thirion *et al.*, 1996). Although this enzyme has been known for many years, its biological function is still in a stage of speculation: for example: (i) the ecto-Ca^{2+}-ATPase might be needed for making adenosine and possibly inorganic phosphate available for the cell uptake (Kortje *et al.*, 1990); (ii) the ecto-Ca^{2+}-ATPase might break down ATP used as a neurotransmitter (Nagy, 1986); (iii) the ecto-Ca^{2+}-ATPase could regulate the relative amount of ATP and adenosine (Barry, 1992); and (iv) the ecto-Ca^{2+}-ATPase activity might be associated with the exocytotic secretion of neuropeptide granules containing ATP (Thirion *et al.*, 1996). From the literature it seems that the cerium precipitated grains, which we observed on the exterior face of the PM during the early stage of dormancy development in poplar (Fig. 20.10), might be the reaction products from the ecto-Ca^{2+}-ATPase activity. If this is the case, the suggestion from Thirion *et al.* (1996) encourages us to speculate that this enzyme might be involved in the exocytosis of cell wall materials, e.g. the secretion of polysaccharides and/or glycoproteins into the cell wall by the fusion of the PM and the transporting vesicles, because, when poplar dormancy was rapidly developing, the cell wall was also markedly thickened (Jian *et al.*, 1997).

References

Ando, T., Fujimoto, K., Mayahura, H., Miyajima, H. and Ogawa, K. (1981) A new one step method for the histochemistry and cytochemistry of Ca^{2+}-ATPase activity. *Acta Histochemica et Cytochemica* 14, 705–715.

Andreeva, N. and Grineva, G.M. (1970) Ultrastructure of the endoplasmic reticulum in maize root cells under anaerobic condition. *Soviet Plant Physiology* 17, 796–801.

Barry, M.J. (1992) Ecto-calcium-dependent ATPase activity of mammalian taste bud cells. *The Journal of Histochemistry and Cytochemistry* 40, 1919–1928.

Bonza, C., Caraelli, A., de Michelis, M.I. and Rasi-Caldogno, F. (1998) Purification of the plasma membrane Ca^{2+}-ATPase from radish seedlings by calmodulin agarose affinity chromatography. *Plant Physiology* 116, 845–851.

Briskin, D.P. (1990) Ca^{2+}-translocating ATPase of the plant plasma membrane. *Plant Physiology* 94, 397–400.

Bush, D.S. (1995) Calcium regulation in plant cells and its role in signaling. *Annual Review of Plant Physiology and Plant Molecular Biology* 46, 95–122.

Cleland, R.E., Fujiwara, T. and Lucas, W.J. (1994) Plasmodesmal mediated cell-to-cell transport in wheat roots is modulated by anaerobic stress. *Protoplasma* 178, 81–85.

Davis, W.L., Jones, R.G. and Goodman, D.B.G. (1987) Electron microscopic cytochemical localization of Ca^{2+}-ATPase in toad urinary bladder. *Journal of Histochemistry and Cytochemistry* 35, 39–48.

Eklund, L. (1991) Relations between indoleacetic acid, calcium ions and ethylene in the regulation of growth and cell wall composition in *Picea abies*. *Journal of Experimental Botany* 42, 785–789.

Eklund, L. and Eliasson, L. (1990) Effects of calcium ions concentration on cell wall synthesis. *Journal of Experimental Botany* 41, 863–867.

Erwee, M.G. and Goodwin, P.B. (1983) Characterization of the *Egeria densa* Planch. Leaf symplast: inhibition of the intercellular movement of fluorescent probes by group II ions. *Planta* 158, 320–328.

Evans, D.E., Briars, S.A. and Williams, L.E. (1991) Active calcium transports by plant cell membranes. *Journal of Experimental Botany* 42, 285–303.

Gilroy, S., Fricker, M.D., Read, N.D. and Trewavas, A.J. (1991) Role of calcium signal transduction of *Commelina* guard cells. *Plant Cell* 3, 333–344.

Gioglio, L., Rapuzzi, G. and Quacci, D. (1991) Ca^{2+}-ATPase and Na^{+}, K^{+}-ATPase activities in the fungi form papilla of the tongue of *Rana esculenta*. *Journal of Morphology* 210, 117–125.

Gunning, B.E.S. and Robards, A.W. (eds) (1976) *Intercellular Communication in Plants: Studies on Plasmodesmata*. Springer, New York.

Gunning, B.E.S. and Steer, M.W. (1996) *Plant Cell Biology*. Jones and Bartlet, Sudbury, Massachusetts.

Hepler, P.K. (1981) The structure of the endoplasmic reticulum revealed by osmium tetroxide-potassium ferricyanide staining. *European Journal of Cell Biology* 26, 102–110.

Hepler P.K., Palvitz, B.A., McCauley, M.M. and Lichtscheidl, I. (1990) Cortical endoplasmic reticulum in plants. *Journal of Cell Science* 96, 355–373.

Jian, L.C., Ching, Y.H. and Chang, P.T. (1973) Cytological studies on the cold resistance of plants: ultrastructure changes of the wheat cells in the winter period. *Acta Botanica Sinica* 15, 22–29.

Jian, L.C., Li, P.H., Sun, L.H. and Chen, T.H.H. (1997) Alterations in ultrastructure and subcellular localization of Ca^{2+} in poplar apical bud cells during the induction of dormancy. *Journal of Experimental Botany* 48, 1195–1207.

Kauffman, S.A. (1993) *The Origins of Order. Self-organisation and Selection in Evolution.* Oxford University Press, New York.

Kiehart, D.P. (1981) Studies on the *in vivo* sensitivity of spindle microtubules to calcium ions and evidence for a vesicular calcium-sequestering system. *Journal of Cell Biology* 88, 604–617.

Kortje, K.H., Freihofer, D. and Rahmann, H. (1990) Cytochemical localization of high-affinity Ca^{2+}-ATPase activity in synaptic terminals. *Journal of Histochemistry and Cytochemistry* 38, 895–900.

Lichtscheidl, I.K., Lancelle, S.A. and Hepler, P.K. (1990) Actin–endoplasmic reticulum complexes in *Drosera*: their structural relationship with the plasmalemma, nucleus, and organelles in cells prepared by high pressure freezing. *Protoplasma* 155, 116–126.

Lucas, W.J., Ding, B. and van der Achoot, C. (1993) Plasmodesmata and the supracellular nature of plants. *New Physiologist* 125, 435–476.

Luft, J.H. (1956) Permanganate – a new fixative for electron microscopy. *Journal of Biophysical and Biochemical Cytology* 2, 799–809.

Maxwell, W.L., McCreath, B.J., Graham, D.I. and Gennarelli, T.A. (1995) Cytochemical evidence for redistribution of membrane pump Ca^{2+}-ATPase and ecto-Ca^{2+}-ATPase activity, and calcium influx in myelinated nerve fibres of the optic nerve after stretch injury. *Journal of Neurocytology* 24, 925–942.

Mughal, S., Cuschieri, A. and Al-Bader, A.A. (1989) Intracellular distribution of Ca^{2+}-Mg^{2+}-ATPase in various tissues. *Journal of Anatomy* 162, 111–125.

Nagy, A.K. (1986) Enzymatic characteristics and possible role of synaptosomal ecto-ATPase from mammalian brain. In: Kreutzberg, G.W., Reddington, M. and Zimmerman, H. (eds) *Cell Biology of Ectoenzymes.* Springer-Verlag, Berlin.

Ogawa, K.S., Fujimoto, K. and Ogawa, K. (1986) Ultracytochemical studies of adenosine nucleotidases in aortic endothelial and smooth muscle cells – Ca^{2+}-ATPase and Na^{+}, K^{+}-ATPase. *Acta Histochemistry and Cytochemistry* 19, 601–610.

Pappas, G.D. and Kriho, V. (1988) Fine structural localization of Ca^{2+}-ATPase activity at frog neuromuscular junction. *Journal of Neurocytology* 17, 417–423.

Penniston, J.T. (1983) Plasma membrane Ca^{2+}-ATPase as active calcium pump. In: Cheung, W.Y. (ed.) *Calcium and Cell Function,* Vol. IV. Academic Press, New York, pp. 99–110.

Plesner, L. (1995) Ecto-ATPase: identities and functions. *International Review of Cytology* 158, 141–156.

Pomeroy, M.K. and Andrews, C.J. (1978) Ultrastructural changes in shoot apex cells of winter wheat seedlings during ice encasement. *Canadian Journal of Botany* 56, 786–794.

Poovaiah, B.W. and Reddy, A.S.N. (1993) Calcium and signal transduction in plants. *Critical Reviews in Plant Sciences* 12, 185–211.

Rasi-Caldogno, F., Pugliarello, M.C., Olivari, C. and de Michelis, M.I. (1989) Identification and characterization of the Ca^{2+}-ATPase which drives active transport of calcium at the plasma membrane of radish seedlings. *Plant Physiology* 90, 1429–1434.

Rasi-Caldogno, F., Carnelli, A. and de Michelis, M.I. (1995) Identification of the plasmalemma Ca^{2+}-ATPase and of its autoinhibitory domain. *Plant Physiology* 108, 105–113.

Rees, A.R. (1981) Concepts of dormancy as illustrated by the tulip and other bulbs. *Annals of Applied Biology* 98, 544–548.

Reuzeau, C., Doolittle, K.W., McNally, J.G. and Pickard, B.G. (1997) Covisualization in living onion cells of putative integrin, spectrin, actin, intermediate filaments and other proteins at the cell membrane and in an endomembrane sheath. *Protoplasma* 199, 173–197.

Robards, A.W. and Lucas, W.J. (1990) Plasmodesma. *Annual Reviews of Plant Physiology and Plant Molecular Biology* 41, 369–419.

Schulz, A. (1995) Plasmodesmal widening accompanies the short-term increase in symplasmic phloem unloading in pea root tips under osmotic stress. *Protoplasma* 188, 22–37.

Shepherd, V.A. and Goodwin, P.B. (1992) Seasonal patterns of cell-to-cell communication in *Chara corallina* Klein ex Wild. II. Cell-to-cell communication in vegetative lateral branches during winter and spring. *Plant, Cell and Environment* 15, 137–150.

Shih, C.Y. and Rappaport, L. (1971) Regulation of rest bud in tubers of potato, *Solanum tuberosum* L, VIII. Early effects of gibberellin A_3 and abscisic acid on ultrastructure. *Plant Physiology* 48, 31–35.

Slocum, R.D. and Roux, S.J. (1982) An improved method for the subcellular localization of calcium using a modification of the antimonate precipitation technique. *Journal of Histochemistry and Cytochemistry* 30, 617–629.

Thirion, S., Troadec, J.-D. and Nicaise, G. (1996) Cytochemical localization of ecto-ATPase in rat neurohypophysis. *Journal of Histochemistry and Cytochemistry* 44, 103–111.

Tucker, E.B. (1988) Inositol bisphosphate and inositol trisphosphate inhibit cell-to-cell passage of carboxyfluorescein in staminal hairs of *Setcreasea purpurea*. *Planta* 174, 358–363.

Tucker, E.B. (1990) Calcium-loaded 1,2-bis-(2-aminophenoxy)-ethane-*N,N,N,N*-tetraacetic acid blocks cell-to-cell diffusion of carboxyfluorescein in staminal hairs of *Setcreasea purpurea*. *Planta* 182, 34–38.

Wang, H. and Jian, L.C. (1994) Changes of level of calcium in the cells of rice seedlings under low temperature stress. *Acta Botanica Sinica* 36, 587–597.

Whaley, W.G., Mollenhauer, H.H. and Leech, J.H. (1960) Some observations on the nuclear envelope. *Journal of Biophysical and Biochemical Cytology* 8, 233–245.

Williams, L.E., Schucler, S.B. and Briskin, D.P. (1990) Further characterization of the red beet plasma membrane Ca^{2+}-ATPase using GTP as an alternative substrate. *Plant Physiology* 92, 747–754.

Wolf, S. and Lucas, W.J. (1994) Virus movement proteins and other molecular probes of plasmodesmata function. *Plant, Cell and Environment* 17, 537–585.

21

The Developmental Basis of Bud Dormancy in 1-year-old *Picea* and *Pseudotsuga* Seedlings

Joanne E. MacDonald

Natural Resources Canada, Canadian Forest Service, PO Box 4000, Fredericton, New Brunswick, Canada, E3B 5P7

Dormancy Defined

Physiologists define dormancy as any case in which a tissue predisposed to elongate does not do so (Doorenbos, 1953). In trees of temperate regions, the cessation of shoot elongation and the appearance of terminal buds is said to indicate bud dormancy (Wareing, 1950). Conifer seedling physiologists use changes in gross bud morphology as an indication of stage of dormancy development (Cleary *et al.*, 1978). Dormancy is verified by placing seedlings under conditions conducive for bud break. If terminal buds fail to break under such conditions, then the buds are dormant (Lavender, 1985). Seedling physiologists have reported that buds that failed to break under such conditions subsequently enlarged. How can dormant buds grow? The answer is found in developmental botany. Although buds have reached the developmental stage after which bud break is not possible, they have not completed development. A definition of bud dormancy – based on lack of mitotic activity in the shoot apical meristem (Owens and Molder, 1973) – may be more meaningful than the physiological definition of bud dormancy.

In the following paragraphs the developmental basis of bud dormancy in 1-year-old *Picea mariana* (Mill.) B.S.P., *Picea glauca* Moench Voss, and *Pseudotsuga menziesii* (Mirb.) Franco var. *menziesii* seedlings will be presented. The observations are from investigations of bud development and mitotic activity in response to a variety of dormancy-induction treatments.

Bud Defined

The terminal bud comprises a bud-scale complex (*sensu* MacDonald and Owens, 1993) and a preformed shoot (*sensu* Hallé *et al.*, 1978) (Fig. 21.1A). The bud-scale complex is composed of outer, scarious and inner, foliaceous bud scales and is borne upon the bud-scale receptacle (Fig. 21.1A). The preformed shoot – the miniature shoot for next year – comprises multiple spirals of leaf primordia and a shoot apical meristem (Fig. 21.1B).

Bud Development

Bud development involves the initiation and differentiation of the bud-scale complex and the preformed shoot. Once a dormancy-induction cue is perceived, the shoot apical meristem ends neoformed-leaf initiation and begins bud-scale initiation. Duration of bud-scale initiation ranges from less than 1 week to 4 weeks, depending on species and dormancy-induction treatment. Upon completion of bud-scale initiation, the shoot apical meristem begins leaf initiation. Leaf initiation continues until the preformed shoot is fully developed. Duration of leaf initiation ranges from

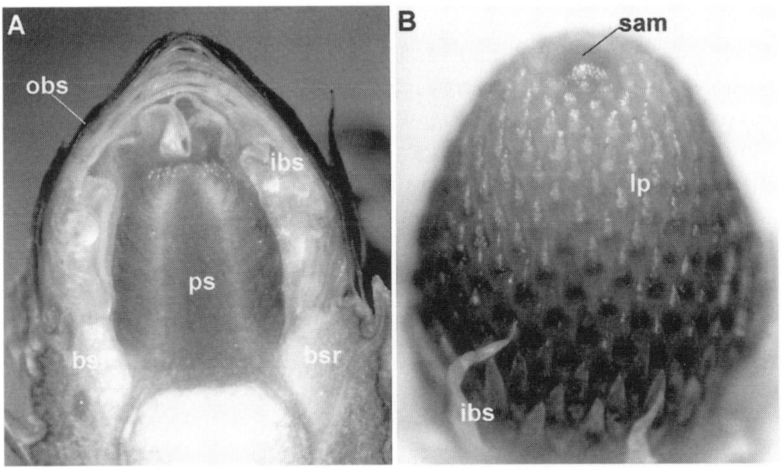

Fig. 21.1. (A) Photograph of a free-hand longitudinal section through a terminal bud of *Picea glauca*. The bud-scale complex encloses the preformed shoot (ps). The bud-scale complex is composed of outer, scarious bud scales (obs) and inner, foliaceous bud scales (ibs) and sits upon the bud-scale receptacle (bsr). (B) Photograph of an intact preformed shoot from a terminal bud of *Picea glauca*. The shoot apical meristem (sam) surmounts the multiple spirals of leaf primordia (lp) that it initiated. A last-formed inner bud scale (ibs) is evident near the base of the preformed shoot.

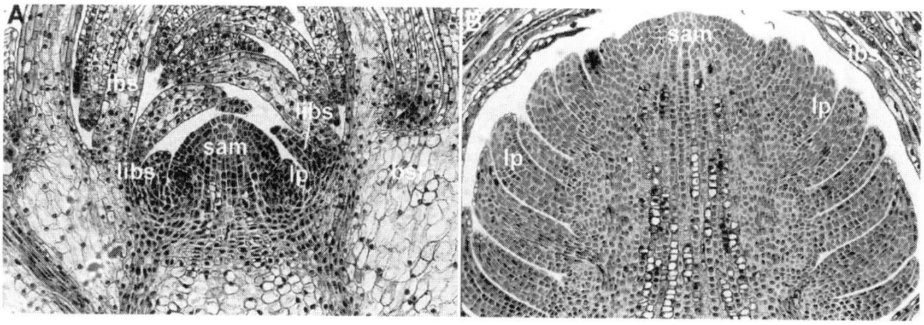

Fig. 21.2. (A) Light micrograph of a median longitudinal section through the developing preformed shoot within a terminal bud of *Picea mariana*. The shoot apical meristem (sam) is just starting leaf initiation. A first-formed leaf primordium (lp) is arising on the flank of the shoot apical meristem. A last-formed inner bud scale (libs) sits on the bud-scale receptacle (bsr). Other inner bud scales (ibs) are more differentiated than last-formed inner bud scales. (B) Light micrograph of a median longitudinal section through a preformed shoot at the end of leaf initiation within a terminal bud of *Picea mariana*. The shoot apical meristem (sam) has completed initiation of leaf primordia (lp). Inner bud scales (ibs) are evident.

5 to 13 weeks depending on species, dormancy-induction treatment and seedling size.

Differentiation of bud-scale primordia into the mature components of the bud-scale complex takes 4–5 weeks. At its completion, rapid leaf initiation is just starting and the preformed shoot has few primordia. A comparison of the size of the developing preformed shoot at this time (Fig. 21.2A) with the size of the preformed shoot at the end of leaf initiation (Fig. 21.2B) illustrates why it is unfortunate that seedling physiologists use changes in gross bud morphology as an indicator of dormancy development.

Mitotic Activity during Bud Development

Median longitudinal sections of shoot apical meristems exhibit cyto-histological zonation (*sensu* Foster, 1938); the apical zone, peripheral zone and rib meristem are discernible (Fig. 21.3A). During bud development mitotic divisions are visible in the apical zone, peripheral zone, and rib meristem of sectioned shoot apical meristems. It is mitotic divisions in the peripheral zone that give rise to primordia (Esau, 1977) (Fig. 21.3B). Mitotic activity and rate of leaf initiation are closely related. Mitotic activity increases as rate of leaf initiation increases, stays high during rapid leaf initiation, and decreases as rate of leaf initiation slows prior to its completion (Fig. 21.4).

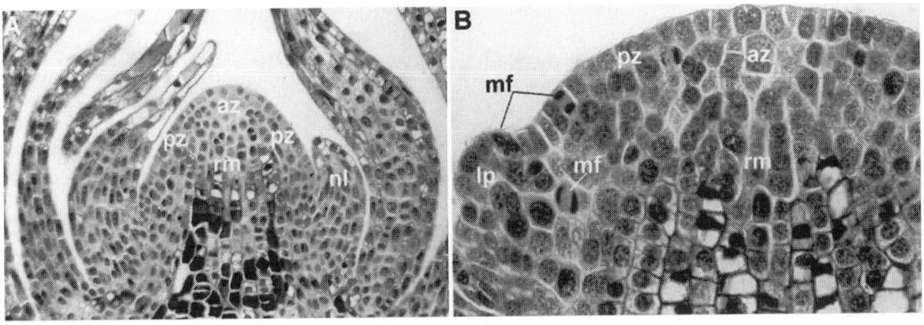

Fig. 21.3. (A) Light micrograph of a median longitudinal section through a shoot tip of *Pseudotsuga menziesii* var. *menziesii*. The shoot tip comprises a shoot apical meristem and neoformed leaves (nl). The apical zone (az), peripheral zone (pz) and rib meristem (rm) within the shoot apical meristem are indicated. (B) Light micrograph of a median longitudinal section through a shoot apical meristem of *Picea mariana*. The shoot apical meristem is nearing completion of leaf initiation. The apical zone (az), peripheral zone (pz), and rib meristem (rm) are indicated. The leaf primordium (lp) is arising from localized mitotic activity in the peripheral zone (pz). Mitotic figures (mf), associated with the primordium and in the position of the next primordium, are indicated.

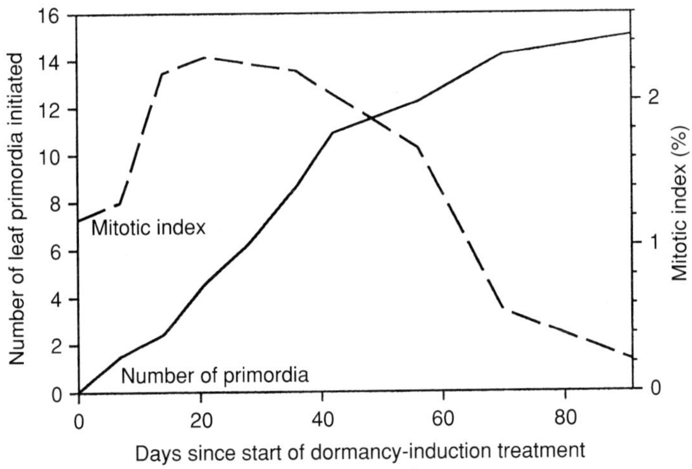

Fig. 21.4. Relationship between rate of leaf initiation by and mitotic index in shoot apical meristems of *Pseudotsuga menziesii* var. *menziesii* during the days since the start of a short-day dormancy-induction treatment.

Mitotic Activity after Bud Development

Mitotic divisions continue throughout all zones of the shoot apical meristem for 1 to several weeks after the completion of leaf initiation. Although mitotic divisions continue in the peripheral zone, primordia will not arise prior to dormancy (Fig. 21.5). Finally, mitotic divisions are limited

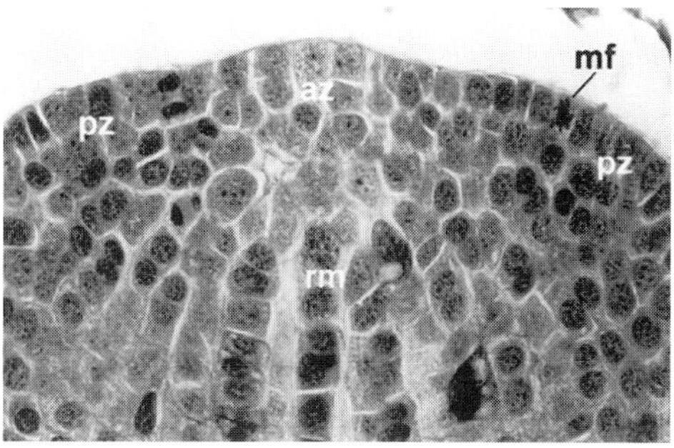

Fig. 21.5. Light micrograph of a median longitudinal section through a shoot apical meristem of *Picea mariana*. The shoot apical meristem has completed leaf initiation. The apical zone (az), peripheral zone (pz) and rib meristem (rm) are indicated. Although a mitotic figure (mf) is evident in the peripheral zone, a primordium would not arise before dormancy.

to the apical zone. Once mitotic division ceases in this zone the shoot apical meristem is dormant.

References

Cleary, B.D., Greaves, R.D. and Owston, P.W. (1978) Seedlings. In: Cleary, B.D., Greaves, R.D. and Hermann, R.K. (eds) *Regenerating Oregon's Forests.* Oregon State University Extension Service, Corvallis, pp. 63–97.

Doorenbos, J. (1953) Review of the literature on dormancy in buds of woody plants. *Medelinger van de Landbouwhogeschool te Wageningen, Nederland* 53, 1–24.

Esau, K. (1977) *Anatomy of Seed Plants*, 2nd edn. John Wiley & Sons, New York.

Foster, A.S. (1938) Structure and growth of the shoot apex in *Ginkgo biloba. Bulletin of the Torrey Botanical Club* 65, 531–556.

Hallé, F., Oldeman, R.A.A. and Tomlinson, P.B. (1978) *Tropical Trees and Forests. An Architectural Analysis.* Springer-Verlag, Berlin.

Lavender, D.P. (1985) Bud dormancy. In: Duryea, M.L. (ed.) *Evaluating Seedling Quality: Principles, Procedures, and Predictive Abilities of Major Tests.* Forest Research Laboratory, Oregon State University, Corvallis, pp. 7–15.

MacDonald, J.E. and Owens, J.N. (1993) Bud development in coastal Douglas-fir seedlings in response to different dormancy-induction treatments. *Canadian Journal of Botany* 71, 1280–1290.

Owens, J.N. and Molder, M. (1973) A study of DNA and mitotic activity in the vegetative apex of Douglas fir during the annual growth cycle. *Canadian Journal of Botany* 51, 1395–1409.

Wareing, P.F. (1950) Growth studies in woody species. I. Photoperiodism in first-year seedlings of *Pinus silvestris. Physiologia Plantarum* 3, 258–276.

Molecular Genetic Analysis of Bud Dormancy-related Traits in *Populus*

22

Tony H.H. Chen,[1], Joel Davis,[1] Barbara E. Frewen,[2] Glenn T. Howe [3] and Harvey D. Bradshaw, Jr[2]

[1]Department of Horticulture, Oregon State University, ALS 4017 Corvallis, OR 97331, USA; [2]College of Forest Resources, University of Washington, Seattle, WA 98195, USA; [3]School of Natural Resources, The Ohio State University, OARDC, Wooster, OH 44691, USA

Introduction

Woody perennial plants native to temperate regions are subjected to large seasonal variations in temperature. Annual growth cycles, involving alternations between active shoot growth and vegetative dormancy (endodormancy), are critical for the long-term survival and growth of these plants. The development of endodormancy is an important adaptive strategy because plants are more resistant to freezing and dehydration stress when they are dormant.

Growth cessation is a prerequisite for the development of endodormancy and substantial cold hardiness. In woody perennial plants, growth cessation is usually associated with terminal bud set or shoot tip abortion. If bud set and cold acclimation do not occur before the first frost of autumn, the tree will be damaged. In the spring, the onset of active shoot growth (bud flush) follows endodormancy release and is generally associated with a decrease in cold hardiness. Therefore, the timing of bud set and bud flush, or bud phenology, represents a trade-off between survival and growth of forest trees. However, if dormancy is released and bud flush occurs too early in spring, the growing tissues may be killed by a late spring frost. On the other hand, trees that stop growing and become dormant too early, or release dormancy and flush too late, have shortened growing seasons that reduce their competitive ability and growth potential. Thus, the genetic controls on bud phenology, as well as the timing of dormancy

(eds J.-D. Viémont and J. Crabbé)

induction and release, have implications for adaptation of trees to their natural environments, and to applied tree breeding.

When grown in a common environment, trees from high latitudes and elevations tend to set bud and break bud early relative to trees from lower latitudes and elevations. Typically, bud phenology is under strong genetic control. In *Populus* clonal field trials, for example, the timing of spring bud flush was among the most highly heritable traits, with up to 98% of the total phenotypic variance of clonal means explained by genetic factors, and only 2% by environmental and other factors (Bradshaw and Stettler, 1995).

Quantitative genetic studies suggest that autumn bud set and spring bud flush are controlled by multiple genes (Pauley and Perry, 1954; Eriksson *et al.*, 1978). Despite the evidence that bud phenology is under strong genetic control, little is known about the identity of the genes responsible for genetic variation in bud phenology. This is because phenological traits are not inherited in a simple Mendelian fashion characteristic of traits under the control of a single gene. Instead, these traits show the continuous phenotypic variation typical of a quantitative trait under the control of several to many genes. The genes that govern quantitative traits are called quantitative trait loci (QTL). Most traits of basic and practical interest in plants are quantitative traits (e.g. Darwinian fitness, adaptation, crop yield and stress tolerance). This explains the intense effort during the past 10 years to genetically map QTL. Although many QTL have been genetically mapped in trees, we know little about these QTL, except their map positions on chromosomes.

Populus as a Model System for Studying Endodormancy

Species and hybrids of *Populus* (poplars, cottonwoods and aspens) are widely accepted as models for studying woody plant biology (Stettler *et al.*, 1996). Their key features are:

1. *Populus* species have wide latitudinal and/or elevational ranges. For example, *Populus trichocarpa* occurs at low elevations from 30° N to 60° N in western North America, and at elevations up to 1000 m in the centre of its range. *Populus deltoides* extends from 30° N to 50° N at low elevations in eastern and central North America.
2. Common garden studies demonstrate that northern and southern ecotypes of *P. trichocarpa* (Pauley and Perry, 1954; Howe *et al.*, 1995) and *P. deltoides* (Farmer, 1996) are strongly differentiated for bud phenology, and that bud phenology is under strong genetic control.
3. The ease with which *Populus* can be cloned (i.e. vegetatively propagated) makes it possible to test individual genotypes in replicated field and greenhouse trials, giving accurate estimates of their phenotypes.
4. Genetic markers and linkage maps have been developed for *Populus* (Bradshaw *et al.*, 1994). QTL mapping studies using F_2 hybrids of

P. trichocarpa and *P. deltoides* demonstrate that bud phenology is under the control of a few loci with modest effects (Bradshaw and Stettler, 1995).

5. Candidate genes with plausible connections to endodormancy-related traits have been cloned and characterized by molecular biologists and physiologists working on *Populus*. Three phytochrome genes (*PHYA*, *PHYB1* and *PHYB2*) have been characterized, for example, that may be involved in sensing photoperiodic signals that trigger growth cessation, terminal bud set, and the onset of endodormancy (Howe *et al.*, 1998). The *Populus* homologues of two *Arabidopsis* genes (*ABI1* and *ABI3*) that affect responses to abscisic acid have also been characterized (Rohde *et al.*, 1998; A. Rohde and W. Boerjan, personal communication).

6. Efficient transformation and regeneration systems are used routinely to produce transgenic *Populus* for basic research and commercial plantations (Han *et al.*, 1996).

Mapping Populations and Phenotyping

Two inbred-like, three-generation mapping pedigrees were used to map QTL for endodormancy-related traits. One pedigree was initiated in 1980 by crossing a female *P. trichocarpa* (93-968) from western Washington (48° N latitude) with a male *P. deltoides* (S7C4) from Texas (31° N). These two parents were chosen to maximize segregating variation for endodormancy-related traits in the F_2 generation. In 1993, two of the resulting F_1 progeny (55-265 and 55-263) were crossed to produce a segregating F_2 mapping population (Family 822).

In 1996 and 1997, replicated clonal trials (i.e. common garden experiments) containing 346 F_2 clones from Family 822, along with their parents and grandparents, were planted outdoors in Corvallis, Oregon, and in St Paul, Minnesota. A second, related F_2 family (Family 331) was used to verify the QTL that we detected in Family 822. The same maternal *P. trichocarpa* (93-968) used to produce Family 822 was crossed to a male *P. deltoides* (ILL-129) from central Illinois (39° N). In 1988, two of their F_1 progeny (53-246 and 53-242) were crossed to produce the second F_2 mapping population (Family 331).

Fifty-five F_2 clones from Family 331 were planted in the spring of 1991 in Puyallup, Washington. The date of bud flush, as well as other growth and form traits, were measured in the spring of 1993 (Bradshaw and Stettler, 1995). Spring bud flush was the trait with the highest heritability ($H^2_c = 98\%$). A total of five bud break QTL were detected, which together accounted for 85% of the total phenotypic variance in the F_2 (Bradshaw and Stettler, 1995). Physiologists, pathologists and geneticists from around the world have now mapped QTL affecting more than 100 different traits in Family 331.

Quantitative Genetic Analyses of Dormancy-related Traits

We studied the quantitative genetics of dormancy-related traits in Family 822. Dormancy-related traits were measured in the field and in a greenhouse. Field traits included the timing of bud set (BS) in Minnesota and Oregon, the timing of bud flush (BF) in Oregon, as well as autumn frost damage (FD) and winter survival (WS) in Minnesota. In a warm greenhouse, we measured BS and the number of new leaves (NL) on trees growing under an 8-h photoperiod (short day, SD), and NL on trees growing under a natural photoperiod (NP) from August to December.

As expected, Family 822 showed substantial phenotypic segregation for endodormancy-related traits (Fig. 22.1). In Oregon, mean bud set dates ranged from 16 August to 25 November (101 days), whereas mean bud flush dates ranged from 4 March to 13 April (40 days). The difference in BS between the 'earliest' and 'latest' F_2 clones was more than twice the corresponding difference in BF, apparently because of differences in the environmental control of these traits. The timing of bud set is strongly influenced by photoperiod, whereas the timing of bud flush is influenced

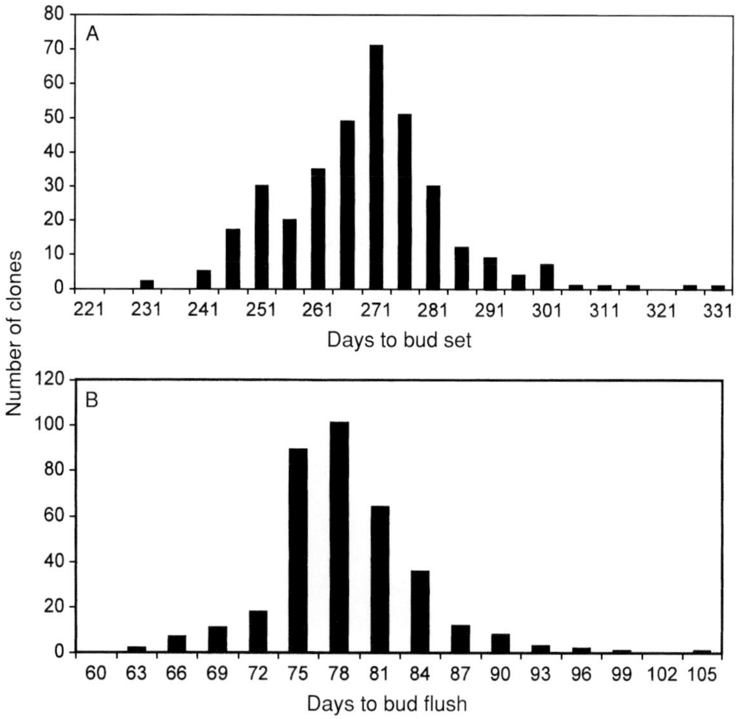

Fig. 22.1. The distribution of number of days to (A) bud set and (B) bud flush (from Frewen *et al.*, 2000).

largely by temperature. Similar differences in the variability of these traits have been observed in pure species of *Populus* (Ernst and Fechner, 1981; Brissette and Barnes, 1984; Farmer, 1993; Dunlap and Stettler, 1996).

The heritability of clonal means (H^2_c) for BS was 91% in Oregon, 75% in Minnesota, 85% in Minnesota and Oregon, and 94% for BF in Oregon (Table 22.1). The heritability of BF was higher than that of any other trait. We did not analyse BF in Minnesota because many of the trees did not survive the harsh winters. These heritabilities indicate that BS and BF are under moderate to strong genetic control. These results are consistent with those reported for poplars collected from natural populations, as well as for other angiosperm and coniferous tree species (Farmer, 1970; Worrall, 1983; Weber *et al.*, 1985; Riemenschneider *et al.*, 1992, 1994; Bradshaw and Stettler, 1995; Aitken and Adams, 1997; Thomas *et al.*, 1997).

FD was measured in Minnesota using a 0–10 score that incorporated information on both the timing and severity of damage. Although many trees were severely damaged or killed, other trees exhibited no frost damage. The heritability of clonal means for FD was moderately low ($H^2_c = 51\%$), which is consistent with heritabilities for autumn cold hardiness in natural populations of Douglas fir (Aitken *et al.*, 1996). WS ranged from 0 to 100% among the F_2 clones. There was a moderately strong negative genetic correlation between WS and FD (–0.70), suggesting that the mortality that occurred during the autumn and winter was caused primarily by autumn frosts, or that genotypes that are susceptible to frost are also prone to being killed by low midwinter temperatures. The genetic correlation between BS and FD ranged from 0.58 to 0.72, indicating that

Table 22.1. Heritability for bud set, bud flush and frost damage in Family 822. H^2_c is the heritability of clonal means.

Environment	Traits	H^2_c (%)
Field		
Corvallis, Oregon	Bud set (BS)	91
	Bud flush (BF)	94
St Paul, Minnesota	Bud set (BS)	75
	Frost damage (FD)	51
Oregon and Minnesota	Bud set (BS)	85
Greenhouse		
SD photoperiod (SD)	Bud set (BS)	80
	New leaves (NL)[a]	86
Natural photoperiod (NP)	New leaves (NL)[b]	83

[a]Number of new leaves produced within 60 days after the imposition of an 8-h photoperiod.
[b]Number of new leaves produced from the middle of August to the middle of December.

trees that grew later into the autumn experienced greater frost damage. These genetic correlations support the conclusion that the timing of bud set is associated with adaptability. Although the genetic correlation between BS and WS ($r_g = -0.25$ to -0.30) was only weakly negative, it does provide evidence that early bud set is associated with trees that are better able to survive the winter.

SDs are typically viewed as the primary signal for bud set because SDs typically induce bud set under warm, non-inductive temperatures, but low night temperatures often fail to cause bud set under non-inductive long-day (LD) photoperiods (but see Håbjørg, 1972). We studied the effect of SDs under non-inductive temperatures by measuring SD-induced bud set under warm greenhouse conditions. SD-induced bud set in the greenhouse (BS-SD) was highly variable among F_2 clones and under moderately strong genetic control ($H^2_c = 80\%$). The 'earliest' F2 clone set bud 16 days after the start of the 8-h photoperiod, whereas other clones were still growing after 2 months. Because leaf production stops when bud set occurs, NL is an alternative measure of bud set that is based on a developmental scale, rather than a chronological scale (Howe *et al.*, 1995). The heritability of clonal means for NL-SD was 86%, and the genetic correlation between NL-SD and BS-SD was 0.93, which demonstrates the close relationship between these traits. Genetic correlations between BS-SD and BS were used to judge the relative impact of photoperiodic responses on the timing of bud set in the field. Surprisingly, these correlations were relatively modest. The genetic correlations between BS-SD and BS ranged from 0.53 to 0.59, indicating that BS-SD explained only about 28–35% of the variation in BS. Slightly lower genetic correlations were observed between NL-SD and BS. These correlations ranged from 0.45 to 0.51.

We also measured bud set and leaf production in a warm greenhouse under a natural photoperiod. Although bud set occurred readily in the field, bud set occurred late, or not at all, when the trees were grown in the greenhouse under a natural photoperiod. In the greenhouse, for example, only 29.5% of the trees had terminal buds by the middle of December, but in the Minnesota plantations, 86.2% of the trees had terminal buds when they were last measured in the beginning of November. The trees in the field which had not set bud by this time had already been damaged or killed by frost, so their bud set dates were never determined. In Oregon, which has a photoperiodic regime similar to that of Minnesota, all of the trees had a terminal bud by 25 November. Although few trees set bud in the greenhouse, the production of new leaves seemed to be inhibited by the naturally decreasing photoperiod. At the end of the experiment, NL-NP varied from 2.9 to 90.9 among the clones, with a clone mean heritability of 83%.

Clearly, the trees in the greenhouse experiments performed differently from the trees in the field. In the SD experiment, this is illustrated by the modest genetic correlation between BS in the two environments. In the NP experiment, this is illustrated by the modest genetic correlation between

NL-NP and BS, as well as by the small number of trees that set bud in the greenhouse compared with the field. Because temperature regimes obviously differed between the field and greenhouse environments, we hypothesize that genetic differences in temperature responses are more important than previously recognized. None the less, differences in soil moisture and nutrients may also be involved. The trees in the greenhouse experiments were well watered and fertilized weekly, and it is known that soil moisture and nutrient deprivation can cause growth cessation and bud set. In judging the NP experiment, it should be noted that light from street lamps and other greenhouses was higher in the greenhouse than in the field during the night. The light levels in the greenhouse were estimated to be about 0.2–0.5 μM, mostly from sodium vapour lamps. Although these levels are very low, contaminating light in the field was below the level of detection.

QTL Mapping

Linkage map construction

We constructed a linkage map for Family 822 consisting of 335 dominant AFLP (315) and RAPD (20) markers (Frewen *et al.*, 2000). Due to the high percentage of dominant markers used (95%) and the paucity of linkage information from dominant markers linked in repulsion (Ott, 1985; Knapp *et al.*, 1995), two linkage maps were constructed, representing the parental genomes of *P. trichocarpa* (T map) and *P. deltoides* (D map). The markers in T map were distributed in 26 linkage groups, and spans a distance of 2002 cM (centiMorgan), covering 77% of the estimated genome length (Bradshaw *et al.*, 1994) with an average marker interval of 13.6 cM. The D map consists of 24 linkage groups with a total length of 1778 cM, covering 68% of the genome with an average marker spacing of 12.3 cM. Because the haploid chromosome number is 19 in *Populus*, some of these linkage groups must be representing different sections of the same chromosome.

QTL analysis

We initially scanned for QTL using MAPMAKER/QTL 1.1 (Lincoln *et al.*, 1992) using a detection threshold of LOD > 2. The QTL that we detected were then examined in detail using Model 6 of QTL CARTOGRAPHER (Zeng, 1994), which tests for the presence of multiple QTL. Empirical statistical significance thresholds (experiment-wise $P < 0.05$) for declaring the presence of a QTL were determined by performing 100 permutations of the dataset (Churchill and Doerge, 1994).

Four QTL affecting the timing of bud set were found, two on the T map and two on the D map. Because one QTL on each map was detected at a similar location on linkage group *F*, these QTL probably represent the same locus. Therefore, a maximum of three unique bud set QTL appear to have been identified. All bud set QTL are on linkage groups that also contain bud flush QTL, although no QTL for bud set are located in the same marker interval. The estimated magnitudes of effects of the bud set QTL ranged from 6 to 12 PVE (percentage of F_2 phenotypic variation explained). The *P. trichocarpa* alleles reduced the number of days to bud set at two of the three unique QTL. Differences in the timing of bud set between homozygous classes at each QTL ranged between 6 and 13 days, which represents 6–14% of the period over which bud set occurred.

Nine bud flush QTL were identified, five on the T map and four on the D map. Because three QTL on the T map (on linkage groups *F*, *J* and *D*) seem to occur at the same positions as QTL on the D map, these QTL probably represent the same loci. Thus, a maximum of six unique bud flush QTL were apparently detected. The estimated magnitudes of effects of the bud flush QTL ranged from 6 to 18 PVE. The *P. trichocarpa* alleles hastened the time to bud flush at four of the six unique QTL. The difference between the two homozygous classes for each bud flush QTL was between 4 and 6 days, representing 9–14% of the period over which bud flush occurred.

Candidate gene mapping

A small set of genes (*PHYB1*, *PHYB2*, *ABI1D*, *ABI1B*, and *ABI3*) were tested as candidates for dormancy-related QTL (Frewen *et al.*, 2000). Each of these genes was mapped to a different linkage group. *PHYB1*, *ABI1D* and *ABI3* mapped to linkage groups on which no bud set or bud flush QTL were detected. *PHYB2* mapped to linkage group *J* on each map, within a marker interval that also contains a bud flush QTL. In addition, a bud set QTL was detected on linkage group *J*. *ABI1B* mapped to linkage group *P*, within an interval that contains a bud flush QTL accounting for 9 PVE. On the D map, *ABI1B* mapped to linkage group *P*.

Analyses of variance were used to test for associations between the timing of bud flush and the candidate genes *PHYB2* and *ABI1B* in Family 331. The effect of *PHYB2* genotype on the timing of bud flush was not significant ($P = 0.32$), but the effect of *ABI1B* genotype was significant ($P = 0.018$). F_2 trees homozygous for the *P. trichocarpa* allele at *ABI1B* flushed an average of 9 days earlier than those homozygous for the *P. deltoides* allele.

In both F_2 families, the *PHYB2* allele from *P. trichocarpa* was associated with delayed bud flush. However, the *ABI1B* allele from *P. trichocarpa* had the opposite effect, causing bud flush in both families to be earlier. For both families, *PHYB2* and *ABI1B* had the same effect on bud set; the *P. trichocarpa* alleles were associated with earlier bud set.

Conclusion

Species and hybrids of *Populus* are excellent models for studying traits related to bud dormancy. Considerable progress has been made towards understanding the genetic controls on dormancy-related traits. First, quantitative genetic analyses demonstrate moderate to high heritabilities for these traits. Second, QTL affecting bud set and bud break have been identified. QTL analyses of other dormancy-related traits (e.g. deep dormancy and chilling requirements) are underway. Third, we have identified two candidate genes whose map positions coincide with QTL for bud set and/or bud flush.

Acknowledgements

This work was supported by a grant to THHC, HDB and GTH from the USDA NRICGP(9600678), a grant to GTH from the National Science Foundation (IBN9421420), and by the members of the Poplar Molecular Genetics Cooperative.

References

Aitken, S.N. and Adams, W.T. (1997) Spring cold hardiness under strong genetic control in Oregon populations of *Pseudotsuga menziesii* var. *menziesii*. *Canadian Journal of Forestry Research* 27, 1773–1780.

Aitken, S.N., Adams, W.T., Schermann, N. and Fuchigami, L.H. (1996) Family variation for fall cold hardiness in two Washington populations of coastal Douglas-fir (*Pseudotsuga menziesii* var. *menziesii* (Mirb.) Franco). *Forest Ecology and Management* 80, 187–195.

Bradshaw, H.D. and Stettler, R.F. (1995) Molecular genetics of growth and development in *Populus*. IV. Mapping QTLs with large effects on growth, form, and phenology traits in a forest tree. *Genetics* 139, 963–973.

Bradshaw, H.D., Villar, M., Watson, B.D., Otto, K.G., Stewart, S. and Stettler, R.F. (1994) Molecular genetics of growth and development in *Populus*. III. A genetic linkage map of a hybrid poplar composed of RFLP, STS, and RAPD markers. *Theoretical and Applied Genetics* 89, 167–178.

Brissette, J.C. and Barnes, B.V. (1984) Comparisons of phenology and growth of Michigan and western North American sources of *Populus tremuloides*. *Canadian Journal of Forestry Research* 14, 789–793.

Churchill, G.A. and Doerge, R.W. (1994) Empirical threshold values for quantitative trait mapping. *Genetics* 138, 963–971.

Dunlap, J.M. and Stettler, R.F. (1996) Genetic variation and productivity of *Populus trichocarpa* and its hybrids. IX. Phenology and Melampsora rust incidence of native black cottonwood clones from four river valleys in Washington. *Forest Ecology and Management* 87, 233–256.

Eriksson, G., Ekberg, I., Dormling, I. and Matern, B. (1978) Inheritance of bud-set and bud flushing in *Picea abies*. *Theoretical and Applied Genetics* 52, 3–19.

Ernst, S.G. and Fechner, G.H. (1981) Variation in rooting and juvenile growth phenology of narrowleaf cottonwood in Colorado. *Proceedings of the North Central Tree Improvement Conference, Madison, Wisconsin,* pp. 111–118.

Farmer, R.E. (1970) Genetic variation among open-pollinated progeny of eastern cottonwood. *Silvae Genetica* 19, 149–151.

Farmer, R.E., Jr (1993) Latitudinal variation in height and phenology of balsam poplar. *Silvae Genetica* 42, 148–153.

Farmer, R.E., Jr (1996) The genecology of *Populus*. In: Stettler, R.F., Bradshaw, H.D., Jr, Heilman, P.E. and Hinckley, T.M. (eds) *Biology of* Populus *and its Implications for Management and Conservation.* NRC Press, Ottawa, pp. 33–55.

Frewen, B.E., Chen, T.H.H., Howe, G.T., Davis, J., Rohde, A., Boerjan, W. and Bradshaw, H.D., Jr (2000) QTL and candidate gene mapping of bud set and bud flush traits in *Populus*. *Genetics* 154, 837–845.

Han, K.-H., Gordon, M.P. and Strauss, S.H. (1996) Cellular and molecular biology of *Agrobacterium*-mediated transformation of plants and its application to genetic engineering of *Populus*. In: Stettler, R.F., Bradshaw, H.D., Jr, Heilman, P.E. and Hinckley, T.M. (eds) *Biology of* Populus *and its Implications for Management and Conservation.* NRC Press, Ottawa, pp. 201–222.

Håbjørg, A. (1972) Effects of photoperiod and temperature on growth and development of three latitudinal and three altitudinal populations of *Betula pubescens* Ehrh. *Meldinger fra Norges Landbrukshøgskole* 51, 1–27.

Howe, G.T., Hackett, W.P., Furnier, G.R. and Klevorn, R.E. (1995) Photoperiodic responses of a northern and southern ecotype of black cottonwood. *Physiologia Plantarum* 93, 695–708.

Howe, G.T., Bucciaglia, P.A., Hackett, W.P., Furnier, G.R., Cordonnier-Pratt, M. and Gardner, G. (1998) Evidence that the phytochrome gene family in black cottonwood has one *PHYA* locus and two phyB loci but lacks members of the *PHYC*/F and *PHYE* subfamilies. *Molecular Biological Evolution* 15, 160–175.

Knapp, S.J., Holloway, J.L., Bridges, W.C. and Liu, B.H. (1995) Mapping dominant markers using F2 matings. *Theoretical and Applied Genetics* 91, 74–81.

Lincoln, S.E., Daly, M.J. and Lander, E.S. (1992) Mapping genes controlling quantitative traits with MAPMAKER/QTL 1.1. Whitehead Institute Technical Report, Cambridge, Massachusetts, pp. 1–73.

Ott, J. (1985) *Analysis of Human Genetic Linkage.* Johns Hopkins University Press, Baltimore, Maryland.

Pauley, S.S. and Perry, T.O. (1954) Ecotypic variation in the photoperiodic response in *Populus*. *Journal of the Arnold Arboretum* 35, 167–188.

Riemenschneider, D.E., McMahon, B.G. and Ostry, M.E. (1992) Use of selection indices to increase tree height and to control damaging agents in 2-year-old balsam poplar. *Canadian Journal of Forestry Research* 22, 561–567.

Riemenschneider, D.E., McMahon, B.G. and Ostry, M.E. (1994) Population-dependent selection strategies needed for 2-year-old black cottonwood clones. *Canadian Journal of Forestry Research* 24, 1704–1710.

Rohde, A., Ardiles-Diaz, W., van Montagu, M. and Boerjan, W. (1998) Isolation and expression analysis of an *ABSCISIC ACID-INSENSITIVE 3* (*ABI3*) homologue from *Populus trichocarpa*. *Journal of Experimental Botany* 49, 1059–1060.

Stettler, R.F., Bradshaw, H.D., Jr, Heilman, P.E. and Hinckley, T.M. (1996) *Biology of* Populus *and its Implications for Management and Conservation*. NRC Press, Ottawa.

Thomas, B.R., MacDonald, S.E. and Dancik, B.P. (1997) Variance components, heritabilities and gain estimates for growth chamber and field performance of *Populus tremuloides*: growth parameters. *Silvae Genetica* 46, 317–326.

Weber, J.C., Stettler, R.F. and Heilman, P.E. (1985) Genetic variation and productivity of *Populus trichocarpa* and its hybrids. I. Morphology and phenology of 50 native clones. *Canadian Journal of Forestry Research* 15, 376–383.

Worrall, J. (1983) Temperature–bud-burst relationships in amabalis and subalpine fir provenance tests replicated at different elevations. *Silvae Genetica* 32, 203–209.

Zeng, Z.B. (1994) Precision mapping of quantitative trait loci. *Genetics* 136, 1457–1468.

Regulation of Growth and Dormancy in Pea Axillary Buds

23

Joel P. Stafstrom

Department of Biological Sciences, Plant Molecular Biology Center, Northern Illinois University, De Kalb, IL 60115, USA

Introduction

'Dormancy' refers to a set of adaptive mechanisms that allows plants and plant organs to cease growing under specific conditions, and then to resume growing when conditions change or improve. Dormancy is frequently linked to seasonal cycles, such as cold temperatures at high latitudes (Powell, 1988) or drought in some tropical regions (Borchert, 1991). The annual growth cycle of temperate perennials includes at least three distinct types or phases of bud dormancy (Fuchigami and Nee, 1987; Borchert, 1991). Following a flush of spring growth, meristems made during the current year or previous years stop growing. Since growth of terminal and axillary buds is regulated by physiological processes occurring throughout the plant, this type of within-season dormancy is called correlative inhibition. The best known example of correlative inhibition is the regulation of axillary bud development by growing terminal buds, which is called apical paradormancy or, more commonly, apical dominance. The onset of winter dormancy is promoted by short photoperiods (photoperiodic endodormancy). Morphological changes necessary for cold acclimation usually include development of protective bud scales. Ensuing cold temperatures fulfil a bud's chilling requirement (cryogenic endodormancy). Buds then become competent to grow (thermal eco-dormancy), and may begin to develop after being exposed to warm temperatures. In general, one phase of the seasonal cycle must be completed before the next can begin (Fuchigami and Nee, 1987). Apical paradormancy may be considered to be a stage in the developmental programme leading to endodormancy (Crabbé and Barnola, 1996). The extent to which various types of bud dormancy may be similar in their hormonal regulation, patterns of gene expression, or biochemical control mechanisms is not well understood. Lessons learned from model systems such as the garden pea may lead to insights that are relevant to dormancy mechanisms in long-lived perennials and other less tractable plants.

Hormonal Regulation of Apical Dominance: an Overview

By definition, correlative inhibition involves a signalling organ (a growing terminal bud or developing fruit) and, some distance away, a responding organ (a dormant axillary bud). All five 'classical' plant hormones, water, nutrients and unidentified graft-transmissible substances have been implicated in regulating apical dominance. The following overview highlights only a few factors that appear to play major roles in controlling bud development. Some of these factors are known or suspected to be transported from one part of the plant to another, and others are thought to act within the bud or in the vicinity of the bud. Recent reviews discuss these topics in much greater depth than is possible here (Cline, 1994; Tamas, 1995; Napoli *et al.*, 1999).

Apical dominance was one of the first developmental phenomena shown to be regulated by a hormone. Auxin (IAA) is synthesized in the terminal bud, transported basipetally and then, in some manner, inhibits the growth of axillary buds (Thimann and Skoog, 1933; Cline, 1996). Despite the early discovery of this phenomenon, the mechanism by which IAA inhibits axillary buds is unknown. According to the direct theory, bud dormancy results from high concentrations of IAA within buds (Fig. 23.1A). This model is difficult to reconcile with the fact that terminal buds contain high levels of IAA and yet are able to grow. Direct measurement of IAA in *Phaseolus* axillary buds indicates that growing buds contain more IAA than dormant buds (Gocal *et al.*, 1991). Furthermore, auxin-stimulated genes are expressed at low levels in dormant buds (Stafstrom, 1993). Therefore, it is more probable that auxin inhibits axillary bud growth by an indirect mechanism, either by promoting the synthesis of a secondary inhibitor or by inhibiting the synthesis of a growth promoter.

ABA (abscisic acid) was first identified based on its ability to induce dormancy in sycamore maple buds. ABA also appears to be important in inhibiting the growth of paradormant buds. ABA content is elevated in dormant *Phaseolus* and *Elytrigia* buds (Gocal *et al.*, 1991; Pearce *et al.*, 1995). ABA is not transported from the inhibiting organ (Tamas *et al.*, 1981), so it is probably synthesized within the bud or in the vicinity of the bud, perhaps in response to IAA within the stem (Fig. 23.1B). Ethylene does not appear to play a role in inhibiting bud growth (Romano *et al.*, 1993).

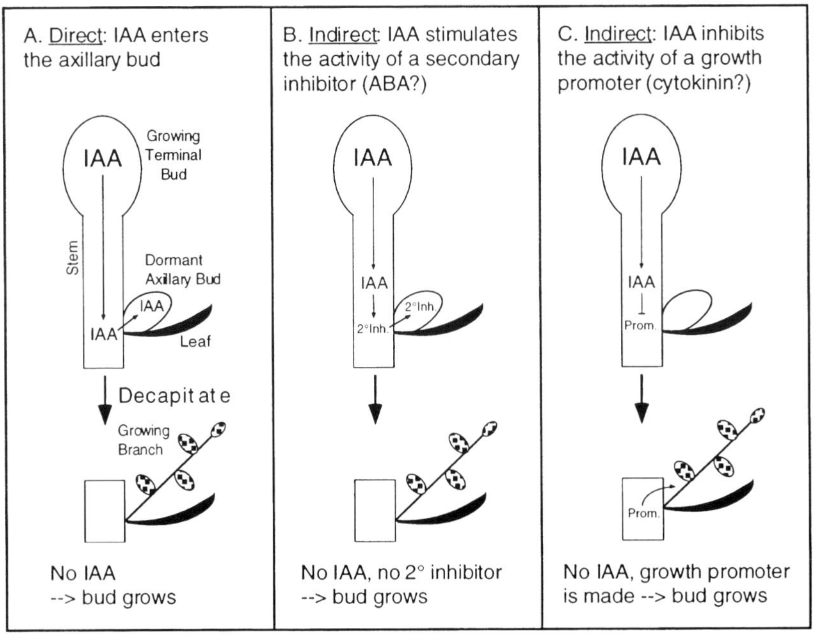

Fig. 23.1. Models for the involvement of IAA derived from the growing terminal bud in regulating axillary bud growth.

Cytokinins (CKs) are good candidates for the promoters of bud growth (Fig. 23.1C). Cytokinins stimulate axillary bud development even in the presence of auxin (Sachs and Thimann, 1967). Transgenic tobacco plants that express the *Agrobacterium ipt* gene, which leads to high levels of CK accumulation, show enhanced bud growth (Medford *et al.*, 1989). Expression of the same gene within small clones of cells promotes adventitious bud development on tobacco leaves (Estruch *et al.*, 1991). However, it is not yet clear whether CKs are the primary promoter of bud growth and development. For example, although the abundance of CKs within *Cicer* buds increases shortly after decapitation, this increase does not occur until after buds have started to grow (Turnbull *et al.*, 1997). It also is not clear whether high levels of CKs within buds are sufficient for maintaining growth once it has begun (Faiss *et al.*, 1997). Although roots synthesize CKs and transport them to the shoot, it now appears that root-derived CKs are not responsible for stimulating the growth of axillary buds (see below).

Branching and Plant Architecture

Plant shoots are constructed from repeating units or modules (also called metamers or phytomers) consisting of a leaf, one or more axillary organs (vegetative buds, floral buds or tendrils) and a section of stem (node and internode). New modules are initiated repeatedly through organogenic activities of the shoot apical meristem or SAM (Sussex, 1989; Kerstetter and Hake, 1997). Vegetative axillary buds have the potential to develop into branches. In turn, buds on primary branches may develop into secondary branches, and so on. Development of all buds on a plant would lead to competition for resources such as photosynthetic light, substrate space or mineral nutrients, and would result in reduced vigour of the plant (Harper, 1985; Givnish, 1995). Most 'reserve meristems' within dormant buds never develop; none the less, they represent a repository of future developmental potential (Stafstrom, 1995).

Shoot-derived auxin and root-derived cytokinins are thought to integrate plant development by inhibiting growth locally (auxin inhibits bud development) and promoting growth at a distance (auxin promotes root development) (Sachs, 1991). Thus buds might be able to perceive their position along the plant axis and, as a consequence, grow or remain dormant based on the relative or absolute level of these two hormones in adjacent stem tissue (Sachs and Thimann, 1967; Stafstrom, 1993). Recent work on branching mutants from several species provides important insights into bud development and plant architecture (reviewed in Napoli *et al.*, 1999; Schmitz and Theres, 1999). The *ramosus1* mutant of pea is illustrative. Whereas wild-type pea plants bear only a few small branches proximal to the flowering node, *rms1* plants have branches at most nodes (Beveridge *et al.*, 1997). Based on the auxin–cytokinin gradient model, it might be expected that *rms1* shoots would have reduced auxin levels, that root sap would be enriched in CKs, or both. Contrary to these expectations, *rms1* plants contain elevated levels of auxin in the shoot and greatly reduced levels of CKs in the root sap. Grafting of *rms1* scions to wild-type root stocks does not promote branching in the scions. Thus, some graft-transmissible factor(s) other than CKs influences bud development (Beveridge *et al.*, 1997; Napoli *et al.*, 1999). Grafts also have been made between tobacco root stocks that over-express the *ipt* gene and wild-type scions. In these experiments, enhanced levels of CKs in roots did not promote bud growth (Faiss *et al.*, 1997). However, CK synthesis within or in the vicinity of the bud can stimulate buds to grow, at least transiently (Faiss *et al.*, 1997).

Growth–Dormancy Cycles in Pea Axillary Buds

Reversibility between growing and dormant states is a hallmark of bud dormancy. Whereas a complete growth–dormancy cycle in buds of temperate perennials typically encompasses an entire year, multiple growth–dormancy cycles can occur during a single season in paradormant buds (Stafstrom, 1995; Crabbé and Barnola, 1996). Axillary buds on Alaska pea plants can be stimulated to undergo more than one complete growth–dormancy cycle (dormant → growing → dormant → growing) over a period of 6 days (Stafstrom and Sussex, 1992). There are four accessory buds at node-2. The largest and oldest bud is referred to as bud-1, the second largest bud is bud-2, and so forth. Bud-1 begins to grow visibly within 8 h of decapitating the terminal bud (Stafstrom and Sussex, 1992). Buds are considered to be in the 'dormancy-to-growth transition stage' between the time of decapitation and 8 h, and to be growing at subsequent times. Decapitating the terminal bud also stimulates the smaller buds to develop. After 4–5 days, small buds become fully dormant due to dominance imposed by bud-1. At intermediate times (2–3 days after decapitation) small buds are in the 'growth-to-dormancy transition stage'. Small buds resume growing within 1 day of removing bud-1.

Two-dimensional SDS-PAGE analysis of silver-stained proteins, *in vivo* labelled proteins and *in vitro* translation products demonstrates that buds in each stable developmental state (growth and dormancy) and each transition state (growth-to-dormancy and dormancy-to-growth) express unique sets of proteins and mRNAs (Stafstrom and Sussex, 1988; Stafstrom, 1993). Dormant pea buds incorporate labelled amino acids into protein at rates equivalent to growing buds (Stafstrom and Sussex, 1988). Based on these results, it appears that dormancy is an active developmental state. Maintenance of the dormant state may be regulated by the expression of dormancy-specific genes, the lack of expression of growth-specific genes, or both.

Origin of Axillary Buds

Axillary organs are derived from a group of meristematic cells in the leaf axil, the axillary meristem. Axillary meristems of ferns, dicots and conifers are usually derived from detached meristems, groups of cells that became separated from the terminal SAM due to the differentiation of intervening cells (Wardlaw, 1965; Sussex, 1989). In addition to vegetative axillary buds, which are indeterminate, the axillary meristem also gives rise to determinate organs such as flowers and thorns. Adventitious buds may arise from nearly any part of the living plant, including roots.

The axillary meristem may give rise to more than one bud. Pea plants contain a characteristic number of accessory buds at each leaf node. Node-2, for example, contains four preformed accessory buds (Fig. 23.2A). *In situ* hybridization analysis indicates that rpL27 mRNA, a growth-associated marker, increases in all four accessory buds following decapitation of the terminal bud (Stafstrom and Sussex, 1992). Thus, all preformed buds are capable of growing. As described above, rapid growth by bud-1 inhibits further growth of the smaller buds, which become dormant again. If bud-1 is removed, bud-2 resumes growing and becomes dominant

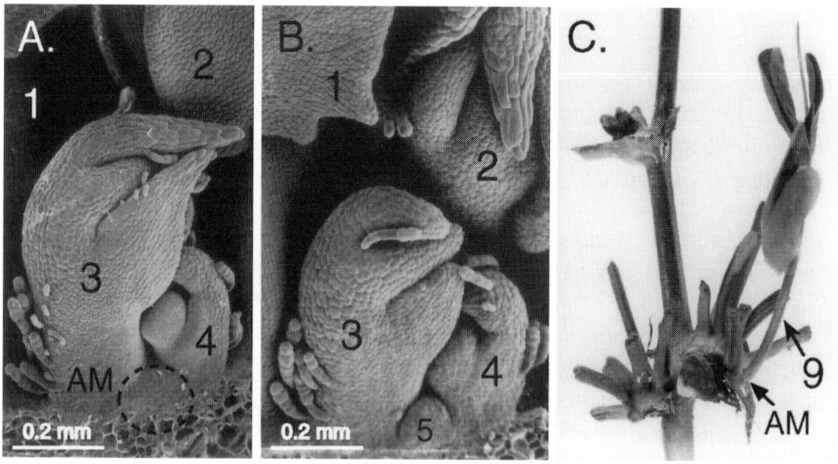

Fig. 23.2. Development of supernumerary buds from the pea axillary meristem. (A) An intact Alaska pea plant contains four dormant accessory buds in the axil of node-2. The approximate position of the axillary meristem (AM) is indicated. (B) One to two days after decapitation, bud-5 (the first supernumerary bud) emerges from the axillary meristem. (C) Preformed and supernumerary buds were removed from each leaf node as soon as they began to grow. Nodes 1, 2 and 3 are shown. Twenty-six days after the terminal bud was decapitated, the stumps of several buds are present at each node. Node-2 contained eight stumps; bud-9, the fifth supernumerary bud, is developing into a shoot. The approximate position of the persistent axillary meristem is indicated. (Adapted from Stafstrom and Sarup, 2000.)

over the other buds (Stafstrom and Sussex, 1992; Stafstrom *et al.*, 1998a). Recently, we explored the effects of removing each preformed axillary bud as soon as it began to grow (Stafstrom and Sarup, 2000). By 1–2 days after decapitation, the apical meristem of the first supernumerary bud (bud-5) begins to emerge from the nodal region between bud-3 and bud-4, which we interpret to be the location of the axillary meristem (Fig. 23.2B). New buds continue to form from the axillary meristem. After 26 days, bud-9 (the fifth supernumerary bud) is developing into a shoot (Fig. 23.2C). All of the nodes that were examined were capable of forming supernumerary buds. The persistence and continued activity of the axillary meristem is not well documented in the literature. Other than pea, persistence of the axillary meristem has been studied experimentally only in *Eucalyptus* spp. (Cremer, 1972) and the *Araucariaceae* (Burrows, 1990). However, the presence of a persistent axillary meristem has been inferred in other species (Fink, 1984). In the tomato mutant *lateral suppressor*, axillary buds do not develop during vegetative development because 'cells in the axils of the leaf primordia fail to retain their meristematic character' (Schumacher *et al.*, 1999).

Molecular and Cellular Activities of Growing Buds

Two-dimensional protein gels have provided a global view of gene expression patterns in developing pea buds. A more precise molecular understanding of bud development has emerged by analysing a few stage-specific genes in greater detail.

Some 80 different proteins comprise the two subunits of the eukaryotic ribosome. Since growth commonly requires enhanced levels of protein synthesis, synthesis of new ribosomes is usually associated with growth. Ribosomal protein L27 gene (*rpL27*) is a useful molecular marker for bud growth during growth–dormancy cycles in bud-2. Accumulation of rpL27 mRNA is several-fold higher in growing buds (1 day after decapitation) than in dormant buds on intact plants (Fig. 23.3). As bud-2 becomes dormant due to inhibition by bud-1, *rpL27* expression declines to basal levels. If bud-1 is removed 5 days after the first decapitation (stage 5+1), bud-2 grows and again expresses high levels of *rpL27*. Northern blotting and *in situ* hybridization experiments indicate that rpL27 mRNA levels increase in all regions of the bud within 1 h of decapitation and reach maximal levels within 6 h (Stafstrom and Sussex, 1992). It appears that regulatory signals from the plant are transported very rapidly into the buds and that all bud cells are competent to respond to these signals.

Histone, MAP kinase, cdc2 (cyclin-dependent kinase) and cyclin B mRNAs accumulate preferentially in growing pea buds (Devitt and Stafstrom, 1995). PsDRG, a novel GTP binding protein, is also preferentially

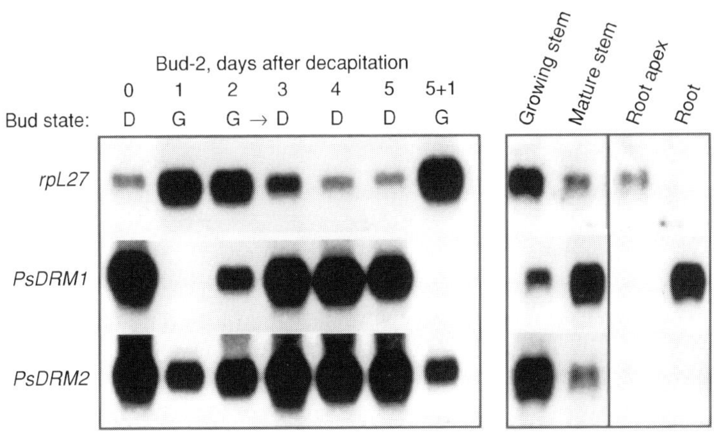

Fig. 23.3. Northern blot analysis of gene expression in pea axillary buds and other organs. The *rpL27* gene is preferentially expressed in growing axillary buds and in other growing organs (root apices and elongating stems). *PsDRM1* expression is much greater in dormant axillary buds and in non-growing stems and roots than in their growing counterparts. *PsDRM2* mRNA accumulated preferentially in dormant buds but was also present in growing buds. (Adapted from Stafstrom *et al.*, 1998a.)

expressed in growing buds (Devitt *et al.*, 1999). Histone mRNAs accumulate almost exclusively during S phase, whereas cyclin B accumulates during G2 and M phases (Jacobs, 1995). Expression of the cyclin D and PCNA genes, which are associated with the G1 checkpoint and S phase, respectively, are also upregulated in growing pea buds (Shimizu and Mori, 1998a). Interestingly, cyclin D and PCNA proteins are present in both dormant and growing buds, but they interact with each other only in dormant buds (Shimizu and Mori, 1998b). The complex formed by these proteins (which may also include cdc2 kinase and a CDK inhibitor), may be involved in blocking entry into S phase in dormant bud cells. Activation of cdc2 and cyclin genes is also correlated with bud growth in *Populus* (Rohde *et al.*, 1997). These results indicate that bud growth and dormancy, which are organ-level phenomena, are closely linked to cell proliferation and quiescence (Nougarède and Rondet, 1978).

Flow cytometry analysis indicates that cells in dormant buds are arrested in G1 and G2 (approx. 3 : 1 ratio; Devitt and Stafstrom, 1995). Based on the following interpretations, both cell populations begin to proliferate very soon after decapitation. Histone mRNAs begin to accumulate within 1 h of decapitation, which indicates that at least some G1 cells are arrested very near to the G1/S boundary. A second peak of histone expression beginning at 9 h may represent a population of G1 cells arrested earlier in G1. Cyclin B expression is tightly linked to late G2 and M phases of the cell cycle, and thus is an excellent marker for mitosis (Jacobs, 1995). A small and transient increase in cyclin B expression occurs in pea buds at 6 h after decapitation and a large and sustained increase begins at 18 h (Devitt and Stafstrom, 1995). The first peak of cyclin B expression is due to proliferation of cells that had been arrested in G2 and the second peak is due to cells that had been arrested in G1 (Devitt and Stafstrom, 1995). In contrast to our results, Shimizu and Mori (1998a) only observed the later peak of cyclin B expression, which began at 14 h after decapitation (we did not analyse time points between 12 and 18 h). To address this discrepancy, we examined the distribution of dividing cells in Feulgen-stained sections of developing pea buds. In dormant buds, dividing cells were present in the apex proper (distal to the youngest leaf primordium) and in the youngest leaf/internode modules (L/I 5 and L/I 4); cell division was quite rare in older, more proximal regions of the bud (Fig. 23.4A). At 6 h after decapitation, the frequency of dividing cells was much higher, particularly in L/I 3 (Fig. 23.4B). Twenty-four hours after decapitation, the frequency of dividing cells was higher than at 6 h, but the spatial distribution of dividing cells was similar at both stages (Fig. 23.4C). The mitotic index was determined for these and additional regions of buds from intact plants and from plants 3, 6 and 24 h after decapitation (Fig. 23.5). Leaf-1 and internode-1 are from the oldest and most basal region of the bud, and together comprise about 90% of the mass of the bud. Thus when buds are being studied by Northern blotting or by some other biochemical technique, it is predominantly these

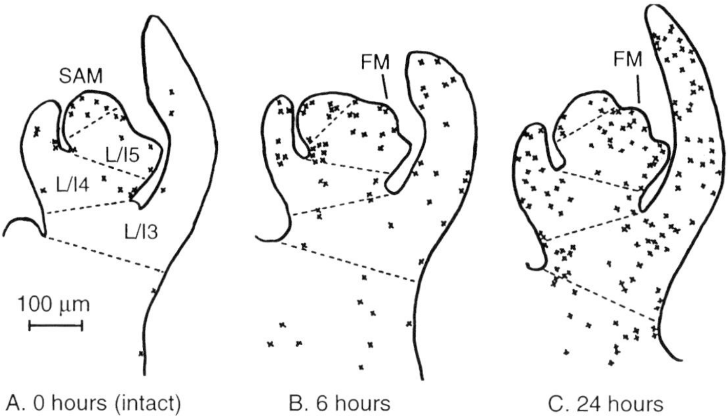

Fig. 23.4. Localization of dividing cells in developing pea buds. The apical region of bud-1 from node-2 was analysed. This bud typically contains five leaf primordia; the three youngest leaf/internode modules are indicated (L/I 3, L/I 4 and L/I 5). Histological sections were stained with Feulgen to highlight dividing cells. For each stage, the locations of mitotic figures (X) from eight serial sections were superimposed. FM, floral meristem; SAM, shoot apical meristem.

cells that are being analysed. In both organs, there is a clear increase in mitotic index at 6 h and an even greater increase at 24 h. The distributions of mitotic cells at each stage are indistinguishable (data not shown). Therefore, it appears that cells arrested in G1 and G2 are rather uniformly distributed throughout the bud and that both populations begin to proliferate soon after decapitation. Cells in the first two internodes of pea cotyledonary buds are also arrested in both G1 and G2; a peak in mitotic activity 4 h after decapitation indicates proliferation of cells arrested in G2 (Nougarède and Rondet, 1978).

ABA inhibits cell division in pea cotyledonary buds (Nougarède *et al.*, 1987). The *Arabidopsis ICK1* gene encodes a protein that interacts with cdc2 kinase and cyclin D3, thereby inhibiting progression through the G1/S checkpoint (Wang *et al.*, 1998). *ICK1* expression is stimulated by ABA, which therefore suggests a mechanism by which ABA may inhibit cell proliferation and promote bud dormancy.

Gene Expression in Dormant Buds

Very little is known about genes that are necessary for promoting or maintaining the dormant state. We recently described two genes, *PsDRM1* and *PsDRM2*, that are preferentially expressed in dormant pea buds (Stafstrom *et al.*, 1998a). Expression of *PsDRM1* is closely linked with the dormant state

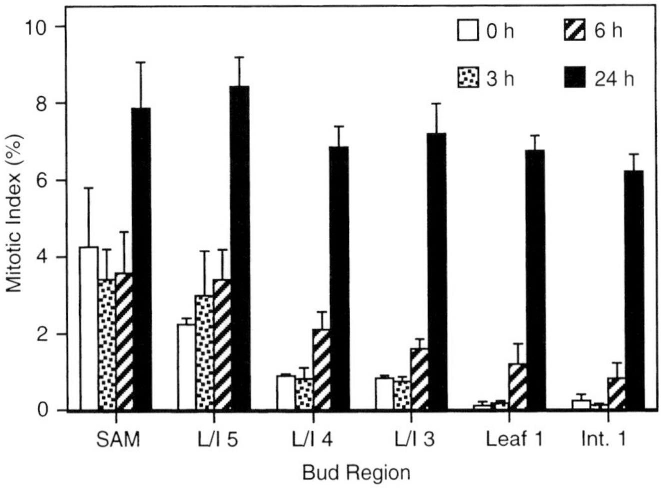

Fig. 23.5. Mitotic index (mean ± SE)in various regions of developing pea buds. Bud-1 from node-2 was studied. Three buds were analysed for each stage, and three serial sections were analysed for each bud. The SAM and leaf/internode modules 3, 4 and 5 are as shown in Fig. 23.4. Leaf-1 and internode-1 are the largest and most basal parts of the bud; together they represent about 90% of the bud mass.

during growth–dormancy cycles, that is, its expression pattern is reciprocal to that of *rpL27* (Fig. 23.3). *PsDRM1* mRNA abundance in axillary buds declines 20-fold within 6 h of decapitation. *PsDRM1* expression is also enhanced in other non-growing organs (fully elongated roots and stems) compared with their growing counterparts. A related gene from *Arabidopsis*, *AtDRM1*, is expressed at relatively high levels in 6-week-old rosette leaves, flower buds, open flowers, roots, siliques and inflorescence stems. This mRNA was not detected in 8-day-old rosette leaves, the only organ assayed that was not yet fully grown (Stafstrom *et al.*, 1998b).

The best-studied homologue of *PsDRM1* is *λSAR5*, which was isolated from strawberry receptacles based on its repression by auxin (Reddy and Poovaiah, 1990). The effects of hormones and other factors on *PsDRM1* expression was examined in seedling roots. Roots normally express *PsDRM1* at high levels, they readily take up exogenously supplied hormones and, most importantly, they are much easier to collect in large quantities than axillary buds. *PsDRM1* expression is repressed by 100 μM 2,4-D, a synthetic auxin, and enhanced by 100 μM ABA (Fig. 23.6). These responses are consistent with the observation that dormant buds contain relatively low levels of auxin and high levels of ABA (Gocal *et al.*, 1991; Fig. 23.1). It should be noted, however, that expression was altered in roots only in response to very high concentrations of these hormones. *PsDRM1* was stimulated by high NaCl and by cold. Transgenic *Arabidopsis* plants expressing

AtDRM1::*GUS* fusion will allow us to analyse how this promoter is regulated in buds and other organs (Stafstrom *et al.*, 1998b).

The expression pattern of *PsDRM2* in pea buds is qualitatively similar to that of *PsDRM1*, but the level of *PsDRM2* mRNA is only about threefold lower in growing buds than in dormant buds (Fig. 23.5). In contrast with *PsDRM1*, *PsDRM2* is expressed at higher levels in growing stems than in mature stems. *PsDRM2* expression is barely detectable in roots and root apices. Thus, *PsDRM2* expression is not strictly associated with non-growing organs. *PsDRM2* encodes a glycine-rich protein whose sequence is related to cold- and ABA-responsive genes from lucerne (Luo *et al.*, 1991; Laberge *et al.*, 1993). *PsDRM2* expression is stimulated by 100 μM ABA but not by cold. Unlike *PsDRM1*, *PsDRM2* expression is stimulated by 100 μM 2,4-D (Fig. 23.6). Glycine-rich dehydrin-like proteins accumulate in cold acclimated peach bark (Arora and Wisniewski, 1994). At this point, it is not known whether these two classes of glycine-rich proteins perform related functions in the dormant tissues in which they accumulate.

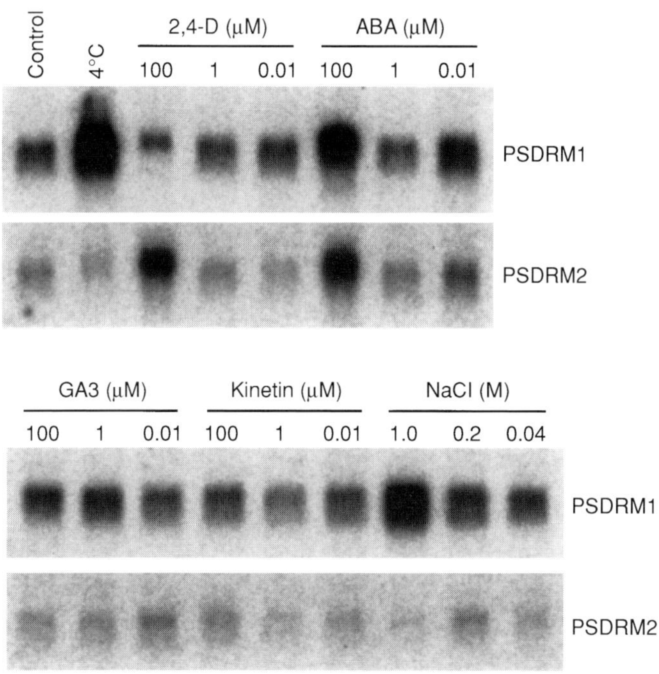

Fig. 23.6. Northern blot analysis of gene expression in pea seedling roots in response to hormones, NaCl and cold. Two-day-old pea seedlings were incubated in water (control) or the indicated treatment for 8 h. RNA was isolated from a region of the root between 1 and 2 cm from the root apex. *PsDRM1* expression is enhanced by cold and 100 μM ABA, and repressed by 100 μM 2,4-D. *PsDRM2* expression is enhanced by 100 μM ABA and 100 μM 2,4-D.

Manipulation of Hormone Levels in Transgenic Plants

Traditionally, hormone responses have been studied by applying a hormone to a tissue, waiting some period of time, and then recording its effect on growth. Molecular approaches have at least two advantages over measuring growth *per se*. First, hormone-induced changes in gene expression occur much more rapidly than most growth responses. In some instances, altered gene expression can be detected within minutes of hormone application (Abel *et al.*, 1994). Second, genetic engineering techniques can be used to alter the activity of genes responsible for the synthesis, breakdown, transport and detection of hormones (Klee and Lanahan, 1995). The 35S promoter leads to very high levels of transcription in essentially every cell of the plant. When this promoter was linked to *iaaH* and *iaaM*, two genes from *Agrobacterium* that are involved in IAA synthesis, plants produced very high levels of IAA in all tissues. These plants exhibited very strong apical dominance; in fact, buds did not grow even when plants were decapitated (Klee and Lanahan, 1995). In a related experiment, the 35S promoter was linked to the *iaaL* gene, which conjugates lysine to IAA and renders the IAA inactive (Romano *et al.*, 1991). Such plants, which contain very low levels of active auxin, are highly branched. Plants that over-expressed the *ipt* gene under the control of a heat-shock promoter are also very bushy (Medford *et al.*, 1989). While these experiments demonstrated the power of genetic engineering, they actually added little to what was known from earlier work. The reason is that hormone levels are seldom uniformly high (or low) within the various tissues and organs of a plant. As was discussed above, high auxin in the stem leads to bud dormancy whereas high auxin within a bud is correlated with bud growth. By using more specific promoters, however, it is possible to alter hormone levels in only specific cells. For example, by fusing the *ipt* gene to a promoter that is de-repressed by tetracycline, it was possible to increase cytokinin accumulation only in roots or only in buds (Faiss *et al.*, 1997). Only CK accumulation within buds stimulated their growth.

Gan and Amasino (1995) developed an elegant method for testing whether the local synthesis of a hormone (cytokinin) is sufficient for effecting a physiological phenomenon (inhibiting leaf senescence). The promoter from *SAG12*, a leaf senescence gene, was fused to *ipt*. The onset of leaf senescence triggered expression of *ipt*, which led to cytokinin accumulation, which then turned off the senescence programme. Similar approaches could be used to better understand the role of various hormones in controlling bud growth and dormancy. Let us suppose, for example, that the synthesis of IAA within an axillary bud is necessary and sufficient for promoting and maintaining growth. If a dormant bud-specific promoter were linked to the *iaaM* and *iaaH* genes, then auxin would accumulate within buds whenever they received signals from the rest of the plant that said 'be dormant'. Instead, these buds would grow. To further test this hypothesis, one could fuse a growing bud-specific promoter to the *iaaL* gene. Plants expressing this gene construct would be expected to respond to the 'grow' signal (e.g. plant decapitation) by inactivating all IAA within the bud. As a result, buds on these plants would not develop. Many related experiments can be envisioned. At present, the biggest hurdle to performing such experiments is the lack of promoters that are expressed exclusively in one stage of bud development or another. For example, PsDRM1 is expressed at very high levels in dormant buds, but it is also expressed at high levels in other non-growing organs (Fig. 23.3).

The complete sequences of the *Arabidopsis* and rice genomes will be available within a couple of years. Also, the ability to simultaneously analyse the expression of thousands of genes using microarrays and other developing technologies will greatly accelerate the pace of discovering genes associated with many developmental processes (Ruan *et al.*, 1998). As a result, researchers may soon have plentiful dormancy genes in hand. The next and much harder challenge will be to determine how these genes function individually and together within the developing plant to regulate dormancy and other processes.

Acknowledgements

This work was supported by NIH grant GM54276-01 and by the Plant Molecular Biology Center, Northern Illinois University.

References

Abel, S., Oeller, P.W. and Theologis, A. (1994) Early auxin-induced genes encode short-lived nuclear proteins. *Proceedings of the National Academy of Sciences, USA* 91, 326–330.

Arora, R. and Wisniewski, M.E. (1994) Cold acclimation in genetically related (sibling) deciduous and evergreen peach (*Prunus persica* [L.] Batsch). II. A 60-kilodalton bark protein in cold-acclimated tissues of peach is heat stable and related to the dehydrin family of proteins. *Plant Physiology* 105, 95–101.

Beveridge, C.A., Symons, G.M., Murfet, I.C., Ross, J.J. and Rameau, C. (1997) The *rms1* mutant of pea has elevated indole-3-acetic acid levels and reduced root-sap zeatin riboside content but increased branching controlled by graft-transmissible signal(s). *Plant Physiology* 115, 1251–1258.

Borchert, R. (1991) Growth periodicity and dormancy. In: Raghavendra, A.S. (ed.) *Physiology of Trees*. John Wiley & Sons, New York, pp. 221–245.

Burrows, G.E. (1990) The role of axillary meristems in coppice and epicormic bud initiation in *Araucaria cunninghamii*. *Botanical Gazette* 151, 293–301.

Cline, M.G. (1994) The role of hormones in apical dominance. New approaches to an old problem in plant development. *Physiologia Plantarum* 90, 230–237.

Cline, M.G. (1996) Exogenous auxin effects on lateral bud outgrowth in decapitated shoots. *Annals of Botany* 78, 255–266.

Crabbé, J. and Barnola, P. (1996) A new conceptual approach to bud dormancy in woody plants. In: Lang, G.A. (ed.) *Plant Dormancy: Physiology, Biochemistry and Molecular Biology*. CAB International, Wallingford, UK, pp. 83–113.

Cremer, K.W. (1972) Morphology and development of the primary and accessory buds of *Eucalyptus regnans. Australian Journal of Botany* 20, 175–195.

Devitt, M.L. and Stafstrom, J.P. (1995) Cell cycle regulation during growth-dormancy cycles in pea axillary buds. *Plant Molecular Biology* 29, 255–265.

Devitt, M.L., Maas, K.J. and Stafstrom, J.P. (1999) Characterization of DRGs, developmentally regulated GTP-binding proteins, from pea and *Arabidopsis. Plant Molecular Biology* 39, 75–82.

Estruch, J.J., Prinsen, E., Van Onckelen, H., Schell, J. and Spena, A. (1991) Viviparous leaves produced by somatic activation of an inactive cytokinin-synthesizing gene. *Science* 254, 1364–1367.

Faiss, M., Zalubìlová, J., Strnad, M. and Schmülling, T. (1997) Conditional transgenic expression of the *ipt* gene indicates a function for cytokinins in paracrine signalling in whole tobacco plants. *Plant Journal* 12, 401–415.

Fink, S. (1984) Some cases of delayed or induced development of axillary buds from persisting detached meristems in conifers. *American Journal of Botany* 71, 44–51.

Fuchigami, L.H. and Nee, C.C. (1987) Degree growth stage model and rest-breaking mechanisms in temperate woody perennials. *HortScience* 22, 836–845.

Gan, S. and Amasino, R.M. (1995) Inhibition of leaf senescence by autoregulated production of cytokinin. *Science* 270, 1986–1988.

Givnish, T.J. (1995) Plant stems: biomechanical adaptation for energy capture and influence on species distributions. In: Gartner, B.L. (ed.) *Plant Stems: Physiology and Functional Morphology*. Academic Press, San Diego, pp. 3–49.

Gocal, G.F.W., Pharis, R.P., Yeung, E.C. and Pearce, D. (1991) Changes after decapitation of indole-3-acetic acid and abscisic acid in the larger axillary bud of *Phaseolus vulgaris* L. cv Tender Green. *Plant Physiology* 95, 344–350.

Harper, J.L. (1985) Modules, branches, and the capture of resources. In: Jackson, J.B.C., Buss, L.W. and Cook, R.E. (eds) *Population Biology and Evolution of Clonal Organisms*. Yale University Press, New Haven, pp. 1–33.

Jacobs, T.W. (1995) Cell cycle control. *Annual Review of Plant Physiology and Plant Molecular Biology* 46, 317–339.

Kerstetter, R.A. and Hake, S. (1997) Shoot meristem formation in vegetative development. *Plant Cell* 9, 1001–1010.

Klee, H.J. and Lanahan, M.B. (1995) Transgenic plants in hormone biology. In: Davies, P.J. (ed.) *Plant Hormones. Physiology, Biochemistry, and Molecular Biology*. Kluwer Academic Publishers, Dordrecht, pp. 340–353.

Laberge, S., Castonguay, Y. and Vezina, L.P. (1993) New cold- and drought-regulated gene from *Medicago sativa. Plant Physiology* 101, 1411–1412.

Luo, M., Lin, L., Hill, R.D. and Mohapatra, S.S. (1991) Primary structure of an environmental stress and abscisic acid-inducible alfalfa protein. *Plant Molecular Biology* 17, 1267–1269.

Medford, J.I., Horgan, R., El-Sawi, Z. and Klee, H.J. (1989) Alterations of endogenous cytokinins in transgenic plants using a chimeric isopentenyl transferase gene. *Plant Cell* 1, 403–413.

Napoli, C.A., Beveridge, C.A. and Snowden, K.C. (1999) Reevaluating concepts of apical dominance and the control of axillary bud outgrowth. *Current Topics in Developmental Biology* 44, 127–169.

Nougarède, A. and Rondet, P. (1978) Evénements structuraux et métaboliques dans les entre-noeuds des bourgeons axillaires du pois, en réponse à la levée de dominance. *Canadian Journal of Botany* 56, 1213–1228.

Nougarède, A., Rondet, P., Landré, P. and Rembur, J. (1987) Effet d'un traitement par l'acide abscisique sur la division cellulaire, les teneurs en ADN et l'élongation du burgeon cotylédonaire de plants de pois décapités. *Canadian Journal of Botany* 65, 907–915.

Pearce, D.W., Taylor, J.S., Robertson, J.M., Harker, N. and Daly, E.J. (1995) Changes in abscisic acid and indole-3-acetic acid in axillary buds of *Elytrigia repens* released from apical dominance. *Physiologia Plantarum* 94, 110–116.

Powell, L.E. (1988) The hormonal control of bud and seed dormancy in woody plants. In: Davies, P.J. (ed.) *Plant Hormones and Their Role in Plant Growth and Development.* Kluwer Academic Publishers, Dordrecht, pp. 539–552.

Reddy, A.S.N. and Poovaiah, B.W. (1990) Molecular cloning and sequencing of a cDNA for an auxin-repressed mRNA: correlation between fruit growth and repression of the auxin-regulated gene. *Plant Molecular Biology* 14, 127–136.

Rohde, A., Van Montagu, M., Inzé, D. and Boerjan, W. (1997) Factors regulating the expression of cell cycle genes in individual buds of *Populus. Planta* 201, 43–52.

Romano, C.P., Hein, M.B. and Klee, H.J. (1991) Inactivation of auxin in tobacco transformed with the indoleacetic acid-lysine synthetase gene of *Pseudomonas savastanoi. Genes and Development* 5, 438–446.

Romano, C.P., Cooper, M.L. and Klee, H.J. (1993) Uncoupling auxin and ethylene effects in transgenic tobacco and *Arabidopsis* plants. *Plant Cell* 5, 181–189.

Ruan, Y., Gilmore, J. and Conner, T. (1998) Towards *Arabidopsis* genome analysis: monitoring expression profiles of 1400 genes using cDNA microarrays. *Plant Journal* 15, 821–833.

Sachs, T. (1991) *Pattern Formation in Plant Tissues.* Cambridge University Press, Cambridge.

Sachs, T. and Thimann, K.V. (1967) The role of auxins and cytokinins in the release of buds from apical dominance. *American Journal of Botany* 54, 136–144.

Schmitz, G. and Theres, K. (1999) Genetic control of branching in *Arabidopsis* and tomato. *Current Opinion in Plant Biology* 2, 51–55.

Schumacher, K., Schmitt, T., Rossberg, M., Schmitz, G. and Theres, K. (1999) The *Lateral suppressor* (*Ls*) gene of tomato encodes a new member of the VHIID protein family. *Proceedings of the National Academy of Sciences, USA* 96, 290–295.

Shimizu, S. and Mori, H. (1998a) Analysis of cycles of dormancy and growth in pea axillary buds based on mRNA accumulation patterns of cell cycle-related genes. *Plant Cell Physiology* 39, 255–262.

Shimizu, S. and Mori, H. (1998b) Changes in protein interactions of cell cycle-related genes during the dormancy-to-growth transition in pea axillary buds. *Plant Cell Physiology* 39, 1073–1079.

Stafstrom, J.P. (1993) Axillary bud development in pea: apical dominance, growth cycles, hormonal regulation and plant architecture. In: Amasino, R.M. (ed.) *Cellular Communication in Plants.* Plenum Press, New York, pp. 75–86.

Stafstrom, J.P. (1995) Developmental potential of shoot buds. In: Gartner, B.L. (ed.) *Plant Stems: Physiology and Functional Morphology.* Academic Press, San Diego, pp. 257–279.

Stafstrom, J.P. and Sarup, V.B. (2000) Development of supernumerary buds from the axillary meristem of pea, *Pisum sativum* (Fabaceae). *Australian Journal of Botany* (in press).

Stafstrom, J.P. and Sussex, I.M. (1988) Patterns of protein synthesis in dormant and growing vegetative buds of pea. *Planta* 176, 497–505.

Stafstrom, J.P. and Sussex, I.M. (1992) Expression of a ribosomal protein gene in axillary buds of pea. *Plant Physiology* 100, 1494–1502.

Stafstrom, J.P., Ripley, B.D., Devitt, M.L. and Drake, B. (1998a) Dormancy-associated gene expression in pea axillary buds. Cloning and expression of PsDRM1 and PsDRM2. *Planta* 205, 547–552.

Stafstrom, J.P., Krueger, M.T. and Stoudt, W. (1998b) Nucleotide sequence of AtDRM1 cDNA and genomic clones (Accession nos. AF053746 and AF053747), an *Arabidopsis* homologue of a pea dormant bud-associated gene. *Plant Physiology* 117, 718.

Sussex, I.M. (1989) Developmental programming of the shoot meristem. *Cell* 56, 225–229.

Tamas, I.A. (1995) Hormonal regulation of apical dominance. In: Davies, P.J. (ed.) *Plant Hormones. Physiology, Biochemistry, and Molecular Biology*. Kluwer Academic Publishers, Dordrecht, pp. 572–597.

Tamas, I.A., Engels, C.J., Kaplan, S.L., Ozbun, J.L. and Wallace, D.H. (1981) Role of indoleacetic acid and abscisic acid in the correlative control by fruits of axillary bud development and leaf senescence. *Plant Physiology* 68, 476–481.

Thimann, K.V. and Skoog, F. (1933) Studies on the growth hormone of plants. III. The inhibiting action of the growth substance on bud development. *Proceedings of the National Academy of Sciences, USA* 19, 714–716.

Turnbull, C.G.N., Raymond, M.A.A., Dodd, I.C. and Morris, S.E. (1997) Rapid increases in cytokinin concentration in lateral buds of chickpea (*Cicer arietinum* L.) during release of apical dominance. *Planta* 202, 271–276.

Wang, H., Qi, Q., Schorr, P., Cutler, A.J., Crosby, W.L. and Fowke, L.C. (1998) ICK1, a cyclin-dependent protein kinase inhibitor from *Arabidopsis thaliana* interacts with both Cdc2a and CycD3, and its expression is induced by abscisic acid. *The Plant Journal* 15, 501–510.

Wardlaw, C.W. (1965) Leaves and buds: Mechanisms of local induction in plant growth. In: Beerman, W., Wardlaw, C.W., Gautheret, R.J., Wigglesworth, V.B., Nieuwkoop, P.D., Wolff, E. and Zeevaart, J.A.D. (eds) *Cell Differentiation and Morphogenesis*. North-Holland Publishing, Amsterdam, pp. 96–119.

Expression of Endo-β-mannanase and SNF-related Protein Kinase Genes in True Potato Seeds in Relation to Dormancy, Gibberellin and Abscisic Acid

24

Veria Alvarado, Hiroyuki Nonogaki and Kent J. Bradford

Department of Vegetable Crops, University of California, One Shields Avenue, Davis, CA 95616-8631, USA

Introduction

In an imbibed but dormant seed, some metabolic processes, such as respiration and repair, can be expected to be similar to the corresponding processes in a non-dormant seed. However, at least two processes will be different between dormant and non-dormant seeds: the growth of the embryo associated with radicle emergence and the mobilization of storage reserves. In dormant seeds, growth of the embryo is prevented and storage reserves must be conserved until germination eventually occurs. Thus, we can expect that differential gene expression between dormant and non-dormant seeds will include genes involved in these processes.

Tomato seeds (*Lycopersicon esculentum* Mill.) have been particularly well studied as a model system for physiological mechanisms involved in germination (reviewed in Hilhorst *et al.*, 1998; Welbaum *et al.*, 1998). Tomato seeds comprise an embryo completely surrounded by a relatively hard and brittle endosperm and a thin testa (Fig. 24.1). In order to complete germination, the embryo must generate sufficient force to penetrate the endosperm cap and testa in the micropylar region opposite the radicle tip. The endosperm tissue enclosing the radicle tip (called the endosperm cap or the micropylar endosperm) can be excised and its strength determined by puncture force measurements using a mechanical force analyser (Groot and Karssen, 1987). These experiments showed that weakening of the endosperm cap is required for radicle emergence and

(eds J.-D. Viémont and J. Crabbé)

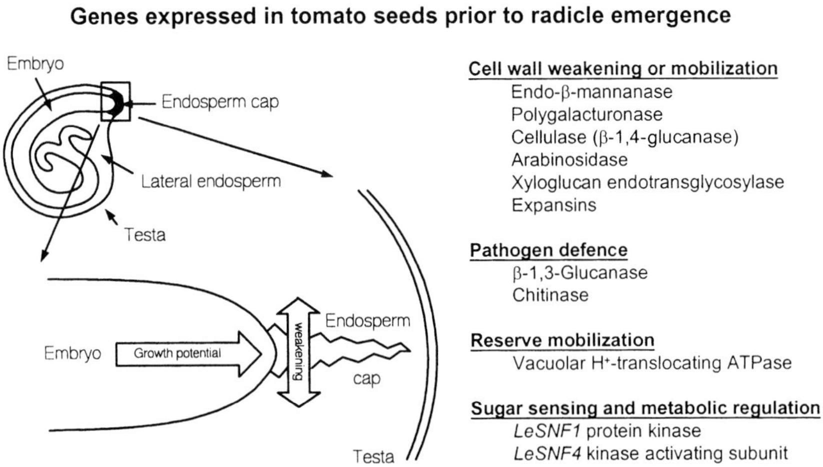

Fig. 24.1. The tomato embryo is surrounded by a rigid endosperm that restricts radicle emergence. Weakening of the endosperm cap, accompanied by the expansive force or growth potential of the embryo, results in radicle emergence and the completion of germination. On the right are listed some of the cloned genes that have been identified as being differentially expressed in germinating versus non-germinating tomato seeds. For details on the individual genes, see Cooley *et al.* (1999), Sitrit *et al.* (1999) and Bradford *et al.* (2000).

that the weakening process is primarily controlled by gibberellin (GA). Thus, studies of the regulation of tomato seed germination have focused on processes occurring in the endosperm cap and the enclosed radicle tip.

As would be expected for a tissue-weakening process, an array of cell wall hydrolases are expressed in the endosperm caps of germinating tomato seeds, including endo-β-mannanase, mannosidase, galactosidase, polygalacturonase, pectin methylesterase, cellulase, arabinosidase, and xyloglucan endotransglycosylase (Groot *et al.*, 1988; Downie *et al.*, 1998; Sitrit *et al.*, 1999; Bradford *et al.*, 2000). We have identified and cloned a number of these hydrolases and other genes that are expressed preferentially in germinating seeds (Fig. 24.1; see Bradford *et al.*, 2000, for details). Expansins are also expressed in both the endosperm cap and the embryo, and may be involved in both weakening and growth. In addition, β-1,3-glucanase and chitinase are expressed specifically in the endosperm cap and could be involved in pathogen defence mechanisms as radicle protrusion exposes the endosperm to the biotic environment. Mobilization of protein and lipid reserves also occurs first in the endosperm cap in association with tissue weakening (Mella *et al.*, 1995; Nonogaki *et al.*, 1998; Toorop, 1998). Vacuolar H^+-translocating ATPase (V-ATPase) expressed in the cap tissue is likely to be involved in the mobilization of protein bodies and possibly in the secretion of hydrolases to the cell wall (Cooley *et al.*, 1999).

Among these genes, endo-β-mannanase has been studied most intensively. The endosperm cap cell walls of tomato contain 60–70% mannose, suggesting that they are primarily composed of mannan polymers, most likely galactomannans (Groot *et al.*, 1988; Dahal *et al.*, 1997). Groot *et al.* (1988) showed that endo-β-mannanase, mannosidase and galactosidase activities were present in GA-treated gibberellin-deficient (*gib-1*) tomato seeds, with endo-β-mannanase showing the most dramatic increase prior to radicle emergence. Multiple isoforms of endo-β-mannanase are present in tomato seeds, with different electrophoretically distinguishable isoforms being expressed first in the endosperm cap, then in the embryo and the lateral endosperm (Nonogaki and Morohashi, 1996; Toorop *et al.*, 1996; Voigt and Bewley, 1996; Nonogaki *et al.*, 1998). Bewley *et al.* (1997) cloned an endo-β-mannanase cDNA from germinated tomato seeds (*LeMAN1*), and Southern blots indicated that there are four or more endo-β-mannanase genes in tomato. We recently isolated a second endo-β-mannanase cDNA (*LeMAN2*) from an imbibed tomato seed cDNA library (H. Nonogaki *et al.*, unpublished results). *LeMAN2* is expressed only in the endosperm cap prior to radicle emergence, and *LeMAN1* is subsequently expressed in the lateral endosperm after radicle emergence. We have used the endosperm cap-specific mannanase cDNA *LeMAN2* to study the regulation of expression of this gene in relation to dormancy and germination.

Another group of genes expressed in germinating tomato seeds may have roles in sugar sensing, energy metabolism, stress responses and reserve mobilization. In a differential cDNA display screen of *gib-1* tomato seeds imbibed in water or in GA, we identified a mRNA that was present in mutant seeds imbibed in water but which disappeared rapidly in response to GA. We subsequently cloned a full-length cDNA corresponding to this mRNA and found that it had significant sequence homology to protein kinase-related proteins known as SNF4 in yeast and AMPK-γ in mammals. The yeast SNF (sucrose non-fermenting) protein kinase and mammalian AMP-activated protein kinase (AMPK) complexes are central components of conserved kinase cascades that act as metabolic sensors of changes in glucose availability and of ATP/AMP ratios, respectively (Hardie *et al.*, 1998). Protein sequence and functional homology exists between the kinase subunits (SNF1/AMPK-α), activating subunits (SNF4/AMPK-γ) and bridging or docking subunits (SIP/GAL83/AMPK-β) that constitute the functional kinase complexes in yeast and mammals.

In plants, a large family of SNF1-related kinases (SnRKs) has been identified that shares sequence and functional homology with the kinase subunit (Halford and Hardie, 1998), and homologues of the yeast docking subunit have been identified in potato (*Solanum tuberosum* L.; Lakatos *et al.*, 1999) and *Arabidopsis* (Bouly *et al.*, 1999). Homologues of the SNF4/AMPK-γ activating subunit of the protein kinase complex have been reported from developing bean (*Phaseolus vulgaris*) seeds (Abe *et al.*, 1996) and *Arabidopsis* (Bouly *et al.*, 1999). We demonstrated that our tomato

SNF4-related cDNA functionally complements a yeast *SNF4* deletion mutant and therefore termed this gene *LeSNF4* (*Lycopersicon esculentum SNF4*). We also cloned a tomato SNF1 kinase subunit homologue (termed *LeSNF1*) that complements the corresponding *SNF1* deletion mutant in yeast (P. Dahal and K.J. Bradford, unpublished results). Given the sequence and functional homology among yeast, plants and mammals, it is likely that these SNF-related kinase (SnRK) genes in tomato also play a role in sugar sensing or metabolic regulation (Halford and Hardie, 1998; Halford *et al.*, 1999).

With respect to germination, *LeSNF4* mRNA rapidly disappears upon imbibition of wild-type seeds in water or of *gib-1* seeds in GA. However, conditions that prevent the completion of germination, including low water potential, ABA, GA deficiency, far-red light or natural dormancy, all maintain *LeSNF4* mRNA abundance (Bradford *et al.*, 2000). An ABA-responsive protein kinase from wheat (PKABA1) has a similar expression pattern (Holappa and Walker-Simmons, 1995) and was recently demonstrated to be involved in the regulation of ABA-suppressed genes in aleurone cells (Gomez-Cadenas *et al.*, 1998). On the other hand, *LeSNF1* mRNA abundance is relatively unresponsive to developmental transitions or stress conditions (Bradford *et al.*, 2000), as is the case with the yeast and animal kinase complexes. In the latter cases, activity is regulated biochemically (as by phosphorylation/dephosphorylation and protein/protein interactions) rather than transcriptionally (Hardie *et al.*, 1998). Our results with tomato suggest that in plants, transcriptional regulation of the SnRK activating subunit (SNF4 homologue) may be an additional mechanism by which metabolic pathways are regulated in response to developmental or environmental cues. For example, sink activity (i.e. the conversion of sucrose to storage compounds) during seed development may be associated with high *LeSNF4* expression, while mobilization of storage reserves and the export of sucrose to the growing seedling during germination may be related to low *LeSNF4* expression.

While tomato has many advantages for the study of germination mechanisms, it generally exhibits only relatively weak primary dormancy. Seed lots with partial dormancy are sometimes produced, but inconsistently, and dormancy is lost quickly during storage. On the other hand, botanical or true seeds of potato have a strong primary dormancy at maturity that prevents germination at warm temperatures. Germination of dormant true potato seeds (TPS) is inhibited at 27°C, but germination will occur at 17°C. This dormancy can be released by prolonged storage under ambient conditions or by rapid ageing treatments at high temperatures (40–45°C) and 4% seed moisture content (Pallais, 1995a,b). Germination of dormant TPS is also stimulated by GA (Bamberg and Hanneman, 1984). Since tomato and potato are closely related, we expect that homologous genes will be present in both species, and that the cDNAs identified in tomato will be applicable to investigate gene expression in TPS. We describe here investigations of the expression of endo-β-mannanase and SnRK genes from tomato in relation to primary dormancy of TPS.

While GA is required for tomato seed germination, ABA can inhibit germination of wild-type seeds (Ni and Bradford, 1992). Seeds of tomato mutants deficient in ABA synthesis (*sit*^{w}) exhibit reduced primary dormancy and rapid germination (Groot and Karssen, 1992; Ni and Bradford, 1993; Hilhorst and Downie, 1995; Downie *et al.*, 1999). As ABA is synthesized from carotenoid precursors (Zeevaart and Creelman, 1988), fluridone, an inhibitor of phytoene desaturation in the pathway of carotenoid biosynthesis (Bartels and Watson, 1978), also blocks ABA synthesis. Fluridone applied during development of sunflower (*Helianthus annuus* L.) and sorghum (*Sorghum bicolor* L. Moench.) seeds resulted in mature seeds having reduced primary dormancy (Le Page-Degivry *et al.*, 1990; Steinbach *et al.*, 1997). Fluridone also stimulated the germination of dormant *Nicotiana plumbaginifolia* seeds (Jullien and Bouinot, 1997) and seeds of lettuce (*Lactuca sativa* L.) imbibed at high temperatures (Yoshioka *et al.*, 1998). Thus, we tested whether fluridone would influence germination of dormant TPS and, if so, whether it also altered expression of endo-β-mannanase and SnRK genes.

Materials and Methods

Plant material

Hybrid botanical or true potato seeds (TPS) were produced in Jambon, Peru (3500 m elevation), in 1997 from parental lines Yungay × 104.12 L.B. Seeds were stored initially at 0°C at a seed moisture content of 4.5% and subsequently at −20°C at UC Davis to preserve primary dormancy.

Germination tests

Four replicates of 25 seeds each were imbibed on blotters in 5 cm diameter Petri dishes and germination was evaluated at 17 and 27°C. Germination tests were conducted in water and in solutions containing 50 μM fluridone, 100 μM GA_{4+7} or 100 μM ABA. Fluridone (1-methyl-3-phenyl-5-[3-trifluoromethyl-(phenyl)]-4-(1*H*)-pyridinone) stock solutions (100 μM) were prepared according to the method of Ober and Sharp (1994).

Mannanase diffusion assays

Endo-β-mannanase activity was determined using a gel diffusion assay (Downie *et al.*, 1994; Still *et al.*, 1997). An agarose gel plate containing 0.5%

(w/v) locust bean galactomannan (Sigma, St Louis, Missouri, USA) was prepared and wells were formed in the solidified gel by scoring with a 3-mm diameter cork borer and removing the plug by suction. The extracts (10 μl) from TPS were applied to the wells and the plate was incubated at 25°C for 24 h. After incubation, the agarose gel plate was stained using 0.5% (w/v) Congo Red dye (Sigma). The diameter of the cleared area on the gel is logarithmically related to the mannanase activity.

Tissue print activity assays

Tissue printing was done as previously described for tomato seeds (Nonogaki *et al.*, 1998) except that seed halves were used instead of tissue sections from the seeds. Briefly, a seed was bisected laterally and the endosperm and the embryo were separately laid with the cut surface down on top of the activity gel (about 150 μm thick) containing 7.5% (w/v) polyacrylamide and 0.5% (w/v) locust bean galactomannan in water. After 15 min incubation at 25°C, the tissue parts were removed from the gel and the gel was stained with periodic acid–Schiff's reagent. The locations where mannanase activity was present were visible as clearing of the underlying gel.

Northern hybridization

Total RNA was prepared from 100 whole TPS by a standard phenol extraction method (Sambrook *et al.*, 1989). Equal quantities of RNA were separated on 1.3% (w/v) agarose gel and transferred to a neutral membrane (Hybond-N, Amersham Pharmacia Biotech, Inc., Piscataway, New Jersey, USA). RNA probes for Northern blots were prepared using a digoxigenin (DIG)-labelled dNTP mixture (Boehringer Mannheim Corp., Indianapolis, Indiana, USA) and T3 or T7 RNA polymerases (Ambion, Inc., Austin, Texas, USA). The following cDNAs were employed: *LeMAN2* (Genbank Accession No. AF184238), *LeSNF1* (Accession No. AF143743) and *LeSNF4* (Accession No. AF143742). Hybridization was performed at 55°C overnight in buffer containing 50% (v/v) deionized formamide, 5× SSC, 0.1% (w/v) sodium-lauroylsarcosine, 0.02% (w/v) SDS, 2% blocking reagent (Boehringer Mannheim) and 0.1 mg ml^{-1} salmon sperm DNA. The membranes were washed once for 25 min at 65°C in 2× SSC, 0.1% (w/v) SDS and twice for 25 min at 65°C in 0.2× SSC, 0.1% (w/v) SDS. After washing, the membranes were rinsed in 0.1 M maleic acid buffer, pH 7.5, including 0.15 M sodium chloride and 0.3% (w/v) Tween 20 (buffer A) and blocked with 5% non-fat milk in buffer A for 1 h prior to detection by anti-DIG antibody conjugated with alkaline phosphatase (Boehringer Mannheim). The membranes were incubated with the antibody (1 : 15,000 dilution) at 25°C for 1 h. After three 10-min washes with buffer A, the signal was detected on film using chemiluminescent substrates for alkaline phosphatase (CDP-Star, Ambion, Austin, Texas, USA, or Lumi-phos 530, Lumigen, Inc., Southfield, Michigan, USA).

Results

Dormant TPS imbibed in water germinated to 40% at 17°C (Fig. 24.2A), but were completely inhibited from germinating when imbibed at 27°C (Fig. 24.2B). This temperature sensitivity is gradually lost during dry afterripening and the range of temperatures at which germination will occur expands (Pallais, 1995a,b). Short exposures to 27°C (e.g. 5 days) do not induce secondary dormancy, and germination will still occur if seeds

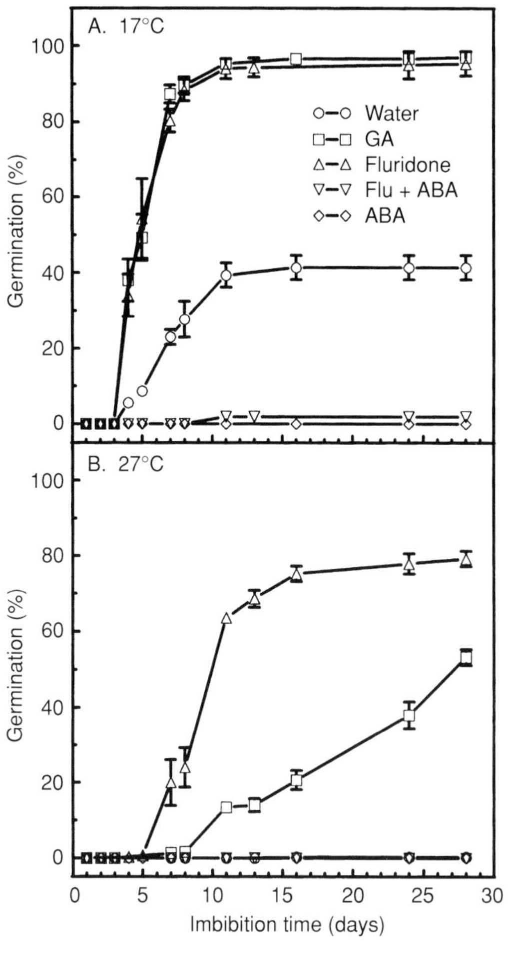

Fig. 24.2. Germination time courses of dormant true potato seeds (TPS) at (A) 17°C and (B) 27°C. At 17°C, 40% of the seeds can germinate in water, and both GA_{4+7} (100 μM) and fluridone (50 μM) can increase this to 100%. Seeds imbibed in ABA (100 μM) or ABA + fluridone (Flu, 50 μM), on the other hand, are prevented from completing germination. At 27°C, only seeds imbibed in GA_{4+7} or in fluridone were able to germinate, with fluridone being more effective than GA.

initially imbibed at 27°C are transferred to 17°C (data not shown). Germination of seeds imbibed at 17°C was stimulated by GA and inhibited by ABA (Fig. 24.2A). Fluridone also promoted germination at 17°C, and this effect was reversed by simultaneous application of ABA (Fig. 24.2A). Both fluridone and GA also stimulated germination at 27°C, with fluridone being more effective than GA (Fig. 24.2B).

As has been shown for tomato seeds, imbibition of TPS at 17°C was accompanied by the appearance of endo-β-mannanase activity in the endosperm cap tissue prior to radicle emergence (Fig. 24.3A). Both the embryo and the endosperm cap exhibited mannanase activity after 96 h of imbibition in water, but no activity was detected in the lateral endosperm at this time in ungerminated seeds (Fig. 24.3B). However, ABA did not prevent the increase in mannanase activity in the endosperm cap (Fig. 24.3A), as is also the case with tomato (Toorop *et al.*, 1996; Dahal *et al.*, 1997; Still and Bradford, 1997).

Dormant TPS initially imbibed at 27°C and maintained at that temperature neither germinated (Fig. 24.2B) nor expressed mannanase activity in the endosperm cap (Fig. 24.4). However, upon transfer of seeds to 17°C after 5 days of imbibition at 27°C, both germination and mannanase activity were induced (Fig. 24.4). Similarly, transfer of TPS from water to either 100 μM GA_{4+7} or 50 μM fluridone at 27°C stimulated mannanase activity and germination, although both processes were slower in fluridone (Fig. 24.4). Curiously, although simultaneous addition of ABA with fluridone completely prevented germination (e.g. Fig. 24.2B), ABA did not prevent the increase in mannanase activity that occurred in the presence of fluridone (Fig. 24.4). While this is consistent with the lack of effect of exogenous ABA on mannanase activity at 17°C (Fig. 24.3A), it is surprising that supplemental ABA does not reverse the effect of the ABA synthesis inhibitor.

Expression of the endosperm cap-specific *LeMAN2* mRNA in TPS imbibed at 17°C was consistent with the presence of mannanase activity. That is, expression increased in seeds imbibed in water, was even greater in the presence of GA and in germinated seeds, and was not blocked by ABA in either control or fluridone-treated seeds (Fig. 24.5). After 96 h imbibition at 27°C, *LeMAN2* mRNA was detected only in the presence of GA (Fig. 24.5), although seeds in fluridone will also germinate at this temperature (Fig. 24.2B). However, germination is delayed by several days at the higher temperature (Fig. 24.2), consistent with the low expression at 96 h. Thus, at least for the endosperm cap, mannanase activity appears to be regulated primarily at the transcriptional level and to be associated closely with whether radicle emergence is imminent or has recently occurred.

Expression of SnRK mRNAs was also assayed under the same conditions. Dry TPS seeds contain mRNAs hybridizing to both *LeSNF4* and *LeSNF1* riboprobes, as in tomato seeds (Fig. 24.5). Abundance of the *LeSNF4* homologue mRNA remained high in ungerminated seeds imbibed

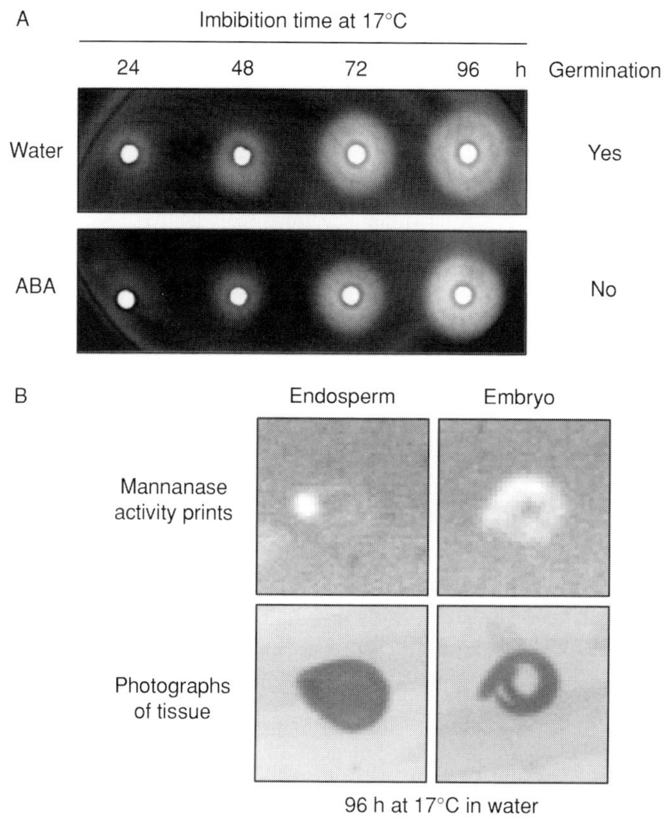

Fig. 24.3. (A) Endo-β-mannanase activity extracted from endosperm caps of TPS at different times after imbibition in water or 100 μM ABA at 17°C. Activity is assayed by the extent of hydrolysis of galactomannan substrate contained in the gel due to the diffusion of the enzyme from the centre well. The diameter of the cleared area is logarithmically related to the mannanase activity. Germination occurred in water but was inhibited in ABA. (B) Tissue prints of endo-β-mannanase activity in the endosperm and embryo of TPS imbibed for 96 h at 17°C. The lower panels show photographs of the tissues on the gels, while the upper panels show the endo-β-mannanase activity as cleared areas. In the endosperm, only the micropylar cap region exhibits activity prior to radicle emergence.

for 96 h at 17°C in water, ABA or fluridone + ABA, but decreased in ungerminated seeds imbibed in GA_{4+7} or fluridone and was not detected in germinated seeds (Fig. 24.5). The signal remained high under all conditions at 27°C, consistent with the slower rate of progress toward germination and low expression of *LeMAN2* mRNA under the same conditions (Fig. 24.5). In contrast, the abundance of *LeSNF1* homologue mRNA in TPS showed relatively little variation due to germination conditions (Fig. 24.5).

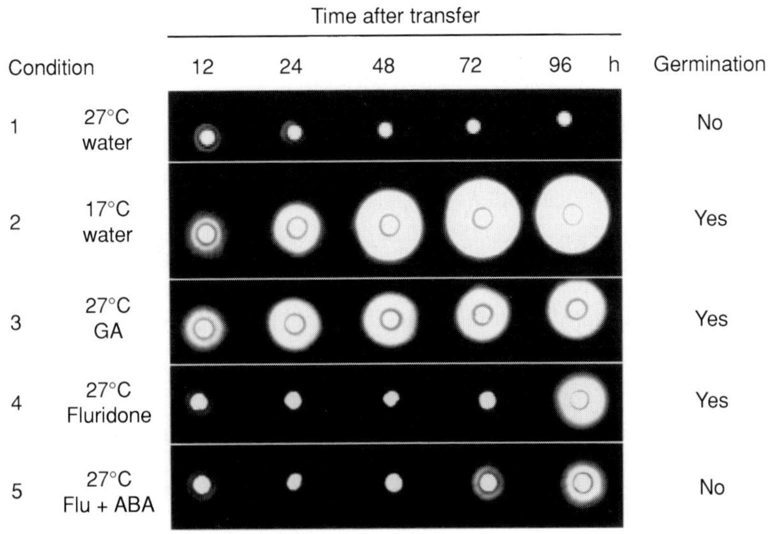

Fig. 24.4. Endo-β-mannanase activity extracted from endosperm caps of TPS at various times after transfer to different conditions following initial imbibition in water for 120 h at 27°C. After imbibition at 27°C to ensure dormancy, seeds were transferred to the following conditions: (1) maintained at 27°C; (2) transferred to 17°C; (3) transferred to 100 μM GA_{4+7} at 27°C; (4) transferred to 50 μM fluridone at 27°C; (5) transferred to 50 μM fluridone plus 100 μM ABA at 27°C. Endosperm caps were excised at the indicated times after transfer and extracts were assayed for endo-β-mannanase activity by gel diffusion. Whether germination occurred after the transfer is indicated to the right of each condition.

Discussion

Similar to previous reports for *N. plumbaginifolia* (Jullien and Buinot, 1997) and lettuce (Yoshioka *et al.*, 1998) seeds, fluridone promoted germination of dormant TPS imbibed at both 17 and 27°C (Fig. 24.2). Germination induced by fluridone can be completely prevented by addition of ABA at both temperatures (Fig. 24.2). Thus, although Karssen (1995) stated that 'ABA is not resynthesized in mature seeds', continued synthesis of ABA appears to be involved in maintaining the temperature-sensitive dormant state in TPS. We do not have data on ABA contents of TPS in the presence and absence of fluridone, but similar studies in *N. plumbaginifolia* and lettuce found that fluridone reduced the ABA content of imbibed dormant seeds (Yoshioka *et al.*, 1998; Grappin *et al.*, 2000). Thus, it is likely that fluridone acted in TPS by blocking ABA synthesis, although the possibility of additional effects cannot be excluded. The lack of dormancy and tendency toward viviparous germination in seeds of the ABA-deficient *sit*^{w} mutant indicated that ABA is essential for the induction of primary dormancy in tomato (Groot and Karssen, 1992; Hilhorst and Downie, 1995;

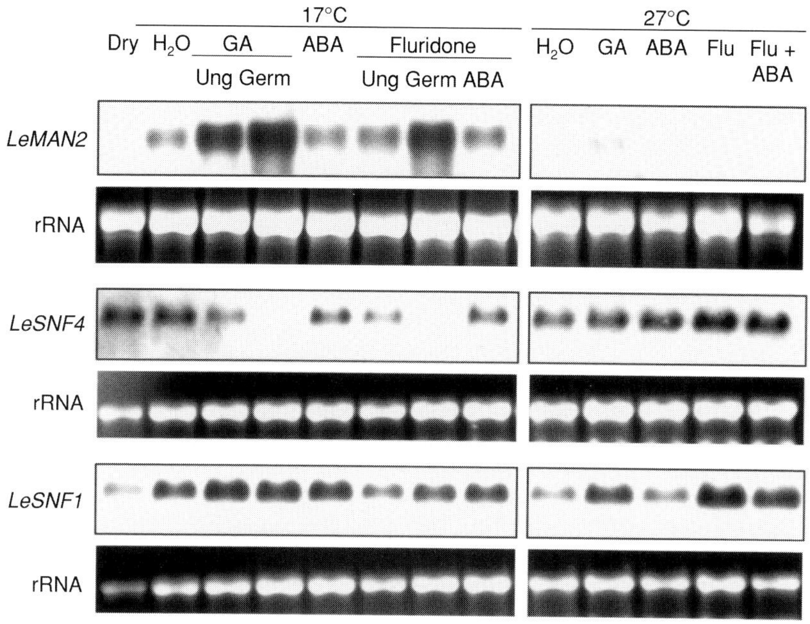

Fig. 24.5. Abundance of mRNAs from TPS imbibed for 96 h at either 17 or 27°C that hybridize to riboprobes for tomato genes expressed during germination: *LeMAN2*, an endosperm cap-specific endo-β-mannanase; *LeSNF4*, the activating subunit of the SNF1-related protein kinase; and *LeSNF1*, the tomato homologue of the SNF1-related protein kinase. Panels labelled rRNA show the ethidium bromide-stained ribosomal RNA bands to indicate the quantity of RNA loaded in each lane. Total RNA was extracted from whole TPS after imbibition for 96 h at 17 or 27°C under the conditions indicated: dry seeds, seeds imbibed in water (H_2O), in 100 μM GA_{4+7}, in 50 μM fluridone (Flu), or in fluridone + ABA (100 μM). In cases where germination had occurred by 96 h (i.e. in GA and fluridone), seeds were separated into ungerminated (Ung) and germinated (Germ) fractions prior to RNA extraction.

Downie *et al.*, 1999), and the present study suggests that continued synthesis of ABA is required for the maintenance of dormancy in imbibed TPS.

As the anatomy of TPS is similar to that of tomato seeds, it is not surprising that their physiological processes also appear to be similar. For example, expression of endo-β-mannanase activity prior to radicle emergence is limited to the endosperm cap and the embryo tissues of TPS (Fig. 24.3B) as in tomato (Toorop *et al.*, 1996; Nonogaki *et al.*, 1998). Gibberellin also stimulates germination and mannanase expression in both tomato and potato seeds (Figs 24.2, 24.4, 24.5; Groot and Karssen, 1987; Groot *et al.*, 1988). Furthermore, while germination is blocked by ABA in both species, the appearance of mannanase mRNA and activity is not affected by ABA in either species (Figs 24.3A and 24.5; Toorop *et al.*, 1996; Dahal *et al.*, 1997; Still and Bradford, 1997). Thus, the inhibition of germination by ABA is not due to an absence of mannanase activity in the endosperm cap. In addition, ABA does not prevent the weakening of tomato endosperm cap tissue as measured by puncture force analysis (Toorop, 1998; F. Chen *et al.*, unpublished results). Thus, ABA apparently exerts its effect on the embryo growth potential (Schopfer and Plachy, 1985; Ni and Bradford, 1992) or on a final phase of endosperm weakening (Karssen *et al.*, 1989; Toorop, 1998). The simpler hypothesis proposed earlier (Ni and Bradford, 1993) that GA and ABA might regulate germination by reciprocally inducing or repressing cell wall hydrolases involved in endosperm weakening must be rejected. In fact, while mRNAs of all of the cell wall hydrolases (and expansin) listed in Fig. 24.1 are upregulated by GA or during germination, only β-1,3-glucanase expression is clearly repressed by ABA (Bradford *et al.*, 2000).

As noted in the Introduction, reserve mobilization occurs in the endosperm cap concomitantly with weakening (Mella *et al.*, 1995; Nonogaki *et al.*, 1998; Toorop, 1998). Indeed, as cell wall mannans constitute the major carbohydrate storage reserve of tomato seeds, it may not be possible to distinguish biochemically between the two processes, both of which involve cell wall disassembly. However, let us assume that the conserved SnRKs of plants are involved in metabolic processes related to the synthesis and/or mobilization of storage reserves, as appears likely based on transgenic experiments that altered SNF1-like kinase activity in potatoes (Purcell *et al.*, 1998). In tomato seeds, *LeSNF4* mRNA is present in mature, dormant or ABA-treated seeds, while its abundance declines rapidly in germinating or GA-treated seeds (Bradford *et al.*, 2000).

Similar results are shown here for TPS, although the changes in mRNA abundance occur more slowly and are more closely linked with whether or not radicle emergence has occurred (Fig. 24.5). As in tomato, abundance of SNF1-like mRNA is relatively constant regardless of germination status (Fig. 24.5). Thus, we can hypothesize that the SNF1-related kinase subunit is constitutively expressed and may have multiple biochemical roles, while expression of the SNF4-related activating subunit is sensitive to hormonal and environmental conditions. Abundance of the SNF4-related transcript appears to be associated with conditions under which reserve mobilization is blocked (i.e. in developing and dormant seeds, in ABA, in high temperature, in far-red light, etc.), while transcript levels decline when reserve mobilization associated with seedling growth occurs. One can envisage that via its interaction with the SNF1-related kinase subunit, the SNF4-like protein modifies the kinase activity to shift metabolic pathways to either store or mobilize seed reserves, depending on the dormancy status of the seeds or the ratio of GA to ABA.

We propose a hypothetical model that can accommodate many, but not all, of the observations to date. As a starting point, we suggest that it may be useful conceptually to separate the processes related to embryo growth, endosperm weakening and reserve mobilization (Fig. 24.6). Gibberellin promotes all three processes, while ABA is able to block some effects of GA

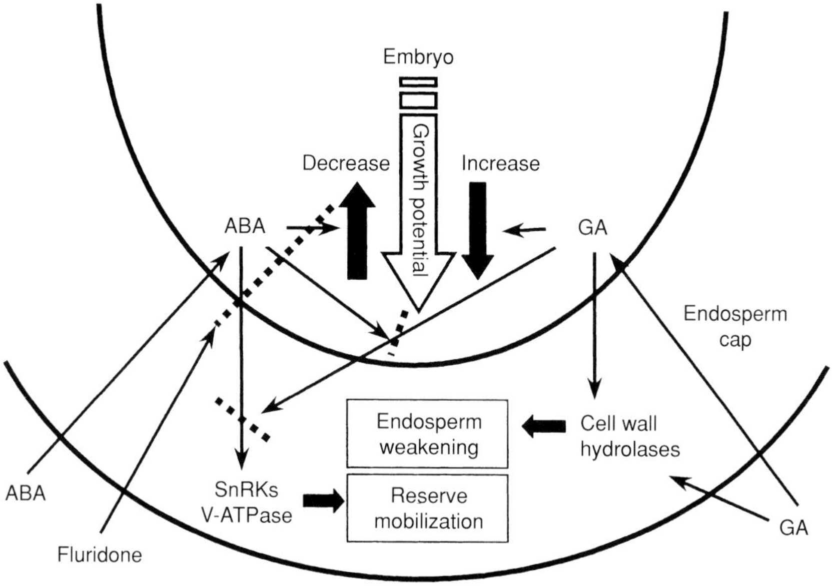

Fig. 24.6. Hypothetical scheme to describe interactions between ABA and GA in regulating dormancy and germination in TPS and tomato. Embryonic ABA (rather than maternal or exogenous ABA) is thought to be critical for the imposition of dormancy during development by decreasing embryo growth potential. We also suggest that it may promote storage reserve accumulation (in both lateral and micropylar endosperm) by influencing the expression and/or activity of SNF1-related protein kinase complexes (SnRKs). GA, on the other hand, increases embryo growth potential and induces cell wall hydrolases that result in endosperm weakening. ABA is not able to block the expression of most of these hydrolases or prevent the majority of endosperm cap weakening from occurring. ABA can maintain SNF4-like gene expression, which we hypothesize may be related to prevention of reserve mobilization, while GA has the opposite effect. The possibility that ABA and GA can each counteract the effect of the other hormone in some pathways is shown, and probably occurs in all parts of the endosperm. Exogenous GA has effects in both the embryo and the endosperm, while exogenous ABA seems to have less effect in the endosperm cap. Fluridone, by blocking ABA synthesis, would be expected to act similarly to GA in most cases, although some effects, such as on cell wall hydrolases or endosperm weakening, may be indirect through GA signals from the embryo. Some cell wall hydrolases, such as mannanase, could be present as multiple genes that are differentially regulated in association with both endosperm weakening and reserve mobilization, making it difficult to interpret function solely from enzyme activity. (Dashed lines indicate inhibition of a given action path.)

but not others. For example, ABA can reduce embryo growth potential, block expression of V-ATPase subunit genes, and maintain expression of SNF4-like genes that may prevent reserve mobilization, but it has little effect on the appearance of cell wall hydrolases or on endosperm weakening (Bradford *et al.*, 2000). It is possible that ABA may act by blocking a GA signalling pathway that represses SNF4-like genes, based upon the ability of the PKABA1 protein kinase to block GA-induced gene expression in barley aleurone cells (Gomez-Cadenas *et al.*, 1998). Fluridone would be expected to have effects opposite to those of ABA by blocking synthesis of the hormone. Thus, it should increase embryo growth potential and stimulate reserve mobilization. The effect of fluridone on dormancy and germination, which is similar to that of applied GA, indicates that at least some physiological components in imbibed dormant seeds are regulated by a balance between ABA and GA. The inability of ABA to reverse the effects of GA on mannanase (and other hydrolase) expression and weakening of the endosperm cap, however, suggests that endosperm cap weakening is regulated in a different way, most likely by GA derived from the embryo (Sánchez and de Miguel, 1997). This is consistent with observations in TPS, where GA can stimulate radicle emergence from dormant seeds (e.g. Fig. 24.2), but the resulting seedlings have low vigour and slow growth compared with seedlings from afterripened seeds (Pallais, 1987, 1989). It is possible that while GA can directly initiate processes leading to radicle protrusion, residual dormancy (possibly due to continued ABA synthesis) represses reserve mobilization pathways and limits seedling vigour. Conversely, fluridone may act primarily in the embryo or on reserve mobilization processes, and only indirectly on endosperm cap weakening, explaining its delayed effect on mannanase expression (Fig. 24.4).

The potential complexity of signalling pathways allowing seeds to sense their environment and transduce those cues into physiological and molecular responses is daunting. At the same time, the tools at our disposal to dissect those pathways and responses are increasing daily. In particular, technologies such as microarrays to assess simultaneously the expression of thousands of genes, coupled with sequence databases containing the identities of those genes, can reveal global gene expression patterns associated with specific dormancy states. Since many of the most interesting examples of seed dormancy do not occur in the cultivated or model system species receiving the most intensive molecular investigation, it is encouraging that tools and insights developed in model plant systems such as tomato can be readily transferred to related species such as potato that exhibit more extreme seed dormancy phenotypes.

Acknowledgements

This research was supported in part by National Science Foundation grant 9722978 to KJB and a UC Davis Katherine Esau Fellowship to HN.

References

Abe, H., Kamiya, Y. and Sakurai, A. (1996) A cDNA clone encoding yeast SNF4-like protein (Accession No. U40713) from *Phaseolus vulgaris* (PGR95-126). *Plant Physiology* 110, 336.

Bamberg, J.B. and Hanneman, R.E., Jr (1984) Hastened germination of slow germinating potato seed lots by the use of gibberellic acid and activated charcoal. *American Potato Journal* 61, 517–518.

Bartels, P.G. and Watson, C.W. (1978) Inhibition of carotenoid synthesis by fluridone and norflurazon. *Weed Science* 26, 198–203.

Bewley, J.D., Burton, R.A., Morohashi, Y. and Fincher, G.B. (1997) Molecular cloning of a cDNA encoding a (1-4)-β-mannan endohydrolase from the seeds of germinated tomato (*Lycopersicon esculentum*). *Planta* 203, 454–459.

Bouly, J.-P., Gissot, L., Lessard, P., Kreis, M. and Thomas, M. (1999) *Arabidopsis thaliana* proteins related to the yeast SIP and SNF4 interact with AKINα1, an SNF1-like protein kinase. *The Plant Journal* 18, 541–550.

Bradford, K.J., Chen, F., Cooley, M.B., Dahal, P., Downie, B., Fukunaga, K.K., Gee, O.H., Gurusinghe, S., Mella, R.A., Nonogaki, H., Wu, C.-T., Yang, H. and Yim K.-O. (2000) Gene expression prior to radicle emergence in imbibed tomato seeds. In: Black, M., Bradford, K.J. and Vasquez-Ramos, J. (eds) *Seed Biology: Advances and Applications.* CAB International, Wallingford, UK, pp. 231–251.

Cooley, M.B., Yang, H., Dahal, P., Mella, R.A., Downie, B., Haigh, A.M. and Bradford, K.J. (1999) Vacuolar H^+-ATPase is expressed in response to gibberellin during tomato seed germination. *Plant Physiology* 121, 1339–1347.

Dahal, P., Nevins, D.J. and Bradford, K.J. (1997) Relationship of endo-β-D-mannanase activity and cell wall hydrolysis in tomato endosperm to germination rates. *Plant Physiology* 113, 1243–1252.

Downie, B., Hilhorst, H.W.M. and Bewley, J.D. (1994) A new assay for quantifying endo-β-mannanase activity using Congo Red dye. *Phytochemistry* 36, 829–835.

Downie, B., Dirk, L.M.A., Hadfield, K.A., Wilkins, T.A., Bennett, A.B. and Bradford, K.J. (1998) A gel diffusion assay for quantification of pectin methylesterase activity. *Analytical Biochemistry* 264, 149–157.

Downie, B., Gurusinghe, S.H. and Bradford, K.J. (1999) Internal anatomy of individual tomato seeds: relationship to abscisic acid and germination physiology. *Seed Science Research* 9, 117–128.

Gomez-Cadenas, A., Verhey, S.D., Holappa, L.D., Shen, Q., Ho, T.-H.D. and Walker-Simmons, M.K. (1998) An abscisic acid-induced protein kinase, PKABA1, mediates abscisic acid-suppressed gene expression in barley aleurone layers. *Proceedings of the National Academy of Sciences, USA* 99, 1767–1772.

Grappin, P., Bouinot, D., Sotta, B., Miginiac, E. and Jullien, M. (2000) Control of seed dormancy in *Nicotiana plumbaginifolia*: post-inbibition abscisic acid synthesis imposes dormancy maintenance. *Planta* 210, 279–285.

Groot, S.P.C. and Karssen, C.M. (1987) Gibberellins regulate seed germination in tomato by endosperm weakening: a study with gibberellin-deficient mutants. *Planta* 171, 525–531.

Groot, S.P.C. and Karssen, C.M. (1992) Dormancy and germination of abscisic acid-deficient tomato seeds. Studies with the *sitiens* mutant. *Plant Physiology* 99, 952–958.

Groot, S.P.C., Kieliszewska-Rokicka, B., Vermeer, E. and Karssen, C.M. (1988) Gibberellin-induced hydrolysis of endosperm cell walls in gibberellin-deficient tomato seeds prior to radicle protrusion. *Planta* 174, 500–504.

Halford, N.G. and Hardie, D.G. (1998) SNF1-related protein kinases: global regulators of carbon metabolism in plants? *Plant Molecular Biology* 37, 735–748.

Halford, N.G., Purcell, P.C. and Hardie, D.G. (1999) Is hexokinase really a sugar sensor in plants? *Trends in Plant Science* 4, 117–120.

Hardie, D.G., Carling, D. and Carlson, M. (1998) The AMP-activated/SNF1 protein kinase subfamily: metabolic sensors of the eukaryotic cell? *Annual Review of Biochemistry* 67, 821–855.

Hilhorst, H.W.M. and Downie, B. (1995) Primary dormancy in tomato (*Lycopersicon esculentum* cv. Moneymaker): studies with the *sitiens* mutant. *Journal of Experimental Botany* 47, 89–97.

Hilhorst, H.W.M., Groot, S.P.C. and Bino, R.J. (1998) The tomato seed as a model system to study seed development and germination. *Acta Botanica Neerlandica* 47, 169–183.

Holappa, L.D. and Walker-Simmons, M.K. (1995) The wheat abscisic acid-responsive protein kinase mRNA, PKABA1, is up-regulated by dehydration, cold temperature, and osmotic stress. *Plant Physiology* 108, 1203–1210.

Jullien, M. and Bouinot, D. (1997) Seed dormancy and responses of seeds to phytohormones in *Nicotiana plumbaginifolia.* In: Ellis, R.H., Black, M., Murdoch, A.J. and Wong, T.D. (eds) *Basic and Applied Aspects of Seed Biology.* Kluwer Academic Publishers, Dordrecht, pp. 203–214.

Karssen, C.M. (1995) Hormonal regulation of seed development, dormancy, and germination studied by genetic control. In: Kigel, J. and Galili, G. (eds) *Seed Development and Germination.* Marcel Dekker, New York, pp. 333–350.

Karssen, C.M., Haigh, A., van der Toorn, P. and Weges, R. (1989) Physiological mechanisms involved in seed priming. In: Taylorson, R.B. (ed.) *Recent Advances in the Development and Germination of Seeds.* Plenum Press, New York, pp. 269–280.

Lakatos, L., Klein, M., Höfgen, R. and Bánfalvi, Z. (1999) Potato StubSNF1 interacts with StubGAL83: a plant protein kinase complex with yeast and mammalian counterparts. *The Plant Journal* 17, 569–574.

Le Page-Degivry, M.T., Barthe P. and Garello, G. (1990) Involvement of endogenous abscisic acid in onset and release of *Helianthus annuus* embryo dormancy. *Plant Physiology* 92, 1164–1168.

Mella, R.A., Maldonaldo, S. and Sánchez, R.A. (1995) Phytochrome-induced structural changes and protein degradation prior to radicle protrusion in *Datura ferox* seeds. *Canadian Journal of Botany* 73, 1371–1378.

Ni, B.-R. and Bradford, K.J. (1992) Quantitative models characterizing seed germination responses to abscisic acid and osmoticum. *Plant Physiology* 98, 1057–1068.

Ni, B.-R. and Bradford, K.J. (1993) Germination and dormancy of abscisic acid- and gibberellin-deficient mutant tomato (*Lycopersicon esculentum*) seeds. Sensitivity of germination to abscisic acid, gibberellin and water potential. *Plant Physiology* 101, 607–617.

Nonogaki, H. and Morohashi, Y. (1996) An endo-β-mannanase develops exclusively in the micropylar endosperm of tomato seeds prior to radicle emergence. *Plant Physiology* 110, 555–559.

Nonogaki, H., Nomaguchi, M., Okumoto, N., Kaneko, Y., Matsushima, H. and Morohashi, Y. (1998) Temporal and spatial pattern of the biochemical activation of the endosperm during and following imbibition of tomato seeds. *Physiologia Plantarum* 102, 236–242.

Ober, E.S. and Sharp, R.E. (1994) Proline accumulation in maize (*Zea mays* L.) primary roots at low water potentials. I. Requirement for increased levels of abscisic acid. *Plant Physiology* 105, 981–987.

Pallais, N. (1987) True potato seed quality. *Theoretical and Applied Genetics* 73, 784–792.

Pallais, N. (1989) Osmotic priming of true potato seed: effects of seed age. *Potato Research* 32, 235–244.

Pallais, N. (1995a) High temperature and low moisture reduce the storage requirement of freshly harvested true potato seeds. *Journal of the American Society for Horticultural Science* 120, 699–702.

Pallais, N. (1995b) Storage factors control germination and seedling establishment of freshly harvested true potato seed. *American Potato Journal* 72, 427–436.

Purcell, P.C., Smith, A.M. and Halford, N.G. (1998) Antisense expression of a sucrose non-fermenting-1-related protein kinase sequence in potato results in decreased expression of sucrose synthase in tubers and loss of sucrose-inducibility of sucrose synthase transcripts in leaves. *The Plant Journal* 14, 195–202.

Sambrook, J., Fritsch, E.F. and Maniatis, T. (1989) *Molecular Cloning: a Laboratory Manual*, 2nd edn. Cold Spring Harbor Laboratory Press, Cold Spring Harbor, New York.

Sánchez, R.A. and de Miguel, L. (1997) Phytochrome promotion of mannan-degrading enzyme activities in the micropylar endosperm of *Datura ferox* seeds requires the presence of the embryo and gibberellin synthesis. *Seed Science Research* 7, 27–33.

Schopfer, P. and Plachy, C. (1985) Control of seed germination by abscisic acid. III. Effect on embryo growth potential (minimum turgor pressure) and growth coefficient (cell wall extensibility) in *Brassica napus* L. *Plant Physiology* 77, 676–686.

Sitrit, Y., Hadfield, K.A., Bennett, A.B., Bradford, K.J. and Downie, B. (1999) Expression of a polygalacturonase associated with tomato seed germination. *Plant Physiology* 121, 419–428.

Steinbach, H.S., Benech-Arnold, R.L. and Sánchez, R.A. (1997) Hormonal regulation of dormancy in developing sorghum seeds. *Plant Physiology* 113, 149–154.

Still, D.W. and Bradford, K.J. (1997) Endo-β-mannanase activity from individual tomato endosperm tissues in relation to germination. *Plant Physiology* 113, 21–29.

Still, D.W., Dahal, P. and Bradford, K.J. (1997) A single-seed assay for endo-β-mannanase activity from tomato endosperm and radicle tissues. *Plant Physiology* 113, 13–20.

Toorop, P.E. (1998) The role of endo-β-mannanase activity in tomato seed germination. PhD Dissertation, Agricultural University, Wageningen, The Netherlands.

Toorop, P.E., Bewley, J.D. and Hilhorst, H.W.M. (1996) Endo-β-mannanase isoforms are present in the endosperm and embryo of tomato seeds but are not essentially linked to the completion of germination. *Planta* 200, 153–158.

Voigt, B. and Bewley, J.D. (1996) Developing tomato seeds when removed from the fruit produce multiple forms of germinative and post-germinative endo-β-mannanase. Responses to desiccation, abscisic acid and osmoticum. *Planta* 200, 71–77.

Welbaum, G.E., Bradford, K.J., Yim, K.-O., Booth, D.T. and Oluoch, M.O. (1998) Biophysical, physiological and biochemical processes regulating seed germination. *Seed Science Research* 8, 161–172.

Yoshioka, T., Takashi E. and Shigueru, S. (1998) Restoration of seed germination at supraoptimal temperatures by fluridone, an inhibitor of abscisic acid biosynthesis. *Plant and Cell Physiology* 39, 307–312.

Zeevaart, J.A.D. and Creelman, R.A. (1988) Metabolism and physiology of abscisic acid. *Annual Review of Plant Physiology and Plant Molecular Biology* 39, 439–473.

The Genetics of Seed Dormancy in *Arabidopsis thaliana*

25

M. Koornneef, C. Alonso-Blanco, L. Bentsink, H. Blankestijn-de Vries, I. Debeaujon, C.J. Hanhart, K.M. Léon-Kloosterziel, A.J.M. Peeters and V. Raz.

Laboratory of Genetics, Wageningen University, Dreijenlaan 2, 6703 HA Wageningen, The Netherlands

Introduction

Genetic variation for seed dormancy has been observed both between and within species. Dormancy is found in crop plants and genetic variation has been observed among cultivars of barley and wheat (Anderson *et al.*, 1993; Li and Foley, 1997). However, it is assumed that domestication of crop plants has led to selection for less dormant genotypes. Many authors (Garbutt and Witcombe, 1986; Platenkamp and Shaw, 1993, and references therein) have observed genetic differences in seed dormancy within and between 'wild' populations. In such genetic studies, the inheritance of dormancy is often described in terms of quantitative inheritance (e.g. Lane and Lawrence, 1995). With the advent of molecular genetics, it became possible to analyse genetic differences at the molecular level, thereby identifying genes that control these differences in molecular terms.

To use genetic variation as a starting point for the molecular identification of genes involved in a developmental or physiological process, it is important that monogenic differences are available and that molecular techniques can be applied efficiently in the species under study. For the combination of genetics and molecular genetics, *Arabidopsis thaliana* is the most suitable plant species (Meinke *et al.*, 1998). To study seed dormancy it is important that clear reproducible phenotypic differences can be monitored for this trait, allowing the detection of monogenic differences under defined conditions. The *Arabidopsis* genotypes Landsberg *erecta* (L*er*) and Columbia (Col), which are mostly used in *Arabidopsis* research, show only a low level of dormancy. This dormancy disappears after approximately

(eds J.-D. Viémont and J. Crabbé)

1 month of afterripening (van der Schaar *et al.*, 1997). However, relatively large germination differences can be found between seed lots of the same genotype (Derkx and Karssen, 1993). Despite such complications, genetic variation for dormancy is available in *Arabidopsis* both among induced mutants and among natural accessions of *Arabidopsis* (Ratcliffe, 1976).

Characteristics of Seed Dormancy and Germination in *Arabidopsis*

Seed development in *Arabidopsis* is completed within 18–21 days after fertilization and is composed of two phases. First, a morphogenetic phase in which the embryo takes its form and size. This phase is completed approximately 10 days after fertilization and is followed by a seed maturation phase. The latter is characterized by the accumulation of nutritive reserves and the induction of desiccation tolerance (Koornneef and Karssen, 1994; Wobus and Weber, 1999). In addition, the endosperm and the testa, together forming the seed envelopes, undergo changes during this phase. The degree of dormancy of the mature seed is determined throughout seed development. The first important stage for dormancy induction is probably the end of the morphogenetic programme, when all tissues present in a mature embryo have been formed and the embryo enters a phase of growth arrest. Secondly, the physiological (including hormonal) changes during seed maturation are assumed to play an important role. This is indicated by the observation that the degree of dormancy can be influenced by the growth conditions of the mother plant (Koornneef and Karssen, 1994, and references therein). A third stage of development affecting dormancy is when mature seeds are imbibed as a preparation for germination. During this process environmental factors such as temperature, light and the degree of afterripening are known to be important controlling factors.

Since dormancy is regulated at different developmental phases, in interaction with environmental factors, it is difficult to distinguish when the genetic and physiological differences are established. This difficulty arises because all assays are based on the germination of seeds, which is the result of the balance between the degree of dormancy and the capacity of the embryo to overcome this dormancy. When germination of immature seeds of *Ler* and Col is tested 1 week after fertilization, either with excised seeds or with excised fruits placed in humid conditions, the seeds do not germinate. Thereafter some germination can be found, which decreases again when the seeds become fully mature. It has been suggested (Karssen *et al.*, 1983) that the latter observation indicates the increase in dormancy during maturation.

Seed germination in *Arabidopsis* is activated by the common germination promoting factors such as afterripening, light, cold treatment and chemicals such as gibberellins and KNO_3 (Derkx and Karssen, 1993). None of these exogenous factors is an absolute requirement for germination because the need for one factor depends on the other factors, as shown for the interaction between light and temperature by Cone and Spruit (1983). This interaction depends also very much on the genotype.

The Genetic Analysis of Seed Dormancy in *Arabidopsis*

Genetic variation for germination can be detected when genotypes are compared in identical environments. This implies that not only the conditions of the germination test must be identical but also growth and storage conditions including the time the seeds are stored must be the same. Furthermore, the test conditions must be discriminative between genotypes. For the standard wild types with low dormancy, germination in light of seeds freshly harvested or stored for a limited time is often used (Léon-Kloosterziel *et al.*, 1996). Other parameters are the germination rate after different periods of cold treatment (Cutler *et al.*, 1996) and germination in darkness. In addition to testing mature seeds, germination of excised immature seeds and of such seeds present in excised fruits can be used to investigate genetic variation during the early stages of seed development.

Since tissues from both maternal (testa) and zygotic origin (embryo and endosperm) contribute to the germination behaviour of the seed, the genetic analyses of seed dormancy have to take into account these different tissue origins. Maternal inheritance can be deduced from the germination of seeds obtained after reciprocal crosses, where the two genotypes are used both as female and as male parent. Although reciprocal differences might also be attributed to cytoplasmic inheritance of the trait, the presence of genetic segregation in the subsequent generations allows the distinction between cytoplasmic effects and maternal inheritance (Léon-Kloosterziel *et al.*, 1994).

Germination tests can be used efficiently for mutant screens, because of the large numbers one can deal with. However, the variability of the germination trait may lead to genetic misclassification of individual seeds and therefore to false-positives in mutant screens. Examples of mutants, selected on the basis of their germination characteristics are gibberellin (GA) and abscisic acid (ABA) mutants (Koornneef *et al.*, 1998). In addition, mutants isolated in different screens can also show differences in dormancy/germination characteristics. An example of these are phytochrome mutants isolated on the basis of their hypocotyl phenotype, but extensively used to investigate the role of the different phytochromes in seed germination (Shinomura *et al.*, 1996).

Mutants Affected in Seed Germination.

The *Arabidopsis* mutants affected in seed dormancy can be subdivided into various groups. The earliest mutants with developmental defects in late embryogenesis and seed maturation are *lec1, fus3* and several additional recently isolated mutants. In these pleiotropic mutants, premature germination is observed very early on when the seeds are excised from the siliques or occurs within the siliques. The analysis of these mutants together with the ABA-insensitive *abi3* mutants and the study of their mutual interactions in double mutants (Parcy *et al.*, 1997) have indicated their crucial role in seed maturation. It appears that these three genes have partially overlapping functions in the overall control of seed maturation leading to mutants defective in many aspects of seed maturation, such as lack of desiccation tolerance, seed storage compounds and dormancy. The subsequent cloning of the genes *ABI3* (Giraudat *et al.*, 1992), *FUS3* (Luerssen *et al.*, 1998) and *LEC1* (Lotan *et al.*, 1998) showed that these three genes encode seed-specific transcription factors.

A second group of mutants specifically affects dormancy. Mutants of this group are those defective in ABA biosynthesis (*aba*) or ABA action (*abi*). The non-dormant phenotype, which is a characteristic of all ABA deficient mutants found in various species, indicates the essential role of this plant hormone in dormancy. ABA accumulates to a maximum in the embryo 10 days after pollination (DAP), after which the ABA levels decrease, until later (14–17 DAP) in seed development when a second peak can be detected. Genetic studies demonstrated that only the second minor ABA peak is clearly correlated with dormancy induction (Karssen *et al.*, 1983) and that this ABA is of embryonic origin. The role of the first peak of maternal ABA is yet to be determined, but might be involved in the inhibition of precocious germination during seed development, thereby acting as an inducer/regulator of growth arrest. It is not clear whether ABA just acts as an inhibitor when it is present or if it induces a dormant state, which does not require a continuous presence as suggested by Karssen *et al.* (1983).

Biochemically well-defined mutants (*ga1, ga2, ga3*) with defects in the early part of GA biosynthesis (Koornneef and van der Veen, 1980) are available in *Arabidopsis,* as well as mutants defective in ethylene perception (*etr, ein4*) and ethylene signal transduction (*ein2, ein3*) (Ecker, 1995). The reduced germination of these mutants indicates the promotive role of the two hormones. GAs are absolutely required for germination, except in some genetic backgrounds, when ABA is lacking or when the testa is mechanically removed or genetically altered (Debeaujon and Koornneef, 2000). However, in ethylene-insensitive mutants, germination is only reduced and cold treatments restore germination effectively, which is not the case in GA-deficient mutants. Testa mutants, including those affected in testa pigmentation or testa structure are a special class of mutants with reduced dormancy in which the germination effect is maternally inherited in agreement with the maternal origin of these tissues. The altered germination behaviour of testa mutations is probably due to the effects of the mutations on the physical and chemical structure of the testa, which influence the permeability of the testa (Debeaujon *et al.*, 2000). In addition to these biochemically defined mutants, a group of at least five reduced dormancy (*rdo*) loci was identified, in which the mutant phenotype seems restricted to seed dormancy (Léon-Kloosterziel *et al.*, 1996 and unpublished data).

The Exploitation of Natural Variation

Arabidopsis is an annual plant for which large differences in dormancy behaviour can be found between genotypes collected from nature (accessions) (Fig. 25.1; Ratcliffe, 1976). Kugler (1951) performed the first genetic study of genetic differences in germination found between *Arabidopsis* accessions. Natural genetic variation can be analysed using quantitative trait locus (QTL) analysis (Alonso-Blanco and Koornneef, 2000). We performed such an analysis of the genetics of the dormancy difference between the *Arabidopsis* accessions L*er* (low level of dormancy) and Cvi (strong dormancy) and identified seven chromosomal regions that affect this trait. One region on chromosome 5 provides the major QTL, and near-isogenic lines (NILs) from only this region have been obtained and will serve as starting material for the molecular cloning of the gene(s). By combining these dormancy loci with mutants affected in this trait, the interaction between these 'natural' dormancy loci with hormones such as ABA can be studied. The use of NILs for this type of analysis is more effective than combining mutants with accessions, because the interactions can be studied per gene region in the same genetic background. Furthermore, these lines can serve as starting material for additional screens for

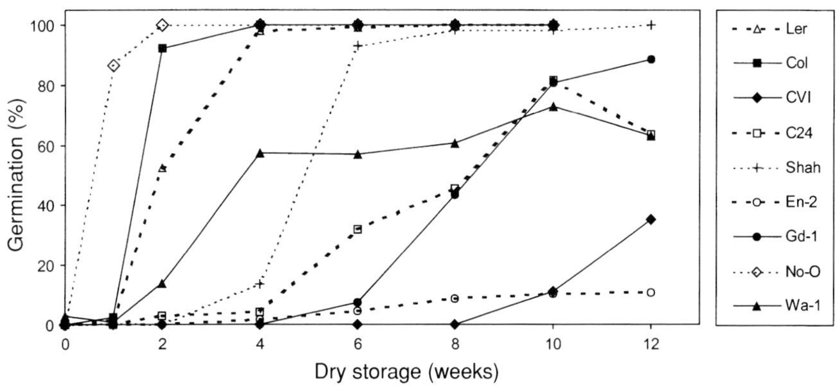

Fig. 25.1. The germination in white light of nine different *Arabidopsis* accessions after different periods of dry storage.

non-dormant mutants. In such stronger non-dormant lines the background of 'false-positives' due to the small fraction of wild type seeds that germinates, as found in screens using L*er* (Léon-Kloosterziel *et al.*, 1996), is expected to be very much reduced or absent.

Small differences in dormancy are also amenable to genetics in *Arabidopsis*, as was shown by the analysis of the genetic differences for seed germination between two commonly used *Arabidopsis thaliana* accessions L*er* and Col (van der Schaar *et al.*, 1997) using QTL analysis. Germination tests were performed with seeds grown from a set of 98 recombinant inbred lines, in three different maternal environments and each seed batch was tested in three different germination environments: in light, in darkness and in the presence of the GA inhibitor paclobutrazol. For 9 loci out of the 14 loci detected, no significant interaction between the locus and environmental factors could be detected. However, three other distinct loci controlling germination behaviour in the presence of the GA inhibitor paclobutrazol had a much lower or no effect when germination was tested in water. Two other loci affecting germination in darkness and/or light had practically no effect on germination in the presence of paclobutrazol. The effects of the individual loci were small in all cases, which makes a thorough molecular analysis very difficult. However, the presence of QTLs in specific regions, which may represent quantitative effects of specific alleles, can be studied, when genes affecting seed dormancy are identified in those regions.

Conclusions

The identification of the molecular nature of genetic differences found for dormancy within species can reveal aspects of the molecular mechanisms underlying this complex trait. Molecular genetic analysis is most efficient in model species such as *Arabidopsis*. In addition to the analysis of mutants, the large natural variation for dormancy in this species can be exploited and is expected to allow an analysis at the molecular level for some of the genes involved. The combination of mutant and natural variation approaches is unique to *Arabidopsis*. From the genetic analyses done thus far, which are still far from saturation, it appears that seed dormancy is a trait that is influenced by many genes, plant hormones and environmental factors. In its genetic complexity seed dormancy may resemble the transition to flowering for which already more than 80 genes have been identified (Levy and Dean, 1998), of which mutants often have only small effects. Probably many of the genes involved in seed dormancy and germination also affect other processes. Mutants of such genes are expected to have adult plant phenotypes or may be lethal, when their function is essential. However, the regulation of these basic processes during seed development and germination may require specific genes, which when mutated will lead to seed or dormancy specific phenotypes. The molecular genetic analysis of seed dormancy should incorporate classical physiological analysis, which has revealed the many environmental influences on this process and which are expected to be translated into the environmental regulation of the expression of many of the genes involved.

The disadvantages of *Arabidopsis* for seed germination studies due to the great variability in seed material coming from the same homozygous genotype and the relatively rapid changes during dry storage in *Arabidopsis* (Derkx and Karssen, 1993) are overruled by the genetic and molecular advantages of this species.

The relevance of comparing genes identified in *Arabidopsis* with those in other plants can be seen in the study of the orthologous genes *ABI3* from *Arabidopsis* and *VP1*, from maize, which seem to have related functions (Holdsworth *et al.*, 1999). This does not imply that there will be no genes that have a major effect only in specific species.

References

Alonso-Blanco, C. and Koornneef, M. (2000) Naturally occurring variation in *Arabidopsis*: an underexploited resource for plant genetics. *Trends in Plant Science* 5, 22–29.

Anderson, J.A., Sorrells, M.E. and Tanksley, S.D. (1993) RFLP analysis of genomic regions associated with resistance to pre-harvest sprouting in wheat. *Crop Science* 33, 453–459.

Cone, J. and Spruit, C.J.P. (1983) Imbibition conditions and seed dormancy in *Arabidopsis thaliana*. *Physiologia Plantarum* 59, 605–612.

Cutler, S., Ghassemian, M., Bonetta, D., Cooney, S. and McCourt, P. (1996) A protein farnesyl transferase involved in abscisic acid signal transduction in *Arabidopsis*. *Science* 273, 1239–1241.

Debeaujon, I. and Koornneef, M. (2000) Gibberellin requirement for *Arabidopsis* seed germination is determined both by testa characteristics and embryonic abscisic acid. *Plant Physiology* 122, 415–424.

Debeaujon, I., Léon-Kloosterziel, K.M. and Koornneef, M. (2000) Influence of the testa on seed dormancy, germination and longevity in *Arabidopsis*. *Plant Physiology* 122,403–414.

Derkx, M.P.M. and Karssen, C.M. (1993) Variability in light-, gibberellin- and nitrate requirement of *Arabidopsis thaliana* seeds due to harvest time and conditions of dry storage. *Journal of Plant Physiology* 141, 574–582.

Ecker, J.R. (1995) The ethylene signal transduction pathway in plants. *Science* 268, 667–675.

Garbut, K. and Witcombe, J.R. (1986) The inheritance of seed dormancy in *Sinapis arvensis* L. *Heredity* 56, 25–31.

Giraudat, J., Hauge, B.M., Valon, C., Smalle, J., Parcy, F. and Goodman, H.M. (1992) Isolation of the *Arabidopsis* ABI3 gene by positional cloning. *Plant Cell* 4, 1251–1261.

Holdsworth, M., Kurup, S. and McKibbin, R. (1999) Molecular and genetic mechanisms regulating the transition from embryo development to germination. *Trends in Plant Science* 4, 275–280.

Karssen, C.M., Brinkhorst-van der Swan, D.L.C., Breekland, A.E. and Koornneef, M. (1983) Induction of dormancy during seed development by endogeneous abscisic acid: studies on abscisic acid deficient genotypes of *Arabidopsis thaliana* (L.) Heynh. *Planta* 157, 158–165.

Koornneef, M. and Karssen, C.M. (1994) Seed dormancy and germination. In: Meyerowitz, E.M. and Somerville, C.R. (eds) *Arabidopsis.* Cold Spring Harbor Laboratory Press, Cold Spring Harbor, New York, pp. 313–334.

Koornneef, M. and van der Veen, J.H. (1980) Induction and analysis of gibberellin sensitive mutants in *Arabidopsis thaliana* (L.) Heynh. *Theoretical and Applied Genetics* 58, 257–263.

Koornneef, M., Léon-Kloosterziel, K.M., Schwartz, S.H. and Zeevaart, J.A.D. (1998) The genetic and molecular dissection of abscisic acid biosynthesis and signal transduction in *Arabidopsis. Plant Physiology and Biochemistry* 36, 83–89.

Kugler, I. (1951) Untersuchungen über das Keimverhalten einiger Rassen von *Arabidopsis thaliana* (L.) Heynh. Ein Beitrag zum Problem der Lichtkeimung. *Beitrage der Biologie der Pflanzen* 28, 211–243.

Lane, M.D. and Lawrence, M.J. (1995) Genetics of seed dormancy in *Papaver rhoeas. Heredity* 75, 84–91.

Léon-Kloosterziel, K.M., Keijzer, C.J. and Koornneef, M. (1994) A seed shape mutant of *Arabidopsis* that is affected in integument development. *Plant Cell* 6, 385–392.

Léon-Kloosterziel, K.M., van der Bunt, G., Zeevaart, J.A.D. and Koornneef, M. (1996) *Arabidopsis* mutants with a reduced seed dormancy. *Plant Physiology* 110, 233–240.

Levy, Y.Y. and Dean, C. (1998) The transition to flowering. *Plant Cell* 10, 1973–1989.

Li, B. and Foley, E. (1997) Genetic and molecular control of seed dormancy. *Trends in Plant Science* 2, 384–389.

Lotan, T., Ohto, M., Yee, K.M., West, M.A.L., Lo, R., Kwong, R.W., Yamagishi, K., Fischer, R.L., Goldberg, R.B. and Harada, J.J. (1998) *Arabidopsis LEAFY COTYLEDON1* is sufficient to induce embryo development in vegetative cells. *Cell* 93, 1195–1205.

Luerssen, H., Kirik, V., Hermann, P. and Misera, S. (1998) *FUSCA3* encodes a protein with a conserved VP1/ABI3-like B3 domain which is of functional importance for the regulation of seed maturation in *Arabidopsis thaliana. The Plant Journal* 15, 755–764.

Meinke, D.W., Cherry, J.M., Dean, C., Rounsley, S.D and Koornneef, M. (1998) *Arabidopsis thaliana*: a model plant for genome analysis. *Science* 282, 662–682.

Parcy, F., Valon, C., Kohara, A., Miséra, S. and Giraudat, G. (1997) The *abscisic insensitive 3 (ABI3), FUSCA 3 (FUS3)* and *LEAFY COTYLEDON 1* (*LEC1*) loci act in concert to control multiple aspects of *Arabidopsis* seed development. *Plant Cell* 9, 1265–1277.

Platenkamp, G.A.J. and Shaw, R.G. (1993) Environmental and genetical maternal effects on seed characters in *Nemophila menziess. Evolution* 47, 540–555.

Ratcliffe, D. (1976) Germination characteristics and their inter-and intra-population variability in *Arabidopsis. Arabidopsis Information Service* 13, 34–45.

Shinomura, T., Nagatani, A., Hanzawa, H., Kubota, M., Watanabe, M. and Furuya, M. (1996) Action spectra for phytochrome A- and B-specific photoinduction of seed germination in *Arabidopsis thaliana. Proceedings of the National Academy of Sciences, USA* 93, 8129–8133.

van der Schaar, W., Alonso-Blanco, C., Léon-Kloosterziel, K.M., Jansen, R.C., van Ooijen, J.W. and Koornneef, M. (1997) QTL analysis of seed dormancy in *Arabidopsis* using recombinant inbred lines and MQM mapping. *Heredity* 79, 190–200.

Wobus, U. and Weber, H. (1999) Seed maturation: genetic programmes and control signals. *Current Opinions in Plant Biology* 2, 33–38.

Index

Page numbers in *italics* refer to figures and tables.

abscisic acid (ABA) 173
 apical dominance 332, *333*
 bud dormancy 339
 embryo growth potential 357–358
 endo-β-mannanase expression 354–355, *356*, 357–360
 induction of gene expression 176–177, 179–180
 potato tuber dormancy 215–216
 seed dormancy 40, *41*
  cereals 183–184
  embryo sensitivity 184, 186–191
  endogenous levels 183–184, 186, 200–204
  GA inhibition of synthesis 205–207
  maintenance 200–201, 206–207
  true potato seed *353*, 354, *359*
 seed lipase activity 48, *49*
 tulip bulbs 7–8, 228–229
  dormancy 229, *230–231*
  stem elongation 233, *234*
  tissue explants 233–237, *238*
Acer platanoides 71–72
Acer pseudoplatanus
 critical fruit moisture content 71–72
 seed pretreatment 72–74, *75–76*
acetone *264–265*, 269
acetyl-CoA 48–49
acid lipase 48
ADP 247–249
Aegilops geniculata 142, *142*
aerial seed banks 140–146
Aesculus hippocastanum see horse chestnut
afterripening
 Arabidopsis thaliana 198–199, *200*
 desert annuals 151–152, *153*
 embryo sensitivity to ABA 188, *189*, 190–191
 Nicotiana plumbaginifolia 198–199
 seed ABA synthesis 196, 203–205
 seed population responses 59–60
Agrobacterium
 iaa genes 342–343
 ipt genes 221, 333, 342
 rol genes 221
alcohol dehydrogenase 267–268, 282

alcohols
 and heat-shock response 281, 285
 seed dormancy breaking 260–261, *264–265*
  chemical lipophilicity 265–266
  leakage of cell components 262
  membrane changes 262–263
  membrane interactions 280–281
  membrane proteins 263
  metabolism of 267–270, 281–282
  pressure reversibility of 261–262
  red light requirement 269, 277
  in soil 282–283
  variability of seed responses 264–267
aleurone layer 176–177
alkaline lipase 48
Amaranthus retroflexus 150–151
aminocyclopropane carboxylic acid 40–41
aminoxyacetic acid 40–41
amphipathic alpha-helices 166
amygdalin 50
anaesthetic compounds 260–261, 280–281
Anastatica hierochuntica 145
annual plants 2–5, *19*
 see also desert annuals
apical dominance 331–332
 hormonal regulation 332–333
 transgenic plant studies 342–343
apical meristem cells
 Ca^{2+}-ATPase 300–303, *304*, 308–309
 intracellular calcium in 298–299, *300–301*, 307
 plasmodesmata/ER channels in 294–298, 305–307
apple seed
 dormancy 47–48
 lipid catabolism 48–49, *52*
 metabolic control of 52–53
 saccharide catabolism 49–52
Arabidopsis thaliana
 AtDRM1 gene 340
 GA deficient mutants 196
 heat-shock proteins 284
 ICK1 gene 339
 seed burial 276
 seed dormancy
  ABA 201–205
  advantages for study 365–366, 371
  characteristics 198–199, *200*, 204
  development 366
  fluoridone 201, *202*, 204
  GA 201, *202*, 204
  genetic analysis 367–368
  germination
   activation 366–367
  membrane receptors 278
  mutants 368–369
  natural genetic variation 365–366, 369–370
  regulation 278
  and soil water potential 276
asparagus 11, *13*
Asteriscus hierochunticus 145
AtDRM1 gene 340
ATP 247–249
ATP/ADP ratio, as marker of endodormancy 249–255
auxin
 apical dominance 332, *333*, 342–343
 branching 334
 potato tuber dormancy 214–215
 transgenic studies 342–343
 tulip bulb
  stem growth 231–233
  tissue explants 233–237, *238*
Avena fatua see oats
axillary bud
 gene expression in 339–342
 growth-dormancy cycles 334–335
 hormonal inhibition of 332–333
 and modular growth 333–334
 molecular/cellular activity 337–339
 origins of 335–336
axillary meristem 335–336

β-inhibitor complex 215, 219–220
bamboo 5, *6*
bark storage proteins (BSPs) 161–162
  in peach 164–169
barley
  ABA-induced gene expression 176–177, 179–180
  embryo sensitivity to ABA 186, *187*, 190
  14-3-3 proteins in 177–181
  seed dormancy 184–185
benzylaminopurine (BAP) 39–42
benzylaminopurine (BAP) riboside 39, *41*
Blepharis spp. 145, 154
blueberry 162, 166–168
Bombax malabaricum 99
branching 10–12, 333–334
Brassica × *napus* 3, *4*
brevideciduous trees *90*, 91, 97, 99, *100*
broad bean 37–38, *39*, *40*
broadleaved trees
  seasonal development 87, *88*, 92–95
  seed pre-treatments 69–80, *81*, 82–83
  storage of pretreated seed 82, *83*
bud, defined 314
bud development 314–315
  correlative control 92–95, 331–333
  mitotic activity following 316–317
  mitotic activity in 315–316
  tropical trees 92–95, 99–105
  xylem sap embolism 109–110, 117, *118*
bud dormancy 313, 319–320
  ATP/ADP ratio 249–255
  bark storage proteins in 161–162, 168–169
  and bud water status 23
  chilling 23–24, 26–31, 163–164, 248–249, 254–255
  control mechanism for 25–26
  correlative inhibition 331–332
  gene expression 339–341
  genetic control 320
  induction and release 319–320
  nucleotide contents 247–249
  nucleotides test 245–246, 250, *251*
  Populus
    candidate genes for 321, 326
    quantitative genetic analysis 322–325
    quantitative trait locus mapping 325–326
  'Tabuenca' test 254
  'whole tree' test 254
  see also axillary bud; floral bud
bud scales 17–18
  initiation 314–315
bulb
  evolution of 9–10
  induction 9–10, 19–20
  suppression of 7–9
  see also tulip bulb
bulbous plants
  evolution 9–10, 121
  growth cycles 7–9
butanol 261, *264*

Ca^{2+}-ATPase 300–303, *304*
  external to plasmalemma 308–309
  methods for localization of activity 305
calcium, intracellular 298–299, *300–301*, 307
Camellia 16–17
Capsicum annuum 18
carbamate 11, *13*
carbonyl cyanide *m*-chlorophenylhydrazone (CCCP) 115–116, *117*
Carrichtera annua 145
cdc2 kinase
  pea buds 337–338
  potato tubers 212–213
cell cycle regulation
  pea axillary bud 337–339
  potato tuber 212–213
cell elongation 35, 37
cell membranes
  dormancy-breaking chemicals 261–263, 280–283, 285
  and gene expression 279–280
  permeability 277
  phase changes 279

phospholipid composition
buds 24, 27–31
seeds 279, 283–284
proteins 277–278, 280–281
and seed dormancy 283–285
cell wall acidification 37–38, *40*
cell wall hydrolases 348–349, 354, *355*
cereals
embryo sensitivity to ABA 184
during afterripening 188, *189*
during thermodormancy 188–190
and water potential 186–187, 190–191
seed dormancy 183–186
seed maturation 185
see also barley
chaperone-like proteins 166, 280
Chenopodium album 151
Chenopodium polyspermum 151
chilling
bud dormancy 23–24, 26–31, 163–164, 248–249, 254–255
cell membranes 24, 26–30, 279–285
chill unit accumulation 162
peach 24, *25–26*, *28*, 163–164
seasonally regulated proteins 161–169
tulip bulbs 121–122, 130–135, 237–240
chitinase 348
chloroform *264–265*, 269–270
Citrus flowering traits 16
cold deserts 139, *140*
conifers
bud development 314–317
seed critical moisture content 71, *74*
seed pretreatment 75
storage of pretreated seed 82
cotyledon 47, *52*, 53
lipid catabolism 48–49
saccharide catabolism 50–53
cryogenic endodormancy 331
Cucumis sativus 11–12, *13*
cyanide 49–53
cyanogenic glycosides 50
cyclin-dependent kinases
dormant potato tuber 212–213
pea axillary bud 337–338
cyclins
dormant potato tubers 212–213
pea axillary bud 337–338
cyclohexylmethanol 269
cytochrome P450 system 268–270, 282
cytochrome P450-dependent monooxygenases 282–283
cytokinins
axillary bud development 333–334, 342
potato tuber dormancy 218–219, 221
seed dormancy release 38–42
cytoskeleton–ER complex 307
cytosolic calcium 298–299, *300–301*, 307

day length 99–105, 331
deciduous stem succulents 89, *90*, 92, 99, *100*
deciduous trees
tropical 89–90
see also brevideciduous trees
dehiscence, delayed 142–144
dehydrin proteins 162, 165–166, 341
in peach bark 164–168
Dendrocalamus giganteus 5, *6*
desert annuals
arable weeds 150–151
germination inhibition in soil 153–154
germination plasticity 147–151, 154
seed dispersal 140–141
aerial seed banks 141–146
soil seed banks 146–147
seed dormancy 139–140
primary 151–152
secondary 152–153
desert soils 153–154
determinate growth 10–12
dimethylnaphthalene 220
dormancy, defined 1–2, 313

dormancy-breaking chemicals 260
  see also alcohols
dormancy–germination transition 35, *36*, 43–44
  dormancy release 38–44
  metabolic activation 35–37
  preparation for cell elongation 37–38, *39*
drought *93–94*, 98–99, 105–106
Dutch iris 7

Echinochloa crus-galli 261–263, 266, 278
ecodormancy 20
  tropical tree periodicity 98–105
ecto-Ca^{2+}-ATPase 308–309
embolism *see* xylem sap embolism
embryo
  ABA and growth potential 357–358
  ABA sensitivity 184, 190–191
    in afterripening 188, *189*
    in thermodormancy 188–189, *190*, 191
    and water potential 186–187, 190–191
  ABA-induced gene expression 176–177, 179–180
  lipid metabolism 48–49
  14-3-3 proteins in 177–181
  saccharide metabolism 50–53
endo-β-mannanase expression
  tomato seed 348–349
  true potato seed 354–355, *356*, 357–360
endoplasmic reticulum (ER)
  apical bud meristem cells
    long day conditions 294–296
    short-day induced dormancy 296–298
  methods of preservation 303–305
  role in bud dormancy development 305–307
endosperm cap 347–348
  cell wall hydrolases 348
  endo-β-mannanase expression 349, 354, *355–356*
  reserve mobilization 348, 358, *359*

ethanol 261, *264*, 282
ether 270
ethyl ether 261, *264–265*, 269
ethylene
  bud growth 332
  potato tuber dormancy 216–217
  seed dormancy release 40–41
evergreen tropical trees 90–91, 97
evolution 2, 9–10, 121
expansins 348

Fagus sylvatica
  seed pretreatment 69, 72–76
  storage of pretreated seed 82, *83*
fatty acids
  β-oxidation in embryo 48–49
  in membrane phospholipids 24, 27–31, 279, 283–284
floral bud
  ADP/ADP ratio 249, *250*, 252–255
  markers of endodormancy 252–255
  nucleotide content 248–249, 253–254
flowering
  carbamate induced 11, *13*
  tropical trees 12–17, *90*, 96
fluoridone
  inhibition of bulb formation 8
  inhibition of thermodormancy 188, 191
  potato tuber 216
  seed dormancy
    ABA-controlled 200–201, *202*, 204–207, 351
    cereal seed 188, 191
    true potato *353*, 354, 356,360
Fraxinus excelsior
  seed dormancy 70, 77
  seed pretreatment 77–78, *79*
fusicoccin (FC) 173–174
  ABA-induced gene expression 176–180
  xylem sap sugar fluxes 115–116, *117*

geophytes
  evolution 9–10, 121

growth cycles 7–9
germination temperature window 72–77, 285
gibberellins (GAs)
Arabidopsis mutants 368
potato tuber dormancy 217–218
seed dormancy
ABA content 205–207
ABA-controlled 196, 199–201, *202*
true potato seed *353*, 354, 358–360
seed lipase activity 48, *49*
tulip bulbs 228–229
auxin-induced stem growth 231–232
dormancy release 229, *230–231*, 237–240
roots 229
tissue explants 233–238
β-1,3-glucanase 348
glucomannans 131–132
glycine-rich proteins 165–168, 341
Golgi body 294, *295*
grasses
dormancy-breaking chemicals 260–261, *264–265*
primary dormancy 183
growth cycles 1, 4–5
and dormancy 18–20
growth habit
annual plants 3–5
bulbous plants 7–10
determinate/indeterminate 10–12
genetic control 10
hormonal control 10
perennial plants 5–7
GTP binding proteins 337–338

heat-shock proteins 279–281, 285
Helianthus annuus 283
Hippeastrum 10–11, *12*, 19
histone expression 337–338
homeophasic adaptation 279
homeoviscous adaptation 279
Hordeum distichium see barley
Hordeum spontaneum 152, *153*
horse chestnut seed
dormancy release 38–43
osmotic solutes *39*
preparation for growth initiation 37–38
Hyacinthus orientalis 7–8
hydrogen cyanide 49–53
13-hydroxyjasmonic acid 220
hygriescens 142, 144–146

iaa genes 342–343
ICK1 gene 339
indeterminate growth 10–12
indole acetic acid (IAA) *see* auxin
intracellular pH 281–282
invertase 51
ipt gene 221, 333, 342
Iris hollandica 7–8
isobutanol 269
isocitrate lyase 49
isopropanol *264*, 265, 269

jasmonic acid (JA)
potato tuber dormancy 220
seed lipase activity 48, *49*

Lablab purpureus 10, *11*
Lactuca sativa 152
Lactuca scariola 152–153
Lactuca serriola 150
late embryogenesis abundant (LEA) proteins 284
light *see* photoperiod; red light
Lilium speciosum 8
linoleate 24, 27–30
linoleate desaturase 24
oxygen requirement 27
temperature optimum 27, 29–30
linolenate 24, 27–30
lipases 48–49
lipids
membrane 24, 27–31, 279
seed reserve mobilization 48–49, *52*
Lycopersicon esculentum see tomato

magnetic resonance imaging (MRI) 122
  tulip bulb
    anatomy and development 125–127
    apparent diffusion coefficient 128–129, *132–133*, 134
    proton density images 127–128, *129*, 130–132
    relaxation times 128, 132–135
Magnolia spp. 13, 15–16
mammalian AMP-activated protein kinase (AMPK) 349
MAP kinase *see* mitogen-activated protein kinase
mean time to bud-break (MTB) 250, *252*
membrane phospholipids 24
  in bud dormancy 24, 27–30
  in cold stress resistance 30–31
  in seeds 279, 283–284
meristems 2
Mesembryanthemum nodiflorum *144*, 146
methanol *264–265*, 269, 281
4-methyl pyrazole 268
microtubule 307
mitochondrion 294, *295*
mitogen-activated protein (MAP) kinase 180
mitosis
  after bud development 316–317
  axillary bud growth 337–339
  during bud development 315–316
modular growth 333–334
MRI *see* magnetic resonance imaging
mucilaginous seed 144–145
multiplicative probability models 58, 66–67
  germination of dormant/non-dormant seed 61–66
  loss of dormancy/loss of viability 59–60
  primary dormancy loss/secondary dormancy induction 60–61

1-naphthalene acetic acid 214–215
naphthylphthalamic acid 233
Negev Desert 139–140, 153–154
Nicotiana plumbaginifolia
  phytochrome-deficient mutant 206
  seed dormancy
    breaking of 282
    characteristics of 198–199, 204
    effects of hormones and fluoridone 199–201
    endogenous ABA content 201–205
nitrate receptor 278
nitrogen storage proteins 161–162, 164–169
non-adenylic tri-phosphate nucleotides (NTP)
  nucleotides test 245–246
  in winter peach buds 248–249, 253–254
non-structural polysaccharides 131–132
norbornadiene 217
nuclear Ca^{2+} 298–299, *300–301*, 307
nuclear envelope 294–296
nucleotides test 245–246, 250, *251*
Nymphaea spp. 5–6

oat 188, *189*, 191, 284
oleate desaturase 24
  oxygen requirement 27
  temperature optimum 27, *28*, 29
oleoyl phosphatidyl choline desaturase 283
one node cuttings test 250
Orobanche aegyptiaca 61–66
Orobanche crenata 60–61, *63–64*
oxygen availability
  bud dormancy 27
  seed germination 186, 275–276, 282

paclobutrazol 233, 370
Panicum capillare 261
Panicum dichotomiflorum 260–261
paradormancy 20

parasitic plants
    germination requirements 58–59
    seed population responses to temperature
        dormancy loss on afterripening 59–60
        induction of secondary dormancy 60–61
parenchyma cells 113–115, 117, *118*
PCNA genes 338
pea axillary buds
    axillary meristem 335–336
    gene expression in dormancy 339–341
    gene expression in growth 337–339
    growth-dormancy cycles 334–335
peach
    chilling requirement 24, *25–26*, *28*, 163–164
    floral bud
        ATP/ADP ratio 254–255
        nucleotide content 253–254
    seasonally regulated proteins 163–169
    sibling genotypes 161–162
    vegetative bud
        ATP/ADP ratio 249–252, 254–255
        nucleotide content 247–249
        nucleotides test 245–246, 250, *251*
pentanol 265, 281
pentose phosphate pathway 50
perennial plants 5–7
phenolic acids 220
phenylacetic acid 214–215
phospholipids 24
photoperiod 99–105, 331
Phyllostachus heterocycla 5, *6*
phytochrome
    and ABA degradation in seed 206
    activity of dormancy-breaking alcohols 269, 277–278
phytochrome genes 321
phytochrome receptor 277
Picea glauca *314*
Picea mariana *315*, *317*
Plantago coronopus 151–152
    seed dispersal 145
    seed germination 153–154
plasmalemma Ca^{2+}-ATPase 300, 302–305, 308–309
plasmalemma H^{+}-ATPase 174
    cell wall acidification 37–38, *40*
    fusicoccin receptor 179–180
    xylem sugar fluxes 115–117
plasmodesmata 291–292
    apical bud meristem cells
        long day conditions 294–296
        SD-induced dormancy 296–298
    EM preservation methods 303–305
    role in bud dormancy development 305–307
polyunsaturated fatty acids 30
poplar
    bud dormancy development
        intracellular calcium 298–299, *300–301*, 307
        plasmalemma Ca^{2+}-ATPase 300, 302–305, 308–309
        plasmodesmata/ER channels 294–298, 305–307
Populus
    endodormancy traits
        candidate genes for 321, 326
        quantitative genetic analysis 322–325
        quantitative trait loci mapping 325–326
    as model for endodormancy 320–321
Populus deltoides see poplar
Portulaca oleraceae 150
potato 211
    see also true potato seed
potato tuber dormancy 211–212
    ABA 215–216
    auxins in 214–215
    cell biology 212–213
    cytokinins 218–219, 221
    endoplasmic reticulum 306
    ethylene 216–217
    factors hindering studies 213–214

GAs 217–218
mutants 221
non-hormonal endogenous factors 219–220
transgenic studies 221
ppdhn1 gene 166–167
propanol 261, *264*, 278
protein kinase C 284
14-3-3 proteins 174, 177–181
Prunus persica see peach
PsDRM genes 339–341
Pseudotsuga menziesii
bud development *315–316*
critical seed moisture content 71, *74*
seed pretreatment 73–76
storage of pretreated seed 82
Pteranthus dichotomus 142, *143*
Pyracantha spp. 13, *14*
pyrazoles 269–270

Rab gene expression 176–177, 180
radicle emergence 35
rain-dispersed seeds 142, 144–146
ramosus 1 pea mutant 334
recalcitrant seed 44
receptor occupancy theory 278
red light
and ABA degradation in seeds 206
dormancy-breaking alcohols 269, 277
rhizomes 5–7
ribosome synthesis 337
rol genes 221
rpL27 gene 337
Rumex spp. 61, *264–265*, 270

saccharide catabolism 49–50, 52–53
cotyledon 51
embryo axis 51
Saccharomyces cerevisiae 285
SAG12 gene 342
Salvia splendens 3
Schismus arabicus 148–150, 152
scutellum 178–179, 181
seasonally regulated proteins 161–162
in peach 163–169
separating dormancy and cold-related proteins 162
sequence homologies *169*
seed banks
aerial 140–146
soil 140, 275–276
seed coat colour 147–148
seed dispersal
aerial seed banks 141–146
soil seed banks 146–151
seed dormancy
apple 47–48
cereals 183–186
control 259–260
cycling of 275–276, 282–286
desert plants 139–140, 150–153
distribution and climatic factors *140–141*
genetic variation 365
hormone balance theory 195–196
population variation 57
role of cell membranes 277–279, 283–286
and thermotolerance 285
see also dormancy-germination transition
seed populations 57
modelling non-linear temperature responses 57–58
dormancy loss on afterripening 59–60
secondary dormancy induction 60–61
seed pre-treatments
control of moisture content 70–71, 82–83
in cold-requiring species 71–72, *73–74*
duration of treatment 72–76
post-treatment storage 80–82
hydrogen cyanide 49–53
problems with 70–71
stratification 69–70
seed reserve mobilization
endosperm cap 348–349, 358–360
lipids 48–49, *52*
saccharides 49–53

seeds 20
  storage 82, *83*
serotiny seed dispersal 142–146
shoot apical meristem 314
  mitotic activity in 315–317
  see also apical meristem cells
sibling genotypes 161–162
silver nitrate 217
Sisymbrium officinale 276, 278
SNF-related kinase (SnRK) genes
  tomato seed 349–350
  true potato seed 354–355, *357*
soil seed banks
  deserts 140
  dormancy cycling in 276
soil water potential 275–276
soils
  desert 153–154
  organic chemicals in 282
  seed dormancy cycling in 275–276
Solanum tuberosum see potato
Spergularia diandra 146–148, 154
stem succulents *90*, 92
storage of seeds 82, *83*
stratification 43, 69–70
Striga hermonthica 59–60
symplast 291–292
  see also plasmodesmata
synaptospermy 141–142
Synechocystis 285

'Tabuenca' test 252–253
temperate trees
  seasonal growth 87, *88*, 92–95
  seed pre-treatment
    in cold-requiring species 71–72, *73–74*
    control of moisture content 70–71, 82–83
    post-treatment storage 80–82
    prechill duration 72–76
    problems with 70–71
    stratification 43, 69–70
    warm/cold treatments 76–80
terminal bud 314
testa mutants 368–369
thermodormancy 188–189, *190*, 191, 331
thermotolerance 285
thylakoid membrane 280
Tilia cordata
  dormancy 79
  seed pretreatment 79–80, *81*
tomato
  growth cycles 18
  lateral suppressor mutant 336
tomato seed 347
  genes expressed in germination 348–350
  hormonal dormancy regulation *359*
topochoric myxospermic seed dispersal 144–145
topochoric seed dispersal 140
topochoric synaptospermic dispersal 141–142
Trachycarpus excelsa 12
tree plants
  annuals 3–5
  growth cycles in 12–18, *19*, 20
  see also temperate trees; tropical trees
triacylglycerol catabolism 48
trifluoroethanol (TFE) 263, 270
  seed dormancy breaking 261, *264–265*
triiodobenzoic acid 233
tropical dry forests
  climatic seasonality 87, *89*
  functional tree types 88–92
tropical trees
  brevideciduous *90*, 91, 97, 99, *100*
  deciduous 89–90
  deciduous stem succulents 89, *90*, 92, 99, *100*
  evergreen 90–91, 97
  periodicity 87–88, *90*, 105–106
    arrest of shoot growth 97–98
    asynchronous crown development 96–97
    bud growth 92–95, 99–105
    and drought *93–94*, 98–99, 105–106
    flowering 12–17, *90*, 96, 101–102, 105

leaf exchange 95–96
and photoperiod 99–105
true potato seed
endo-β-mannanase expression 354, *355*, 357
hormonal dormancy regulation *359*
primary dormancy 350–351
SNF-related kinase expression 354–355, *357*
tuberonic acid 220
tulip bulb
auxin-induced stem growth 231–233, *234*
dormancy 227–229
endogenous hormones 228–229
exogenous hormones
dormancy release 229–231
leaf explants 234–235, *236*
pistil explants 237, *238*
stem explants 233–234, *235*
growth cycles 8–9
hormonal control of growth 237–240
MRI
anatomy and development 125–127
apparent diffusion coefficients 128–129, *132–133*, 134
proton density images 127–128, *129*, 130–132
relaxation times 128, 132–135

Vaccinium section *Cyanococcus see* blueberry
vacuolar H⁺-ATPase 348, 359
vegetative storage proteins *see* bark storage proteins
vessel-associated cells 113–115, 117, *118*
Vicia faba minor see broad bean
volatile compounds 220

walnut, embolism 109–110
and bud break 116–117
role of plasmalemma H⁺-ATPase 115–117
sap hydraulic conductivity 111–112
sap sugar fluxes 113–115
sap water status 112–113, *114*
wheat
embryo sensitivity to ABA 186, *187*
seed cell sap
pH 37–38, *40*
solutes 37, *39*
seed maturation 185
'whole tree' test 254
wind dispersed seed 146–147

xylem sap embolism 109–110
and bud break 116–117
plasmalemma H⁺-ATPase 115–117
sap hydraulic conductivity 111–112
sap sugar fluxes 113–115
sap water status 112–113
xylem tissue
cold hardiness 163
seasonally regulated proteins 164–165, *166*

yeast SNF protein kinase 349

zeatin 219
zeatin-riboside 39, *41*
Zoysia 6–7